Pacing the Void

Tejaprabha Buddha and the deities of the Five Planets, from a mural collected by Aurel Stein now in the British Museum. Original colors would have shown (counter-clockwise, from the left of the Buddha): Mercury in black; Jupiter in blue; Saturn in yellow; Venus in white; and Mars in red.

Pacing the Void

T'ANG APPROACHES TO THE STARS

Edward H. Schafer

FLOATING
WORLD
EDITIONS

First Floating World edition, 2005

LCC 76-48363
ISBN 1-891640-14-3

Published by Floating World Editions
1-860-868-0890
FloatingWorldEditions.com

Distributed byAntique Collectors' Club
1-800-252-5231
AntiqueCollectorsClub.com

to Phyllis

Contents

Illustrations and Tables

Acknowledgments

My thanks are due to Nathan Sivin and Michel Strickmann for needed corrections and scholarly suggestions.

I wish also to thank Georgianna Greenwood for her artistic drawings of "The Polar Asterisms" and "The Twenty-eight Lunar Lodgings", and Yoshimi Nakamura for the Chinese calligraphy accompanying "The Stars of the Northern Dipper", the "Table of Important Starry Chronograms," and inserted in the "The Polar Asterisms".

I am grateful to the National Endowment for the Humanities for a Summer Stipend and to the American Council of Learned Societies for a research grant, both of which were important in making the completion of this book possible.

We demand the stoning of the court astrologers.

J. M. Cameron (title of poem in *The Times Literary Supplement*, 7 November 1975)

1 Introduction

Whatever this book achieves, it is a modest achievement. It makes no significant contribution to the history of science, and very little to our understanding of medieval folklore or to literary criticism. It seems to me to be mostly about images. The most I can hope for is that it will serve as a set of stepping stones to the now almost unimaginable shore where the imagery of the poets of T'ang will be clearly visible in all of its cunning and fantastic workmanship. The title "Pacing the Void" undertranslates the phrase *pu hsü*, common in the titles of Taoist poems of the late T'ang. It means, more exactly, "pacing the barren wastes of space, beyond even the stars, where subjective and objective are indistinguishable." I chose it partly to suggest the exploration of space by astronomers and astronauts, and partly to suggest the obscurity of the conundrums yet to be solved—or even to be recog-

1

nized—before we can achieve a really mature understanding of the symbols and metaphors of medieval Chinese literature.

Although I have drawn up a few charts and tables for the systematization of data which could best be treated in that manner, the general trend of my text is far from "systematic" in the best sense of the word. It is based on a selection of ingredients—literary, folkloristic, and astrological—from which I could blend a reasonably rich goulash whose flavor might be distinguished as *à la T'ang*. I have not tried to write a narrative or to expound a doctrine, or even simply to describe. If a choice must be made, I would say that the language of the volume resembles descriptive prose more than expository prose. But its outstanding stylistic feature is, I think, a somewhat personal blend of selection, suggestion, and analogies. Indeed, my treatment is sometimes less linear than reticulate—occasionally even kaleidoscopic. To make another metaphor, it has been my aim to illuminate a certain area of the stage of T'ang civilization by selective lighting—not neglecting such invisible rays as the ultra-violet and the infra-red—for the purpose of producing surprising but correctly placed highlights, appropriate to the mood of the masque or pageant of T'ang star-lore. Doubtless a secondary and perhaps inevitable result has been the darkening of already perplexing shadows.

While trying not to deviate from verifiable evidence, and labelling guesses and hypotheses for what they are, I have also tried to avoid the repetition of the familiar, except for purposes of orientation or contradiction. In particular, I have wished to invest my chosen "facts" (read "raw data") with an appropriate aura of mystery, or fantasy, or humor, or delight—whichever seemed appropriate to me. But I hope that my readers will be persuaded that the tints I have brushed on the cold marble of respectable philological reconstructions are not "merely subjective"—the glosses produced by a hyperactive imagination deserving no place in serious scholarship. In short, I hope that they will not find the tone of my exegesis unnecessarily frivolous, whimsical, or waggish. I have, I assure them, made a serious attempt to capture in my mind and to re-animate in my prose the true visions of educated or intelligent men of medieval China. I have tried to imagine the moon, for instance, as they imagined it, and to celebrate it as they sometimes celebrated it. To put it differently I have not tried to compile a bare catalogue of T'ang lunar folklore or a handy compendium of "moon images in T'ang poetry." Although my selections and emphases are entirely my own, they are not, I believe, uninformed, and would not, I hope, surprise some literate persons of the T'ang age—an age that is still little *understood*.

The tone of my presentations and interpretations will be regarded, if I am lucky, as appropriate to a defensible style of criticism. The shifts in manner and mood that characterize my study are intended to be cautiously

Introduction

empathetic and concordant with genuine aspects of the imaginings and visions of the men of T'ang, not as arbitrary deformations. If anything, I propose to correct, in some degree, prevailing tendencies to see the T'ang world through European or American eyes—the fallacy of stressing what we stress, of painting with the colors available on our palettes, of fusing only those amalgams which we regard as reasonable or useful.

An indication of the reasons behind some of my choices may be detected in the way I have exploited mythological materials. All readers of Northrop Frye will be aware of the importance of knowing the mythic substructure—perhaps "phylogeny" is a better word—of literary compositions and especially of poems. Some scholars prefer to select or abstract from the mythological content of literature—in what might be called a Lévi-Straussian manner—in such a way as to exclude or minimize the specifics of the costume in which the theme is clad. It is exactly this that I have tried to avoid. For me it is not the theme that makes the poetry, it is the trappings. I am interested in the counterpoint of images, not in archetypal myths put into rhymed verse. The way this preference works out here is that I have written very little about patterns of star-gazing and quite a bit about the magic of star-gazing, with an accent on sensuous details and a tendency to yawn when faced with plot structures.

A similar prejudice will be observed in my treatment of the "scientific" aspect of celestial affairs. It is quite well known that the study of t'ien wen "sky patterns" in pre-modern China was—as the expression clearly suggests—the study of constellations and other figures in the sky, regarded as esoteric diagrams and expressive symbols representing supernatural powers lurking invisibly behind them. Indeed I shall attempt to show that they are very much more than mere symbols. For the Chinese, in short, astronomy was inseparable from astrology and religion, even when temporarily abstracted—as psychology was from philosophy in western Europe—for limited purposes, such as the construction of an accurate calendar. In this book I have paid little attention to such abstract and useful matters. Where the historian of Chinese science tends to be concerned with the accuracy of such ratios and measurements as the obliquity of the ecliptic and the length of the solar years as determined by the astronomer-astrologers of T'ang, I am much more concerned with the benign or alarming visions that signalled to them from the black vault—not only those connected with the art of prediction from omens, in which the Chinese were greatly interested, but also with questions that we might style "ontological," which concerned them relatively little. These are the problems about which there seems to have been no consensus among the medieval Chinese, even within quite small and culturally homogeneous groups—questions about the substance of com-

3

ets, the interaction of sun and moon during solar eclipses, the medium through which the stars pass during their transit below the horizon. For most men such questions were as mysterious and perhaps as frivolous as "what song the Syrens sang, or what name Achilles assumed when he hid himself among women," which, as Sir Thomas Browne concluded, "although puzzling questions, are not beyond all conjecture." They engaged the *imaginations* of some of the men of T'ang, as they did those of other ages, but little intellectual effort was expended on them. Cosmology for instance—for us a serious affair of coagulating galaxies and stellar genesis —attracted little attention from the educated gentry of ancient China, although the arcana of the Taoists are full of dogmatic accounts of theophanies and revelations out of primordial clouds. It follows that modern historians of science have done little with Chinese cosmological speculation, probably wisely. I myself am as fascinated by visions of philosophically-minded goddesses in icy sky-palaces as I am by the evolution of red giants and white dwarfs out of the interstellar dust.

Translations

"I am acquainted with districts in which the young men prostrate themselves before books and barbarously kiss their pages—but do not know how to decipher a single letter."[1] These are the words of Jorge Luis Borges. My translations from medieval Chinese can be regarded as decodings, to which an infinite number of alternatives exist on the shelves of Borges' infinite Library of Babel. Unlike his credulous youths, however, I have little automatic reverence of "masterpieces," and regard my translations as nothing more than aspects of explication—instruments which may help wise men to detect masterpieces. I am certainly not trying to write English poetry—to make pleasing constructs in lieu of hidden Chinese originals—a task for which I am ill suited. In particular I have tried to avoid dressing medieval Chinese verse in twentieth-century American garb. My aim is solely to facilitate the sharing of the visions of the men of T'ang and to reveal, in some degree, their sensibility. If it should be urged that my translations and interpretations are over-imaginative I would answer that if anything they are over-literal. Subjectivity is precisely what I am trying to avoid. I think of myself as a nominalist—as far from Chomsky's mystic universalism as it is possible to be. Sometimes I come close to Nabokov's view of translation, which—as George Steiner conceives it—sees ". . . all but the most rudimentary of interlinear translations as a fraud."[2] I regard almost all approved translations of T'ang poetry as malignant growths. They tell us approximately as much about the real meaning and character of T'ang poems as the works

4

Introduction

of Winkelmann and Canova tell us about the art of ancient Hellas—probably much less. How dreary Tu Fu and Li Po appear in the drab robes laid over them by contemporary taste. How many other poets have written gems which remain undetected, disguised or unnoticed by conventional taste. Serious efforts to discover the excellence of the craftsmanship of Chinese writers are distinguished by their great rarity. The impossible ideal to which I aspire is an imagistic precision to which there are no alternatives. I am outraged by gists and approximations, however palatable and entertaining—the work of men whom Nabokov styles "pitiless and irresponsible paraphrasts." This brings me full circle to the infinite number of translations in cosmic libraries, some differing from others only by the placement of a comma. The unique magic of the original may be obliterated, or (at best) fractionally transmitted, by that very comma. In the end, the aim of my stolid prose renderings is the subversion of most sinological gospel.

Spelling Medieval Chinese

The system of spelling "Middle Chinese" (M.C.) used in this book is a simplification of Karlgren's "Ancient Chinese" romanization. I have explained its meaning and purpose on pages 2–4 of my book *The Vermilion Bird* and tabulated its symbols on pages 267–268 of the same volume.

Chronogram

I have used the word "chronogram" to translate the awkward word *ch'en* which, most commonly, represents the asterisms recognized by the Chinese along the paths of the planets and the moon across the sky, that is, the "Twelve Stations" (*tz'u*) of Jupiter on the ecliptic, analogous to the twelve signs of our zodiac, and the Twenty-eight Lunar Lodgings (*hsiu*) or nakshatras. The idea of time-keeping, hence the element "chrono-," derives easily from this usage. The "-gram" was suggested by the expression *t'ien wen* "patterns or graphic designs in the sky". Accordingly, I treat the common phrase *hsing ch'en* as attributive, and translate it regularly "starry chronograms." (In some instances it seems also to have been a coordinate pair, in the sense of "chronograms and other asterisms," referring to the constellations collectively.)

Counterpart

This is an occasional translation, alternating with "simulacrum," "effigy," "analogue," "doppelgänger," "equivalent," and "other-identity," for *hsiang*, as in the phrase *hsüan hsiang* "occult counterpart," used regularly in medieval literature to refer to the stars and asterisms thought of as harmonized with earthly creatures and institutions, in a universe conceived as a resonant system of celestial and terrestrial correspondences. This is, of course, the basis of Chinese judicial astrology and of much else in medieval belief. The word is discussed further on pages 55–56.

Marchmount

"Marchmount" represents *yüeh* (M.C. **ngauk*), often translated "sacred mountain," a somewhat awkward equivalent, especially when translating poetry. My version is based on the ancient belief that these numinous mountains stood at the four extremities of the habitable world, the marches of man's proper domain, the limits of the ritual tour of the Son of Heaven. There was, of course, a fifth—a kind of axial mount in the center of the world. The concept shows a certain affinity with "landmark," "march," "term[inus]," and "herma."

Numinous

The standardized use of "numen" and "numinous" for *ling* was proposed on page 8 of *The Vermilion Bird*.

Phosphor

This is a word for sky-lights of exceptional luminosity, suggesting a divine message or purpose of particular intensity, normally with auspicious overtones. In Chinese they were called *ching*, which I have translated "phosphor" (*cf.* "phosphor" as "lucifer," the conspicuous star of morning). The common phrase *erh ching* "the two phosphors," refers to the sun and the moon. The reader will find more on this usage on page 182.

Protopsyche

This coinage is used as an equivalent of *p'o* (M.C. **p'āk*), the carnal or residual soul formed in the human embryo by the third day, to which the thin crescent "white-soul" (another possible translation) of the moon on the third day after its monthly rebirth is analogous. It is an entity composed of cool *yin*-light, embryonic lunar light, as opposed to the hot *yang*-light of the sun and to the ethereal soul *hun* (cognate to *yün* "cloud"). Sometimes I have substituted "moon-soul" for "protopsyche." More information is provided on pages 176–182.

Triaster

For this name, assigned to an asterism composed of stars in Orion, see the discussion under "Disastrous Geography" in Chapter 6, "Astrology."

Introduction

A final word. Sprinkled through these pages—not just in the epigraphs—the reader will encounter tags and fragments quoted from European literature. He will not, I hope, regard these as mere displays of frivolous erudition or feeble wit, but accept them, as I hope he will accept such unusual words as protopsyche and Triaster, as bright foils, partial reflections, and surprising analogies, intended to illustrate, either by accent or by contrast, the unique quality—the oddness, if you please—of medieval Chinese images, whose special flavor some writer of Spanish, or Greek, or Old English may, by chance, have captured very nicely.

The seventeenth-century man saw the sky as "a vast cerulean Shield, on which skilful Nature draws what she meditates: forming heroical Devices, and mysterious and witty Symbols of her Secrets."
—E. Tesauro, *Il Cannocchiale Aristotelico*[1]

2 The T'ang Astronomers

The cloudless winter skies over the Yellow River Plain were convenient for the astronomers of ancient China.[2] Nearly perfect conditions for observation were marred only by the waves of fine yellow dust that were driven from the northwest by the dry winter monsoon. These at least had the advantage of providing spectral variations on the main halcyon theme—the orange stars and coppery moons that perplexed the old astrologers. It might seem to follow that the high development of observational astronomy, for which the Chinese are so well known, was the natural outcome of this happy situation. But then one would have to accept that the same results would follow for any observers of clear, dry skies. Observational astronomy did develop in such areas—for example, among the Babylonians—but not inevitably. For premathematical peoples the stunning stellar displays over arid lands are more conducive to star worship than to star study. Indeed the early Bedouins—the Nabataeans are a prime example—were notorious for their star gods. For that matter, despite their development of what might be termed rational astronomy, the Chinese themselves never neglected the sacred powers that showed themselves to human eyes as concentrations of light. This is attested by both official and popular astrology, and by the cults of star deities supported by the government—and the subtler ones adhered to by the Taoists.

For most early Chinese, even for the most advanced authorities on events in the sky, astronomy was indistinguishable from astrology. (In our own tradition we have the great comparable instance of Kepler.) Even as understanding of stellar motions was refined, and more and more aspects of the starry firmament were removed from the realm of conjecture, doubt, and fear—the domain of astrological divination—into the realm of the known and predictable, a major objective of the old Chinese "astronomers",[3] little was changed. Although it is easy to see that comets, meteors, and supernovae might remain forever terrible signals from the powers in space, it would still be wrong to suppose that the inclusion of quite reliable ephemerides in a medieval Chinese almanac meant that the movements of the moon and planets (for instance) were suddenly accepted as merely physical transits of the sky. There were certainly skeptics, but it appears that most men, even well-educated men, continued to believe that a predictable Jupiter remained an awful Jupiter.

Moreover, Chinese astronomy remained mathematical and observational. Little energy was expended on trying to construct a coherent model of the physical universe that would account for the observations and computations.[4] Indeed cosmology languished close to the borderlands of mythology, and for many, perhaps most people, the two were identical. No Chinese poets lamented that "Science" had sullied the lovely moon, or taken the magic out of the stars. On the contrary, the poets found fine metaphors for the simple conjectures of the cosmologists, and saw no intellectual difficulty in retaining the moon in fairyland. Even such significant achievements as the early discovery of the obliquity of the ecliptic, the precession of the equinoxes, and the true length of the tropical year, did not banish the gods from the skies or send the diviners to the bread-lines.[5] Neither did the publication of star maps, whose designs in T'ang times remained, it seems, pretty much the same as they had been in antiquity[6]—compositions of neat little patterns of circles and connecting lines, which failed to exorcise the snowy features and glittering costumes of the gods from the blue fields of Heaven.

A remarkable feature of T'ang astronomy—and astrology—was its syncretist character, or, to be more specific, the extent of Indian influences on it.

These were both theoretical and practical. (A similar condition obtained centuries later, during the Mongol domination of China, when Islamic science prevailed in the office of the Astronomer Royal at Peking.) The extent of western influences on Chinese astronomical and cosmological thought in early antiquity is uncertain. Speculation on the matter has in the past tended to resemble the lush growth of the hot-house or the tropical forest: jungly tangles of colorful lianes and rattans whose stems are confused and whose roots are doubtful. A sober hypothesis by a professional Assyriologist of our own century seems as fair as any other: native Chinese astronomy/astrology was probably modified by the Babylonian by at least the sixth century B.C. This presumably explains why Mesopotamian and Chinese visions of the sky share a number of constellations. The relations of the moon and planets with certain asterisms are similar; so is the significance of the brilliance and color of Venus; and certain omens, especially those with an emphasis on war, victory, civil strife, drought, and rain, are much alike.[7] A specific example is the Near Eastern belief that when Mars lurks near Scorpio war is at hand, and it is a time of anxiety for kings.[8] As late as the year A.D. 675, when Mars was in the lunar asterism "Chamber," close to Antares, the laconic note of the official T'ang astrologer was "There will be anxiety for the sovereign."[9] Of course, many of these parallels and similarities can be attributed to observations and analogies natural to mankind everywhere. In any case, some of them circulated freely throughout Asia during the first pre-Christian millenium.

The later infiltration of Indian conceptions can be traced with somewhat more precision than the hypothetical Chaldean ones. It began with a vengeance in about the first century of our era, if it had not already been taking place. The first Chinese translation of a foreign astrological book of which we have knowledge is of the *Mātānga-avadāna Sūtra* (*Mo-teng-chia ching*), done in A.D. 250. This work refers to the twenty-eight lunar lodgings—which became known also in Persia and Arabia and quickly found a firm bed in the Chinese system of astrology.[10] It has been established from internal evidence that the observational data provided in this treatise were taken in approximately the latitude of Samarkand.[11]

By the sixth century the western zodiac was known in China, but it apparently did not survive beyond medieval times.[12] A comparable set of twelve divisions of the ecliptic, related to the annual position of Jupiter, was already in use, and possibly the new scheme seemed redundant.

Western star wisdom was much honored in T'ang times—much of it patently of Indian origin, but some of it based on Iranian, especially Sogdian, astrology.[13] "The Sutra of the Seven Planetoids" (*Ch'i yao ching*), equipped with ephemerides, which was officially adopted in the mid-eighth century and became very influential throughout T'ang, was of western origin, probably Indian.[14] (The "seven planetoids" are the sun, moon, and five planets, corresponding to the days of the occidental week.) Best known of all these exotic treatises was "The Sutra of Lodgings and Planetoids" (*Hsiu yao ching*), translated from an Indian language in 759 by Amogha (Pu-k'ung). It contained western horoscope procedures conjoined with the twenty-eight lunar lodgings and the seven-day week.[15] "Secrets for Shunning Disasters through the Seven Planetoids" (*Ch'i yao jang tsai chüeh*), with ephemerides from 794, shows Indian influence, as in giving the positions of the invisible planets Rahu and Ketu—which, plus the basic seven, make up the "Nine Plane-toids"—and in describing a solar calendar beginning early in February. It describes celestial movements in Chinese terms, and can actually be used for computing horoscopes, but it is not clear whether or not it was an original production of T'ang.[16] An astrological work with a foreign name, translated into Chinese at the beginning of the ninth century, was the *Tu-li Yü-szu ching*—"The Sutra of *Tu-li Ywĕt-siĕ*." It displays many Greek ideas, apparently from Ptolemy's *Tetrabiblos*, and was probably transmitted to China by way of India through the mediation of the Arabs.[17] A calendrical treatise, titled "Betokened Heaven" (*Fu t'ien*), finished between 780 and 793, was widely but unofficially circulated, becoming especially popular in tenth-century China and Japan. It was based on an epoch beginning with a conjunction of sun and moon in the constellation "Rain Water" (i.e., at the equinox in Pisces) which occurred on a Sunday in A.D. 660. This is of particular interest in that the religious importance of Sunday seems to have been brought to China by Manichaean Sogdians, and the Sogdian name of that day, Mihr (Sun)—Chinese *miet*—has persisted in conservative alma-nacs on Taiwan until recent times.[18] Evidently the name came to T'ang by way of the Uighur Turks in Central Asia.[19] The extent of the influence not only of Indian astronomy but of Indian astronomers at the T'ang court is demonstrated by the fact that the highest available position, that of Astronomer Royal (a paraphrase of *T'ai shih ling*) at the court of Li Lung-chi was attained by an Indian with the inspiring name of Gautama Siddharta. He put together a large compendium of stellar omens titled "Prognostic Canon of [reign] 'Opened Epoch' of Great T'ang" (*Ta T'ang k'ai yüan chan ching*).[20] He is better known now for his translation of the "*Navagrāha* Almanac" (*Chiu chih li*), done in A.D. 718. This was a Graeco-Indian treatise, and was valued for its superiority in the prediction of eclipses, but although it contained such alien usages as the dot for zero and a table of sine functions, these novelties never caught on in China, presumably because they were thought to be impractical.[21]

Astronomical affairs and the formulation of the calendar—two inevi-tably related disciplines whose intimacy had been dislocated but not fatally disrupted by empirical discoveries such as the precession of the equinoxes

—were a monopoly of the court. This was because of the ritualistic and religious component in astronomical activities, which inevitably led back to the sovereign, the Son of Heaven, who was the nexus of the inflow of celestial energy from above and of terrestrial responsibility from below. Nathan Sivin has aptly characterized his position this way: "The Chinese theory of the natural order and the political order as resonating systems, with the ruler as a sort of vibrating dipole between them . . . "[22] Only the Son of Heaven, thus favorably situated, could own true *knowledge* of the stars. Prying into such delicate affairs was threatening and treasonable. To command the data yielded by armillary sphere and star chart was to approach dangerously close to the arcana of spiritual sovereignty. It followed that the possession of such sources of information could only be for treasonable purposes and was tantamount to rebellion against the established order. Accordingly the citizens of the T'ang empire were forbidden to dabble in these matters. The officially pronounced reason for the taboo was benign concern lest inexpert interpreters and other mountebanks, looking only for profit, mislead the ignorant masses and create needless alarm and confusion in their minds. The T'ang code called for stringent penalties for the possession and use of implements and books, with a few trivial exceptions, that led to exact astrological knowledge—that is (as the T'ang code puts it), of our "occult counterparts" (*hsüan hsiang*) in the sky: "All implements and other objects pertaining to the Occult Counterparts; charts and texts about the Sky Patterns; oracular texts and military texts; 'Seven Planetoid Almanacs'; diagrams of the 'Grand Unity' and the 'Thunder Lord'—private households may not possess them. For infringement: two years at state labor."[23]

Despite these restraints there was a certain body of popular astrological lore that was easily accessible to the literate classes. This was in main outline, if not in every detail, essentially the same as that which the privileged court astrologers were perpetually engaged in refining. The old omen books of Han, for instance, were not proscribed, and, since astrology is a conservative science, the basics of Han and T'ang astrology were very much alike. Hence, when reading the poetry of the T'ang era, we are not surprised to find versifiers making homely predictions for their friends and colleagues. A planetary conjunction, visible to all men, plainly spelled victory for a scholar-general whom the poet wished to compliment at a farewell party.

The servants of the Son of Heaven who made the actual observations and computations that were ultimately incorporated into the official almanacs were employed in the court office that I have styled that of the Astronomer Royal. Its actual Chinese name changed several times during the T'ang. Its history may be sketched out roughly as follows:

The T'ang Astronomers

Inspectorate of the Grand Notary (*T'ai shih chien*)	A.D. 618–621
Board of the Grand Notary (*T'ai shih chü*)—attached to imperial library	621–662
Board of the Gallery of Secret Writing (*Pi shu ko chü*)	662–684
Inspectorate of the Spherical Sky (*Hun t'ien chien*); then, shortly, Inspectorate of the Spherical Gauge (*Hun i chien*) —referring to the Armillary Sphere	684–702
Board or Inspectorate of the Grand Notary	702–711
Inspectorate of the Spherical Gauge	711–714
Board or Inspectorate of the Grand Notary	714–758
Estrade for Scrutinizing the Sky (*Szu t'ien t'ai*)— headed by an Inspector	758– [24]

Under whatever name, this important office employed a great assortment of functionaries both major and minor. Among them were archivists, scribes, astrologers, sky-watchers, ritualists, altar-attendants, clock-attendants, and the like. Each of the five senior officers had a colored gem sewn to his hat, the gem displaying the same color as his official costume. These were the heraldic hues of the five planets—red for Mars, white for Venus, and so on.[25] The duties of these dignitaries, who were informally styled "star officers" (*hsing kuan*),[26] extended far beyond what we would regard as the field of "astronomy." Meteorology fell within their province. So they were charged to study "changes in color of the sun, moon, starry chronograms, winds, clouds, and pneumas," and to prepare appropriate divinatory reports. The "Estrade" contained a "Close for Communication with the Occult" (*T'ung hsüan yüan*), manned by scholars who were responsible for the preparation of well-phrased quarterly statements of the omens. These reports were forwarded to the ministries for incorporation into the historical archives.[27]

Also within the purview of the Estrade for Scrutinizing the Sky was the astronomical observatory—still known by a venerable name, "Estrade of the Numina" (*ling t'ai*). Tradition puts an establishment with this title back in the reign of the founder of the Chou dynasty. Another observatory with this name, as always closely affiliated with the divinity of the Son of Heaven, was erected in Han times. Then and later its function was to provide the divine king with an accurate statement of changes in the upper air and what they portended. Its formal charge was "to calculate the verified evidence of stellar measures, the auspicious responses of the Six Pneumas, the permutations and transformations of the divine illuminates," all with the purpose of foreseeing the sources of good and evil fortune, and ultimately of promoting the welfare of the country.[28]

Shih Hu, the infamous but fascinating ruler of Later Chao in the fourth century—a man obsessed equally with magic and with technology—is also remembered for promoting practical and "manly" arts among the women of his palace. He had them instructed in archery, for instance, and in divination by the stars, and established a lady as Grand Notary in the Estrade of the Numina "to look above and watch for disasters and happy omens."[29] This excellent precedent seems not to have been followed by the monarchs of T'ang, but otherwise their instructions to their astrologers followed ancient precedents, with the additional requirement in the eighth century that their prognostications should be based on the most accurate data—that provided by the armillary sphere.[30] (It has been suggested that the stone observatory which still stands at Kyungju in Korea, erected in the seventh century by Queen Sŏndŏk of Silla, might be very like a T'ang Estrade of the Numina.)[31]

The name of a second official T'ang observatory is known. This is the Estrade for Measuring the Shadow (*Ts'e ching t'ai*), at Yang-ch'eng near Lo-yang.[32] Here the location of the winter solstice was calculated by measuring the shadow of a great gnomon. A stele bearing an inscription composed by Li Lung-chi's Astronomer Royal Nan-kung Yüeh in 723 still stands there. The old building and gnomon are gone, but the shell of a Ming replacement is still extant.[33]

One of the most noteworthy achievements of the Chinese in the field of scientific engineering was the building of an armillary sphere, the mechanical heart of their sophisticated observational systems, and the source of many important discoveries. It was a nest of concentric rings representing great reference circles imagined on the inner surface of the celestial sphere. Among the possible armils were those representing the equinoctial colure, the horizon circle, and so on. But for the early Chinese the most important rings represented the Red Road (the celestial equator), the White Road (the lunar orbit), and the Yellow Road (the ecliptic—"the starry girdle of the year").[34] Some kind of armillary sphere was in use already in Han times, and by the fourth Christian century it had made possible the tremendous discovery of the precession of the equinoxes—a discovery that permanently dislocated the traditional belief in cosmic equilibrium by detaching the calendar from the fixed field of stars.

Two celebrated armillary spheres were built during the T'ang period. One was designed by the Astronomer Royal Li Ch'un-feng for Li Shih-min (T'ai Tsung) to prove discrepancies in ancient calendars.[35] Li also wrote the astronomical chapters, models of their kind, in two dynastic histories, those of the Chin and the Sui dynasties,[36] and a book explaining the theory of armillary spheres.[37] The second and more significant of the great T'ang spheres was designed by the Buddhist monk I-hsing, who had a wonderful reputation as a mage and magician, and has deservedly been elevated to the status of a national hero in modern times. His prime purpose was to refine astronomical observations on behalf of his patron, Li Lung-chi, to make possible more accurate prediction of eclipses.[38] His machine was called "Yellow Road Travelling Gauge" (*huang tao yu i*). The design derived from an armillary sphere contrived by Hsieh Lan during the Northern Wei period. It was built in 721 of bronze and iron, and its primary rings were more than four and a half Chinese feet in diameter. (The T'ang foot was about twelve and a quarter of our inches.) It contained a sighting tube something more than four and a quarter Chinese feet long, decorated with jade. The rings themselves were ornamented with silver.[39] Most significant of all, it was driven by water power, operating through an escapement chain that made smooth and regular motion possible, and so it is—as far as can be told now—the ultimate ancestor of all modern clocks and clockwork mechanisms.[40]

The major responsibility of the court astronomers was the compilation of a dependable almanac, whose promulgation was a prerogative of the Son of Heaven. Accordingly the official calendar was a kind of manifesto of the current state of the cosmos as it reflected the ritualistic position of the reigning sovereign at that moment. This quality of the almanac was a very ancient one. Paul Wheatley, writing of the Shang period, might be speaking for all of Chinese antiquity:

But, like all other early calendars, the Shang year count bears the impress of officialdom. It was not concerned with the needs of the farmer—who continued to regulate his activities by the onset of floods, the coming of the rains, the heliacal rising of a star, or some similar phenomenon—but rather was one of a set of accounting devices fashioned to facilitate the ritualistic and managerial functions of sacrally oriented elites.[41]

The system exemplified in an official almanac (*li*) was virtually sacrosanct. New or revised editions often served ritual ends, especially when they were published simultaneously with the inauguration of a new royal era, symbolizing the beginning of a lesser cosmic cycle. Indeed the almanacs of the T'ang Sons of Heaven often took the name of the new era. Otherwise the current almanac was superseded only when a markedly superior scheme had been devised for predicting eclipses and plotting ephemerides. The cosmic system whose operations the almanac aimed to represent may be conceived as a set of mutually engaged cycles. The measurements of actual phenomena made at the Estrade of Numina were, ideally, congruent with the turning of these sets of space-time wheels.[42] Among the nine almanacs promulgated during the T'ang period, eight contained the name of a royal era in their titles. Examples are "Unicorn Virtue" (under Li Chih, Kao Tsung), "Treasure Response" (under Li Heng[a], Su Tsung), "Enduring Felicity" (under Li Heng[b], Mu Tsung).[43]

15

Pacing the Void

Although astronomy and astrology were but different aspects of the same activity, a certain distinction between them was made in the official history of T'ang. The chapters on the calendrical systems used during that era are restricted to observational data and the ephemerides extrapolated from them, while the omen lore that was so significant to the welfare of the state appears in the chapters on the "Patterns of the Sky" (*t'ien wen*). The latter contain not only such reasonably predictable events as occultations of planets by the moon but also such prodigies as supernovae and solar haloes. The planets appear in both treatises: in the first purely as apparitions moving across the sky, in the second (as during conjunctions) as momentous signals of the divine intent. *T'ien wen* (an expression equated with "astronomy" in modern times) means "portent astrology" (as Nakayama has aptly put it) to the men of T'ang.[44]

Yet although astronomical savants and mathematical scholars were not rare at the T'ang court, they were not universally admired by men of sense. The celebrated magistrate and writer Po Chü-i, in one of his poems in the "New *Yüeh-fu*" style, took a censorious tone towards them. The poem is entitled "The Estrade for Scrutinizing the Sky." It has as its published moral: "Draw on the Past to Alert the Present!" In it he laments the failure of the scientists in social responsibility, in not taking their duties as prophets seriously, being concerned only with their "art." Although he takes as his example a display of celestial fireworks which terrified the people long ago in Han times, he is clearly referring to the court astrologers of his own ninth century: these men know what the signs portend, but they approach their sovereign only with flattery and misleading talk about felicitous clouds and longevity stars, then, serenity guaranteed, back to their precious instruments.[45] Perhaps there were such men, but it appears to me that Po Chü-i had little regard for state-supported enterprises whose programs were not primarily and plainly aimed at the public weal: science for science's sake was as abhorrent to him as art for art's sake.

Before leaving the court and its cluster of astrological elite we cannot ignore an institution whose origins were in remote antiquity, and which survived as an only occasionally realized dream. Official relations with the supernatural world were by their nature conservative; it followed that in the new age of T'ang it was possible to give the very modern building in which solstices were determined and latitudes computed the ancient name of "Estrade of Numina." Religious awe persisted even with the advance of scientific technology. This was even more the case with the institution, and the building which was its phenomenal manifestation, which has been called "The Hall of Light" (*ming t'ang*)—a translation that I shall borrow here.[46] "The Hall of Light is the means whereby one communicates with the divine illuminates, responds to Heaven and Earth, adjusts the four seasons, issues instructions and commands, honors those possessed of virtue, emblazons those who hold to the Way, invests those of good conduct."[47] It was, in short, the sacred and ceremonial house of the ancient Son of Heaven. Soothill has aptly compared it to the Regia—the sacred house of the Pontifex Maximus in Rome, where the rituals essential to the welfare of the people were conducted, and the official calendar was designed. (Julius Caesar, as all men know, once held this office—hence the Julian reform of our calendar.) No one was certain about the proper construction and correct appearance of the Hall of Light which, tradition tells us, was first the secret hall of Wen Wang, the great founder of Chou. One Han dynasty source tells us that it was eighty-one feet long (east-west) and sixty-three feet broad (north-south)— "hence it was called the Great House." It was roofed simply with thatch.[48] By T'ang times, on the few occasions when the antiquarians could agree sufficiently to risk a reconstruction of the venerable building, it was chiefly important for housing the high rites of the state cult, presided over by the Son of Heaven. Its astronomical and calendrical functions—other than symbolic ones—had been turned over to the Estrade for Scrutinizing the Sky. In 631 Li Shih-min contemplated building a Hall of Light, and many inconclusive discussions were held among the knowledgeable pedagogues about its authentic characteristics.[49] In the end they led to nothing. In 668 and 669 Li Chih (Kao Tsung) in his turn revived the daring enterprise and even went so far as to designate the eastern half of the capital city as "Township of the Hall of Light" (Ming-t'ang hsien),[50] but a famine in the land during the spring of 669 was considered sufficient warning and the project was discontinued.[51]

The Empress Wu had enough confidence to carry out what her predecessors had been too cautious to achieve—she built a Hall of Light in her Divine Metropolis, Lo-yang. The great lady was fond of monumental architecture in metal, designed for divine purposes. In 695, to celebrate the restoration of the Chou dynasty, she built an octagonal pillar of cast iron 105 feet high, and styled it "Celestial Axis." It was topped by a "fire orb" of polished bronze supported by four dragon-men, inscribed with the names of her great vassals and tributary princes, and by the figure of a divine being holding up a bowl. The orb gathered in the fiery yang power, while the bowl received the beneficent dew of yin.[52] As for the new Hall of Light, considered as a feat of engineering it was not so ambitious, but certainly a magnificent construction—an optimistic attempt to fuse antique precedent with modern splendor. It all began when the monarch asked her magnates how she might "harmonize the Primal Pneuma." Ch'en Tzu-ang, whom we now remember chiefly as a poet, advocated the construction of a Hall of Light. But a

Confucian pedant gave a different view: "Your vassal has heard this from his teacher: 'The Primal Pneuma is the onset of heaven and earth, the grandparent of the myriad creatures, the great initial of the royal administration. In heaven and earth nothing is greater than *yin* and *yang*; among the myriad creatures none is more numinous than man; for the royal administration nothing comes ahead of making mankind secure. Therefore when men are secure, *yin* and *yang* are harmonious; when *yin* and *yang* are harmonious, heaven and earth are tranquil; when heaven and earth are tranquil, the Primal Pneuma will be correct.' "[53] It follows that *man* is the key to good government, and the proposed Hall of Light a cosmological folly. There was more sententious wisdom of this kind purveyed by the conservative scholars. Moreover, should the thing actually come to pass despite their arguments, they wanted the hall built in a geomantically favorable place, rather remote from Lo-yang. The queenly lady would have none of this. She desired the hall to be in her city, and had no patience with nitpicking. Accordingly, in 687, the Basilica of the Generative Prime (*Ch'ien yüan tien*) was razed, and construction of the new Hall of Light began on the site. The Buddhist monk Huai-i was placed in charge of the undertaking, and "several myriads of men" were employed in its achievement. On 14 June 688 it was finished. A great sacrifice to Heaven was offered in the southern suburb and the Lady Wu proceeded to her new sanctum to receive her court.[54] On 28 December of that year all of the surviving scions of the House of T'ang were put to death and their children sent off to languish in tropical exile.[55]

So the restored Chou was firmly established both in sky and on earth. On 23 January 689 the "Hall of Light" was renamed "Divine Palace of the Myriad [Starry] Simulacra" (*Wan hsiang shen kung*) and a great amnesty declared.[56]

Empress Wu's Hall of Light was a three-tiered building 294 feet high and 300 feet square. On the bottom floor were representations of the four seasons, each in its appropriate color. On the second storey the twelve great asterisms of the Jupiter cycle were displayed on the sky canopy, which was furnished with a basin supported by nine dragons. On the upper storey there were simulacra of the twenty-four pneumas and another round canopy. On the roof was a great iron bird, made splendid with gold. It was twenty feet tall, and shown poised as if ready to take flight.[57] This handsome fowl was later destroyed by a great wind and replaced by a magic "fire orb."[58] The court poets, as expected of them, composed elaborate effusions to celebrate this ornament of the new dynasty. Most productive of all, it appears, was Liu Yün-chi, five of whose *fu* ("rhapsodies on a set theme") pertinent to the occasion are still extant.[59] Typical of the rich language of these efforts is the following sample:

The T'ang Astronomers

> Great indeed is the Generative Counterpart—
>> where purple tenuity is diffused through the palace of the Supreme Deity.
> How profound is the Quiescent Carriage—
>> Whose cinnabar pylons open into the precincts of the Unique Person.[60]

With this exordium establishing the cosmic dualism in terms of the harmonic sympathy between the stellar simulacra above and their telluric counterparts below—especially between the great polar deity and the T'ang sovereign—the poet goes on to declare in even more splendid detail the intimate relationship between the divine fires and human affairs.

Whatever the merits of the Hall of Light as cosmic magnet, its form was not universally applauded. In the very year of its construction a courtier memorialized the throne to this effect: "In antiquity the Hall of Light had a thatch of floss-grass that went unclipped, and painted rafters untrimmed. But this one now is decorated with pearl and jade, and painted with vermilion and azure." He went on in this vein, to the point of objecting to the holy iron bird on the roof and the golden dragons within.[61] Evidently he envisaged a simple rustic fane, rather like the wood-and-thatch shrine hidden away behind the splendid vermilion façade of the Heian Jingū in modern Kyoto. These sentiments availed him nothing, but his words were echoed three decades later when an officer of rituals in the palace of Li Lung-chi protested: "The Hall of Light erected by the Heaven-Patterned [Queen] did not match the ancient pattern. Moreover, in a Hall of Light simplicity should come first—but she exhausted extravagance and went the limit in prodigality." Hsüan Tsung was swayed by this argument, and ordered that the divine building be degraded and given the title of its predecessor on the site, the Basilica of the Generative Prime. This was done on 6 September 717.[62] For motives which remain unclear, the edifice became a Hall of Light once more on 28 November of 722, but with curtailed and simplified rites. But on 17 November 738 a Master Architect was sent from the Ch'ang-an palace to Lo-yang to pull it down.[63] Nonetheless in the following year a new Hall of Light was built in the Eastern Capital, and despite unnerving rumors that a child had been buried alive under its foundations, in order that its spirit might guard the holy precincts against evil spirits, the building was finished on 20 December.[64] (It appears that this was not an entirely new construction; one source states that the upper storey was eliminated, and renovation done in the lower.[65]) At any rate, in 740 the building was damaged by a fire that spread to it from a neighboring Buddhist temple, and it was once more degraded to the status of a "basilica."[66] After that we hear no more about it.

Such, in brief, were the professionals, the apparatus, and the housing which occupy the forefront of the astronomical-astrological scene as it is presented by the formal histories of T'ang. Few glimmerings of mythology, of fine fancy, or of starry visions illuminate these rather carefully orthodox presentations, stiffened by official pedantry and stripped of virtually every-thing that might delight the imagination. An important aim of the present book is to correct, if only partially, this traditional blending of conservative statecraft with the lofty abstractions of the history of science. Let us turn, gradually at first, to the mythographers and the poets—the men who were always conscious of the lurking presence of the afreet in the astrolabe, the elf in the orrery.

Till the slow sea rise and the sheer cliff
    crumble,
Till terrace and meadow the deep gulfs
    drink . . .
    —A. C. Swinburne, *"A Forsaken Garden"*

3 Cosmogony

Cosmology, in its preferred sense, was not a science that intrigued Chinese thinkers greatly. Possibly "cosmography" is a more fitting label for their phase-theories, lore of correspondences, and universal mapping of ontological cogwheels. But a sober, analytic, and coherent account of the nature of the present universe was beyond their ambitions. This relative indifference was, it appears, markedly enhanced by the discovery that celestial movements and the calendar did not match. If a choice had to be made, it was in favor of the refinement of the calendar as an admirable abstraction well suited to the ingrained Chinese need for an orderly scheme of things. Empirical crudities had best be blinked at. Accordingly by T'ang times, as one competent authority has pointed out, "the sense of cosmos almost completely dropped out."[1]

Pacing the Void

This is not to say that there was not a great deal of interest in spatial matters—in the world of extension. Indeed there were all kinds of speculations about the distances of the sun and moon from the earth, the nature and size of the solar orb, the manner of the progression of the planets across the celestial dome, and what not. But there was little or no effort to devise a consistent theory that would harmonize all of these guesses, or lead to a succession of better guesses. Admirable as observational astronomy and some kinds of computation, measurement, and prediction were in China, there were no developments that can be compared to the sequences whose end members are Ptolemy-Copernicus, Newton-Einstein. Even the sets of lovely concentric crystalline globes imagined by Eudoxus and Aristotle, seemingly so compatible with Chinese predilections, had no counterpart or rival there.

Although the structure of the universe was neglected, some interest and energy was invested in speculation about cosmic evolution—in imagining the universe in time—however vaguely its physical nature and dimensions might be conceived. In what follows I shall sketch briefly, as supplementary background to the chief preoccupations of this study, some of the more interesting early Chinese notions about cosmic process, with a few notes, of interest mainly to connoisseurs of the curious, about the character of the sky and the color of transient sublunary/superterrestrial phenomena.

I shall begin by dismissing the grand concept of TAO out of hand. For me, at least, it defies discussion, and therefore should not even be accorded the dignity of "concept." It can be spoken of only in metaphors: a womb, a boundless matrix and therefore no matrix, a cloud of non-constituents, an unexplained explanation which does not explain, a cosmic dung, nothing, everything, a four-sided triangle, a gigantic glass of beer, a turtle. Let a reincarnate Chuang Chou deal with the matter.

I have little more to say about the much admired notion of cosmic SPONTANEITY (*tzu jan* in its most elevated sense). (I do not mean to underestimate the vast reverence accorded the word in Chinese literature, both didactic and imaginative; but however much it may have inspired reverence, I do not feel it has inspired much useful thought or attractive poetry.) A few pleasant examples of the treatment of Spontaneity may serve as substitutes for serious meditations on the term—a feat of which I am incapable, anyhow. In Han times Wang Ch'ung, who often managed to conjoin the commonplace with the cosmic (I will not say "the philosophical") provided this interesting analogy: "Heaven and Earth unite their pneumas: the Myriad Creatures are born spontaneously. Husband and wife unite their pneumas: children are born spontaneously."[2] This comparison has a certain homely charm which even I can recognize. Beyond that, the statement—like

22

most assertions using *tzu jan*—is tautological: what happens, happens. Perhaps more sophisticated, certainly more refined, is the Taoist view presented by the T'ang poet Wu Yün, who defines *tzu jan* as "what becomes so without knowing so."[3] (I hold no brief for the intelligibility of either the Chinese or the English phrasing: the sentence may imply "but *we* do not know how it is so," but I prefer the interpretation "it is a blind, unmotivated tendency in the nature of things.") On the theme of mold, or molded form (*hsing*) and energizer, or activator of form (*shen*), he writes as follows:

> Spontaneity bears Empty Nothing;
> Empty Nothing bears Great Tao;
> Great Tao bears Pregnant Cloud;[4]
> Pregnant Cloud bears Heaven and Earth;
> Heaven and Earth bear Myriad Creatures.[5]

Nothing could be clearer than this. Spontaneity is so abstract as to precede even something as far from concretion as vacancy. The Tao is Grossness by comparison; Pregnant Cloud a wormy pullulation.

The idea of an entity or principle in nature whose role is to *fashion* (as opposed to "creating" *ex nihilo*) is not prominent in Chinese metaphysical thought, but is not unknown either. A rather primitive version of a demiurge survived in the form of the venerable goddess Nü Kua, who had affinities with frogs and snails and other quiet, moist creatures and their environment, but in T'ang times she was little remembered except by the poets, sometimes as a phantom in folktales and as a minor figure in local cults.[6] Two aspects of her history and nature were well remembered: she was the repairer of the blue dome of the sky—a feat she accomplished in remote antiquity through the use of suitably colored stones; she was also a kind of female Jehovah who, in an idle moment, made man out of the yellow earth,[7] a feat reported by P'i Jih-hsiu in his verses:

> Nü kua made straight her cord and rope,
> Sketched in the mud—and formed lowly man.[8]

Less stimulating to the imagination perhaps was the abstraction called "Fashioner of Creatures" (*tsao wu che*), or sometimes "Mutator of Creatures" (*hua wu che*). It was a coagulating, specifying principle, also called *tsao hua* (*che*) "Fashioner of Mutations," the generator of the perceptible world. This being (if so indeterminate an entity can be called a "being") had been allotted a whole chapter in the book of *Chuang tzu*,[9] where it exhibits a shadowy kind of life as it ranges the cosmos. In later literature, however, it is an undefinable essence which exercises supernatural arts throughout eternity, to produce the familiar world of concrete and discrete objects revealed by our senses. It is the great Selector, the Synthesizer, the master of determination and individuation.[10] In the western world it has its analogue in the form-giving "Soul of the World" of Giordano Bruno, "continually actualizing all possible forms of being."[11] In the first century of our era Wang Ch'ung gave it a metaphorical identity as a divine craftsman:

> Heaven and Earth are the furnace;
> The Myriad Creatures are the bronze;
> Yin and Yang are the fire;
> The Fashioner of Mutations (*tsao hua*) is the artisan.[12]

In this view—which Wang Ch'ung, a vigorous proponent of "spontaneity," churlishly opposed—the frame of the physical world is imaged as an enclosing crucible, the physical beings of past, present, and future as potential in the plastic medium to be worked, the *yin* and *yang* energies as the softening fire, and the Fashioner himself as the producer of mutants, the originator of actual creatures.[13] Some writers allowed themselves to take a personal and subjective view of the Fashioner. One of these, the late ninth-century poet Su Cheng,[14] took the anthropormorphic liberty of providing him with hands:

> Crow and hare move straight through night and day
> While men wheel round from decay to fruition.
> Do well—abnormal waning;
> Do ill—anomalous waxing!
>
> Heaven's Way lacks party bias,
> Human hearts turn readily upside down—
> Hence so many shifts and changes:
> Best to match feelings with Heaven and Earth.
>
> I wish that the hand of the Shaping Mutator (*tsao hua*)
> Had never set fox to course after hare,
> Nor freed the sea to bear treasures and rarities,
> Nor indulged the lands in growing flowers and willows;
>
> Were Beautiful uniformly beautiful,
> Were Ugly uniformly ugly,
> Men's hearts would return to the Great Unhewn:
> Battle and strife—could they exist either?[15]

("Crow" and "hare" are the emblematic animals of the sun and moon respectively.) Here the Fashioner is a kind of First Cause, but more often he is like a blind potter or whimsical metallurgist—a continual meddler in the world. He also appears as the supreme Magician and Illusionist.

In Taoist poetry of the T'ang period the Fashioner regularly takes on the quality of a mystic state of comprehension or enlightenment—a kind of key to wisdom that the initiate may some day hope to turn, or an empyreal realm beyond space and time where the Meaning of it All will finally become clear.

Cosmogony

Lü Yen, for instance, in two untitled poems, wrote of a kind of "hypophenomenal" (ch'ien)—to coin, I think, a word—access to the Fashioner.[16] He speaks of the sovereignty or authority (ch'üan, "holding the balance of power") of the Fashioner. The Taoist adept Li Hao used the same language in telling us that once one has achieved the secret distilling of the Elixir, "He may wrest from Potent (ch'ien) and Latent (k'un) all the authority of the Fashioner of Mutations."[17] Unlike Lü Yen, for whom the approach to the Shaper was secret and covert, Wu Yün announces, as the climax of one of his saltatory poems, a vision of "the font of the Fashioner of Mutations" at the terminus of a saunter into the void.[18] Metaphorically, however, the secret tunnel and the blinding abyss were equally adequate for the Taoist mind, in which the walk along the Dipper and the dark transit of a sacred grotto lead alike to a higher and more splendid heaven than the one we know.

The more systematic accounts of the creation of the cosmos do not show it as a single event but as a sequence of distinct phases in a continuous process—this is in marked contrast to the endless and unpredictable shapings of the Fashioner in the evolved phenomenal world. Unhappily these sources do not provide a mutually consistent scheme of these stages in the evolution of the universe. It is possible, however, to construct a kind of synthetic scheme that will not differ radically from the specific ones proposed by the several authorities. A convenient way of achieving this aim is to study the taxonomy provided by the rubrics in the first section of the T'ai p'ing yü lan. This yields a scheme that was at least acceptable to the editors of that great anthology, and may be regarded as a fair average of educated opinion and usage in the tenth century. Admittedly my glosses on the sequence are somewhat arbitrary, especially in trying to find a pigeonhole that will not violate the sense of any of the classical sources quoted, but they should serve well enough. The scheme works out like this (beginning at the beginning of time):

1. Yüan ch'i "Primal Pneuma" (the undifferentiated generator of the world).
2. T'ai i "Grand Interchangeability" (the potentiality of formal differentiation).
3. T'ai ch'u "Grand Antecedence" (the germination of formal differentiation).
4. T'ai shih "Grand Initiation" (the realization of formal differentiation).
5. T'ai su "Grand Simplicity" (undifferentiated substance).
6. T'ai chi "Grand Culmination" (material differentiation).

With this idealized medieval arrangement, compare a shorter one that occurs in the book Lieh tzu, much older—though not as old as was once believed:

1. "Grand Interchangeability"—Pneuma not yet visible.
2. "Grand Antecedence"—the beginnings of Pneuma.
3. "Grand Initiation"—the beginnings of Form.
4. "Grand Simplicity"—the beginnings of Substance.[19]

The obvious differences here are the absence of an initial stage in which the primordial breath dwells, imperturbable and aloof. For *Lieh tzu* the Pneuma is merely a secondary product of the Second Age. Moreover, in *Lieh tzu* the stamp of form appears abruptly in "Grand Initiation" without the hesitant preparatory steps provided in the *T'ai p'ing yü lan* system. Finally "substance" (or "stuff" or "matter") is allotted only one phase in the *Lieh tzu* but two in *T'ai p'ing yü lan*. The *Lieh tzu* scheme, then, has fewer refinements, but greater symmetry.

Either system, however, like all similar systems, assumes the existence in time of unformed matter, uninspired form, and so on, as a precondition of the emergence of actual creatures. This is no mere timeless hierarchy of kinds of existence, but a temporal succession—and we must always remember how obsessed the Chinese were with time—of bare ontological concepts. To an aggressive imagination the expression "Primal Pneuma" might suggest an amorphous cloud of faintly visible gas, distinguishable by a vague turbulence, somewhat as we imagine the early condition of our own galaxy to have been. But if we think that this was the case we deceive ourselves. The names "Primal Pneuma", "Grand Antecedence" and the rest cannot properly evoke even that minimal degree of concreteness. They are usually—but not always—treated as abstract markers of the successive epochs in the grand cycle of primordial time. In trying to conceive the *character* of any of these eons, we are attempting a feat as impossible as that of describing the texture of a triangle, the resilience of a cycloid, or the fragrance of a logarithm.

Let us take a somewhat closer look at each of the phases of the sixfold system.

PRIMAL PNEUMA: In the earliest texts the time interval between solitary Pneuma and the emergence of the phenomenal world is not so complex as it appears in the systems we have taken as examples. Sometimes the Pneuma was said to divide directly into the two complementary aspects of reality, like the mitosis of a cosmic zygote: "When the Primal Pneuma first divides, the light and clear is *yang* and is Heaven; the heavy and turbid is *yin* and becomes Earth."[20] In some accounts the Pneuma anticipates the energized material reality that is to succeed it by showing a certain spissitude which is not congruent with its priority over Substance—it is a palpable fog on an enormous scale. In the description which follows, it is not even primordial,

26

but oozes out of the infinite globe, its source and cause: "Before there was Heaven and Earth, the aspect of the cosmic sphere (*hun-t'un*) was like that of an egg. A boundless haze began to sprout—an all-encompassing mist budded and proliferated."[21] The expression I have translated as "all-encompassing mist" (*meng-hung*) represents the character of the Pneuma before its coagulation into nodes. This rather crass quality, which allied the archaic nebula with phenomenal mists and clouds, made it detectable to human eyes through such secondary manifestations. Accordingly, it could become the object of attention of royal sky watchers in their endless search for tokens of the health of heaven and earth. "To ascend the Estrade of the Numina and watch the Primal Pneuma afar"[22]: such was part of the charge to the royal astrologers of Han. Not surprisingly, the potent world-mountain and the "marchmounts" (*yüeh*) of the Middle Kingdom itself, served as spouts and funnels for the noumenal plasma: "The Primal Pneuma of K'un-lun took on substantiality to form the Felicitous Clouds."[23] Among all of the cosmogonic terms, "prime breath" is the most common in poetry, and there it often assumes humble functions through metaphor and analogy. For instance, Meng Chiao has it rising from the mountains just south of the capital city, stimulated, as it were, by the holy presence of a Buddhist monastery: "The foremost hill is the womb of the Primal Pneuma."[24] Sometimes the term was used in an entirely figurative way to characterize the inspired words of an eminent poet, as when Li Shang-yin wrote of an inscription from the hand of Han Yü as "resembling the Primal Pneuma."[25] (The compliment could hardly be overlooked even if the intent was satirical.)

GRAND INTERCHANGEABILITY: Among the set of six phases set forth in the *T'ai p'ing yü lan*, this is the most ambiguous, the most subject to variant interpretations, and the least noticed in literature. We have seen that the *Lieh tzu* awarded it primary status, ahead of the appearance of the cosmic breath, which is primary in the system of the *T'ai p'ing yü lan*. Elsewhere,[26] it has been described laconically as the stage "before the separation of Heaven and Earth"—a description which would apply to all of these phases equally. It is a condition which has even less character than any of the conditions that subsist without form, if that is possible.

GRAND ANTECEDENCE: The view of the *Lieh tzu*, which puts the first appearance of the primitive pneuma at this level of generation, is given elsewhere.[27] This may indicate that it was commonly accepted in preference to the one in which the Pneuma itself was prior to everything. The Grand Antecedence enjoyed special status in some Taoist thought as the eon during which Lao-tzu, in his magnified form as a cosmic, eternal presence looming

27

out of the bottomless abyss of time, separated Heaven from Earth, and ultimately gave form to all the living beings of the world. This account of the Creation may be found expressed most admirably in the "Canon of the Opening of Heaven by the All-Highest Lord Lao," which must have been composed before the year A.D. 1000.[28] This great story shows certain resemblances to those others which, in the Ling-pao tradition of Taoism, tell how the Heaven-Honored Ones (*T'ien tsun*), under various names and aspects, dictated the conditions which shall prevail at each stage of the genesis of the universe. Our text begins as follows:

In the time of the Vast Prime (*hung yüan*) there was still neither Heaven nor Earth. Empty Space had not divided, nor had Clear and Turbid been separated within that obscure and empty desolation. The whole order of the Vast Prime extended into a myriad of kalpas. After the separation of the Vast Prime there was the Composite Prime (*hun yüan*), and the whole order of the Composite Prime was for a myriad of kalpas and extended into a hundred fulfillments, and the hundred fulfillments in turn were for eighty-one myriads of years—and then there was Grand Antecedence. At the time of Grand Antecedence Lord Lao descended from empty space and made himself Master of Grand Antecedence.

Then began the acts of material creation. In view of the preeminent status assigned to the stage of Grand Antecedence in this account, it is not surprising to discover in another canonical book of Taoism the lofty statement "By TAO we mean Grand Antecedence. Grand Antecedence is the antecedence of the TAO."[29] With this stage, the birth of the world is at hand.

GRAND INITIATION: "The initiation of plasm (energizing pneuma, *ch'i*) and structure (impressing matrix, *hsing*) we speak of as Grand Initiation."[30] So says one source; another emphasizes the formative aspect of this stage: "Grand Initiation is the initiation of structure (*hsing*)."[31] Still another says: "The clear became [unformed] germinal essence (*ching*); the turbid became [impressing] form (*hsing*)."[32] A Taoist cosmogonic text is even more explicit—and more theological: "At the time of Grand Initiation, Lord Lao descended to become its Master, and from his mouth he emitted the 'Sutra of Grand Initiation' . . . 'Grand Initiation' is the initiation of the Myriad Creatures."[33]

GRAND SIMPLICITY: The semantics of this term are hard to put adequately into English. "Simplicity" yields only one facet of the many-valued word *su*, which connotes also "innocent," "immaculate," "candid," "untarnished," "unstained," and "unmarked"—even "blank". In the cosmogonic sense it refers to material substance in its undifferentiated condition: "Grand Simplicity" is the inauguration of substance (*chih*). What exists is simple and unshaped ("unhewn" *p'u*), not yet distributed.[34] However, for the Taoist mythographers this implied a separate epiphany of Lord Lao with a new cabala; in this era all creatures lived in utter simplicity and purity, and men were immortal, subsisting on sweet dew and natural fountains of wine.[35] This kind of Golden Age simplicity is a far cry from the cosmic homogeneity called by the same name in the orthodox system. The Grand Simplicity was not unknown to the T'ang poets. Sometimes they used the name simply as a synonym of Divine Philosophy, as when Wang Wei complimented a Taoist alchemist in these terms:

> Naturally she possesses the Technique of Returned Cinnabar;
> At times she discourses on the priority of Grand Simplicity.[36]

Or the name might be applied to the virgin whiteness of new heaven-born snow, as when, after a dangerous drought late in the eighth century, Ch'ang Kun congratulated his sovereign for longed-for precipitation, in the form of snow: "The Grand Simplicity coalesced, unrestricted over a myriad miles!"[37]

GRAND CULMINATION: This stage, it was generally agreed, was marked by the differentiation of the primitive stuff into a polarized pair (yin and yang), and ultimately, by further germination, into the whole variety of nature.[38] It was the culmination of the cosmogonic crisis.

It is worth remarking that some of these names, as the markers of the initiation of eons and of the creation of new cosmic orders, were appropriated by human sovereigns as grandiose labels for their own imperfect reigns. So Liu Ch'e (Han Wu Ti) styled two of his many regnal eras "Grand Antecedence" and "Grand Initiation," while Li Tan (T'ang Jui Tsung) adopted the reign title "Grand Culmination."

The scheme of the T'ai p'ing yü lan provides no special niches for two rather common cosmological terms. The omission of one of these is natural: it belongs to cosmography rather than to cosmogony. It is "Grand Emptiness," an expression with overtones of "waste," "desert," "barrens" in the sense that these words are used in such phrases as "watery waste" and "antarctic desert," but without any unpleasant overtones. The T'ang classical scholiast Li Shan provides us with an interesting but not philosophically useful pair of Equations: "Great Emptiness means Heaven; Spontaneity means Tao."[39] This suggests something like "outer space"—perhaps beyond the illusory sky dome. Some such interpretation is reinforced by the equivocal couplets of Liu Tsung-yüan:

> Sun and Moon—to what are they attached?
>   The array of stars—on what are they displayed?
> The Circumscribed Furnace—the White-souled Vortex—
>   Grand Emptiness is what they are attached to!
> The Myriad Dazzlers on the expanse of the Board-game—
>   Every one of them relies on it.[40]

Unless the expression "Grand Emptiness" is being used paradoxically, it can only be taken to be the matrix in which the celestial bodies are imbedded—a kind of circumambient aether. Whatever its true nature, it carried considerable appeal for poets, who often treated it as a kind of generator of weak phenomena in the upper air—of nebulosities that come close to the limits of perception. Meng Hao-jan, for instance, wrote how "Grand Emptiness gave birth to the lunar nimbus,"[41] while Li Ch'ün-yü was the composer of the following pleasant couplet:

> With footsteps like a flying bird that perches in a high tree;
> With heart resembling idle clouds out in the Grand Emptiness.[42]

Which is to say, walking on air, buoyant of heart—in Seventh Heaven. Otherwise, that high domain may be penetrated by a great Taoist master, such as the Perfected One described by Tu Kuang-t'ing in a fairy fantasy illuminated with the soft colors and clearcut outlines of a painting by Maxfield Parrish. The adept dwells among jade congealed from the clouds and jewel-like magic mushrooms; above all, "His pleasure is to harness the metallic toad and enter the Grand Emptiness."[43] (The metallic toad is, of course, the silver batrachian in the moon.)

The other term that needs some discussion is no such permanent fixture in the sky, but finds a place in many systems of universal genesis, even if it is not prominent in the *T'ai p'ing yü lan* layout. It is the binomial expression *hun-t'un*, which implies both "enveloping" and "integral"—it is a kind of cosmic raviolo or won-ton. (Indeed, the last of these tasty words is a mere dialectal variant of *hun-t'un*, a doughy microcosm.) *Hun-t'un* is the Spherical Totality, the seamless globe of potential being, the primordial egg. At the same time, the word connotes partly a sort of swirling turbulence like the spiral galaxies of the universe we know. The word itself underlies a number of graphic variants,[44] and has at least one important phonetic variant, *hun-lun*. In Taoist literature it has the apocopate and hybrid form *hun i*, "The Integral/Enveloping Monad." In the fable of *Chuang tzu* we have an anthropomorphic Hun-t'un, a dignified deity playing the host between the Gods of the North and South Seas. These graceless goblins despise the seamless monstrosity, and puncture him to provide humanoid orifices—a fatal barbarity as it turn out.[45] For the *Huai nan tzu* it is the primitive unhewn wood, creating but not created;[46] in the *Lun heng* it is characterized as the Unity that preceded the division of the Primary Pneuma;[47] for the *Lieh tzu* it is the stage of cosmogenesis that follows the successive appearances of plasm, form, and substance; the three coexist unsegregated in the integral sphere (here called *hun-lun*).[48] In the Taoist system from which we have already quoted, the era of *Hun-t'un* follows Grand Simplicity (as it does in *Lieh tzu*) as an age of creation marked by the formation of such epiphenomena as mountains and rivers—a far cry from the inchoate ball of mainstream opinion.[49]

The successive stages of creation have no corresponding stages of universal dissolution—much less a cataclysmic Götterdämmerung. Once started, the world keeps going. Obsessed with the processes of time, the Chinese looked eagerly for the evidence of rhythms and progressions within this final stage of cosmic evolution. What they discovered was cycles—and nests of cycles. But there was no glimpse of an "eternal return." The cycles are repeated, but are never identical. It might be said that they are formally the same, but differently colored. Each chimes a different celestial tune, but the tunes are all set to the same mode. Possibly we can think of cosmic history since Grand Culmination as a journey through an infinite piece of neon tubing bent into a sine curve. Amplitude and frequency are constant from alpha to omega, but each phase—the formal duplicate of his predecessor—glows somewhat differently. Hue, saturation, intensity, luster, iridescence, brilliance never recur in identical combinations, even through an infinity of time. A spirit journeying through the endless grotto with its subtly changing glass walls would arrive over and over again at a place where his past and future course were exactly what they had been before, and he would soon learn to predict the direction of his journey from any particular point, since he had traversed identical paths frequently—but he would never be able to guess the *quality* of the stage ahead, however familiar its contours.

One such set of cycles was familiar to many peoples of the world—the one measured out by the precession of the equinoxes, which shows as a gradual shift of the zodiacal position of the sun at the equinoxes. The rate of change is about one sign of the zodiac for each 2,200 years, and the inevitable passage through all twelve brings cosmic time back to zero, marked in many lands as a time of the destruction and regeneration of the world.[50] But although the Chinese were aware of the fact of precession reasonably early, they created no significant doctrine of beginnings and endings from it.

A world-cycle of greater astral significance was that initiated by a conjunction of the Five Planets.[51] Combined with other rare astrological events, such as the addition of sun and moon to the spectacular grouping, and the beginnings of 60-year and 60-day cycles, this juncture marked, in one Han system, the onset of a truly great era, lasting 23,639,040 solar years.[52] Such a spectacular concatenation of celestial cycles was impressive, but it had little to offer common humanity, either for agriculture, or in astrology, or even as food for the imagination. The mind merely boggles at this meshing of numberless cogs in a kind of invisible, icy clockwork.[53]

Eclipse cycles got a fair share of the attention of cosmologists, since the prediction of both solar and lunar eclipses was, for reasons of state policy, a major preoccupation of learned sky-gazers. Unfortunately, until T'ang times at least, their procedures were oversimplified because of the requirements of a neat cosmic numerology. The same was true of the Jupiter cycle, which played a great role in popular astrology.[54]

Taoist philosophers had their own vast cycles of time, to which they applied the Indian and Buddhist name of kalpas (chieh, M.C. kyắp), lasting googols of years. Abstract as they were, they at least formed the skeletal structure of more palpable time phases—the ages of the Primal Pneuma and its successors.

Awesome as these vast temporal frames were, they held little appeal for Chinese writers—or even for most Chinese "thinkers." Literary men preferred a much more primitive system which, so to speak, got one's feet muddy. This was an old mythological topos that finds its *locus classicus* in a tale about the lovely bird-clawed goddess Ma-ku, preserved in the "Biographies of the Divine Transcendents."[55] In it, the lady remarks to a fellow immortal that since they last met she had "seen the Eastern Sea thrice become mulberry fields"—a catastrophe which made it possible to walk dryshod to the holy island of P'eng-lai. In medieval times the phrase "mulberry fields" came to refer not only to topographic but to political convulsions, and from mid-T'ang times it became a metaphor for the great grey-green sea in the east.[56] But above all it became an expression signifying the passage of eons of time, comparable to our geologic periods which we have named "Cretaceous," "Permian," "Carboniferous," and the like. Indeed, at least one T'ang writer, Yen Chen-ch'ing, who lived in the eighth century, made an unqualified connection between the mythic fields and observed geological fact. He observed that rock formations near a Taoist sanctuary on Mount Ma-ku, where the goddess had achieved divine powers, contained the shells of gastropods and oysters. He took these to be relics of "the alteration of the mulberry fields."[57]

"Mulberry fields" was an expression most commonly heard from the lips of the perfected beings of the Tao, who could look with equanimity upon cyclical catastrophes—mere episodes in the lives of beings who have learned the secret of eternal life. This usage is common in poetry:

> He tolerates those mulberry fields that are transformed into the watchet sea;
> A single pill of the cinnabar medicine has fixed a thousand springs for him.

These words tell of a silver-fleshed Taoist adept whose face is as pink as a peach blossom—signs that his physical frame had been profoundly altered to ensure its permanence.[58] The poem from which this couplet was extracted was written by the ninth-century Taoist Lü Yen, now best remembered as one of the so-called "Eight Immortals." He wrote elsewhere along the same lines:

> Concealing himself in the Occult Metropolis he keeps no track of springtimes;
> How many times has the Watchet Sea been transformed into dust?[59]

32

Cosmogony

The periodic inundations of the fields of the sea merely punctuated the changeless existence of the Immortals. Ideally these cycles could be treated as minor esthetic distractions, as spring flowers or autumn leaves appear to ordinary mortals.

To escape from the flux and uncertainties of life in our world was the ambition of Lan Ts'ai-ho. *Lan* means "indigo," and this man, who is said to have lived on earth in T'ang times (if he ever really existed), was named for his blue-dyed shirt. He sang and begged in markets and other public places, hopping about like a madman. Eventually this holy innocent achieved his dream and leaped up into the sky. He, like Lü Yen, is now numbered among the "Eight Immortals," and celebrated in modern folklore and popular art. A single poem is attributed to him: it is his vision of Eternity.

> Stamp and sing, stamp and sing, Lan Ts'ai-ho!
>   What spell—what space you can!
>   A ruddy face—a third-spring tree,
>   The year slips by with one toss of the shuttle!
> The men of old have gone pellmell, never to return;
> While new men come flocking in ever greater numbers.
> But to mount a simurgh—a phoenix at morning—go off into
>   the cyan depths;
> See mulberry tracts at evening—generating whitening waves.
> The enduring sky-lights, luminous flashing—off at the ends of the void;
> Then: gold and silver palace gates—tall thrusting towers![60]

("Third-spring tree" is a flowering tree in the third month of spring—a metaphor for the finest bloom of youth.)

Po Chü-i used the same term in a versified meditation on the sense of history and transience:

> A great army leader in the days of Han—
>   When young, a boy begging for food.
> Once a ranking peer in the time of Ch'in—
>   When old, he became a gentleman hoeing squash.
>
> How the spring flowers blaze and flash—
>   Peaches and plums burst out in the gardens:
> The autumn wind brings sudden desolation;
>   Up in the hall the brambles and boxthorns grow.
>
> Deep valleys are transformed into banks,
> Mulberry fields become sea water.
>   Things tend away—no need for grief;
>   Times to come—why worth our grief?
> For us who put bloom and decay into words
> All is shifting, vacillation without end.[61]

And the likeness of the firmament
upon the living creature
was as the colour of the terrible crystal,
stretched forth over their heads above.

Ezekiel, I, 22

4 The Sky

The blue vault of heaven was believed in antiquity to be a substantial firmament, as attested by the old myth of the repair work done by Nü Kua, and by numberless poetic allusions. Probably this was the opinion of common men at all times. The philosophers, however, held more subtle opinions about its substance, although they generally agreed that it presented a smooth and concave surface to viewers on earth. Orthodox opinion was that the sky was composed of energetic pneuma (*ch'i*)—something like a gigantic bubble with a skin sufficiently tense to maintain its shape indefinitely, and even to support other entities. Probably the most accepted view was that this filmy firmament had, like the earth, congealed directly from the Primal Pneuma,[1] although there was a rival opinion which held that the celestial dome was generated by a pneuma exhaled by the soil and water of the earth itself.[2] Superficially, it would seem difficult to confute these views, but—expectedly—Wang Ch'ung was the man to try it. He argued that Pneuma could not be the stuff of a Sky to which the stars visibly adhere. (He compares the twenty-eight lunar stations to post-stations adhering to terra firma.) Therefore a tenuous vapor was out of the question.[3]

Wang Ch'ung's skepticism extended also to the apparent shape of the sky. It was universally believed that it was bowl-shaped, just as our senses tell us. Not so says the imperturbable critic, staunchly defying both reason and experience. The sky, he wrote, is as flat as the earth. Its apparent curvature is an illusion—the effect of perspective. The sun does not, as it seems to do, climb up the inner surface towards the zenith and then slide down the opposite slope. The sun appears out of nowhere, becomes visible when it is close enough, and vanishes again in the far distance.[4]

Still, the best minds remained obdurate. For them, the important issue was not the shape of the sky but its extent: was it a simple hemisphere, or was it a complete sphere enveloping the inhabited world both above and below? The former picture was the commonly accepted one, and doubtless the oldest. It followed from naïve observation: "Observing the shape of the sky from below, it is like the canopy of a carriage, and the stars are interlinked like strings of beads."[5] This hypothesis was commonly called the "Canopy Heaven" (kai t'ien) theory and alternatively the "Chou Gnomon" theory.[6] Among three competing doctrines it was the most conservative, but had the greatest influence on language and metaphor.[7] The fourth-century authority Yü Hsi expressed this common-sense point of view thus:

They take it that the sky resembles an overturned bowl, with the Dipper and Pole, presumably, in its center. While the center is high, the four margins fall low, and the sun and moon move sidewise around it. The sun is visible when it is near: this makes it daytime; the sun is invisible when it is far: this makes it night.[8]

A close competitor to this doctrine was the "Enveloping Sky" (hun t'ien) theory, congenial alike to the cosmogonists in that it suggested the primordial egg, and to the professional astronomers in that its network of star-paths could conveniently be imitated with an armillary sphere. The name is related to that of the integral globe (hun-t'un) regarded as a stage in the evolution of the universe. Although often expressed in homely language, this theory was in the main progressive and given respectful attention even by the skeptical. Some put it that the sky was "like a pellet or pill."[9] Others, with a better cosmogonic analogy, compared it to a hen's egg. One such authority adds: "The sky is great and the earth is small. There is water both outside and inside the sky, and the earth is stabilized by the pneuma it receives therefrom; it floats buoyed up by the water. The rotation of the sky

35

is like the movement of a wheel on its axle, and water is the envelope of both sky and earth."[10] It did not follow that the earth, too, was spherical. As Yü Hsi put it, "They take it that the earth is at its center, with the sky going round outside. Sun and moon begin by climbing up the sky, and later go into the earth. Daytime, then, is when the sun is above the earth; night is when the sun goes into the earth."[11] Yü Hsi himself had doubts about this doctrine: "I would say, in my stupidity, that if we must have the sky swaddling the earth, like an eggshell enclosing the yolk, then the earth must be an object within the sky. Why then did our master minds distinguish them with separate names?"[12] It remained for Yang Chiung, during the reign of Li Chih (T'ang Kao Tsung) in the seventh century, to attempt a serious reconciliation between the astronomers and the sages. He became a collator of texts at the court, but his heart was among the stars, and he spent his days and nights at the Estrade of Numina studying the figures on the bronze armillary sphere—doubtless the same one that had crowned the work of Li Ch'un-feng. Later in life, after reflecting on the various proposals made to account for the sky, he wrote the celebrated "Rhapsody on the Enveloping Sky." In this elegant composition he lays before us a grand panorama of all the powerful asterisms and describes their orderly and responsible movements, with the great aim of reconciling "enveloping sky" theory with the orthodox systems of morals, divination, and religion.[13] It does not appear that he had great success, although no doubt some more advanced astronomical thinkers thought well of his effort.

I know of no systematic Chinese effort to explain the color of the sky, and must perforce be content with the opinion expressed in the ancient book of *Chuang tzu* which holds that the sky's blue (*ts'ang-ts'ang*, which connotes a pallid shade of blue) is caused by extreme distance.[14] This reminds us of Wang Ch'ung's explanation of the appearance and disappearance of the sun. Doubtless the phenomenon was regarded as similar to the blue cast of distant mountains.

The distance of the sky, on the other hand, was a riddle that engaged many minds and induced as many estimates, ranging from the ten billions of *li* in the *Huai nan tzu*,[15] down to the "more than sixty thousand" of Wang Ch'ung.[16] In the ninth century, Tuan Ch'eng-shih gave an approximation, based on Taoist sources, of 409,000 *li*.[17] Figures provided by the Han dynasty "wefts" on the classical books average about 178,500 *li*, with a corresponding circumference of over a million *li*.[18]

Whether spherical or hemispherical, the star-studded sky plainly rotated above the earth around the polar axis, its motion being to the left, that is, westward. Careful observers noted that the sun and moon do not rotate at the same speed as the sky, the difference in the case of the moon being especially conspicuous since it slips back through a full circuit in twenty-eight days. Mathematical astronomers described this as a contrary motion to the right (east). Wang Ch'ung, used the analogy of a millstone, with ants loosely "attached" to its surface but walking in a direction opposite to the motion of the wheel. Similarly, he thought, the sun and moon do not adhere tightly enough to the surface of the sky to be inhibited in their perverse march eastward.[19]

The physical constitution of the firmament, however, mattered much less to the Chinese than its spiritual component, which Wang Ch'ung minimized. Debate was endless on the question of heaven as a source of moral authority, example, and judgment—matters that had been bruited for centuries on every level, ranging from the interpretation of omens as signifying the responses of heaven to events on earth, or the ancient concept of heaven as a dynastic ancestor, to generalized philosophical opinion about heaven as an impersonal source of both life and ethical standards. Wang Ch'ung, who always had something to say on such matters, took the simple organic view that heaven in motion generates vital energy ("pneuma," ch'i), which induces the birth of all creatures, but that it does it blindly and automatically, without a purpose (the "principle" of Spontaneity). For him there was no science of teleology.[20]

Of particular interest in the eternal, fretful arguments about the nature of Heaven is a set of T'ang dynasty documents in which are recorded the opinion of three intimates of considerable intellectual acumen and literary talent. The men were Han Yü, Liu Tsung-yüan, and Liu Yü-hsi. The reader is invited to make a private study of the contributions of all three men to the debate, but here, since my purpose is not philosophical, I will limit myself to a few remarks on the essay of Liu Yü-hsi, entitled "Discourses of Heaven" (T'ien lun). This was written as a critical amplification of an earlier essay on the same subject by his friend Liu Tsung-yüan, with whom Liu Yü-hsi disagrees on a number of points. He is chiefly concerned with moral-spiritual matters—for instance, with the relations between heaven and mankind on a more elevated plane than the merely astrological one. He holds that the sky is no mere aggregation of abstract power, displayed in the stars, but is immediately concerned with the role of man in nature and his proper spiritual development. T'ien ("heaven"), it seems to me, is here presented as a kind of pantheistic entity, much like "Heaven" in some late Christian literature.[21]

Speculation on what lay behind the shell of the sky seems not to have attracted many early Chinese minds. The Taoists were exceptions. They had a whole hierarchy of celestial realms. In one of their formulations, the outermost realm, beyond the triad of Clarities (san ch'ing, namely Great

Scarlet [*Ta ch'ih*], Yü's Leavings [*Yü yü*], and Clear Tenuity [*Ch'ing wei*]), where all polarization is obliterated, is a kind of Empyrean called "Great Envelope" (*Ta lo*).[22] But here we begin to depart from physical reality and enter unlocalized realms known only to the mystic. On the other hand, the expression "beyond the sky" (*t'ien wai*) is common enough in poetry, and in some contexts suggests that the dome is transparent with the stars shining through, like candles through the glass shell of a lantern:

> Outside the sky the Luminous Ho throws up waves of jade.[23]

This verse, by the remarkable lady named "Flower Stamen" (Hua-jui fu-jen), shows us the candent sky river tossing crystalline white-caps on the far side of the tense film of heaven. Presumably only the very exceptional mortal could expect to trim his sails to meet the gusts of ether in this supercelestial space—but we shall see that it sometimes happened.

Nevertheless, conjecture of a serious sort on these matters was not entirely barren. It produced a third system, often mentioned along with the "Canopy Sky" and "Integral Sky" theories as a radical, minority opinion. It is the doctrine of "Unrestricted Night" (*hsüan yeh*).[24] The antiquity of this cosmological system is unknown, but it was a dying tradition, its books and masters lost beyond retrieval by the end of the Han dynasty, and it apparently lacked important advocates in the fourth Christian century.[25] Nonetheless the incredible notion of infinite space seems to have survived as a kind of eccentric vision, worth discussing in professional circles as an interesting old hypothesis but one without practical application. The cosmos of "Unrestricted Night" was very like the one produced by modern European astronomy: "The School of Unrestricted Night avers that heaven is quite insubstantial, and that the sun, moon, and the host of stars float freely in the midst of empty space."[26] In short, what men see above them is not a stretch of dark firmament but a sky that "is high and distant without limit."[27] In T'ang times Yang Chiung put the doctrine in the mouth of a rather heavy-hearted spokesman (who got laughter for his pains) in this form:

> Sun and moon are held up by the Primal Pneuma
>   so that they are sometimes centered, sometimes aslant;
> Starry chronograms float in the Grand Void
>   so that they now have motion, now have rest.[28]

In T'ang times, then, the faint message was not unheard, but was not taken seriously by established astronomers, who could not relate their computations to it, or make bronze models of it. The poets and the Taoist theologians kept it somewhat in mind. But for most men a taut blue canopy remained posed over earth and ocean.

The Sky

We occidentals have long recognized an affinity between some blue minerals and the blue vault of the sky—a kinship petrified in such modern words as azurite, lazurite, lazulite, and celestite (none of which would have puzzled the early Chinese), to say nothing of a black manganese spinel named galaxite (which might have troubled them). Is the sky indeed made of the same material as the earth? Pablo Neruda has gone so far as to assert that the great firmament was derived from the dry land beneath:

> De endurecer la tierra
> se encargaron las piedras:
> pronto
> tuvieron alas.[29]

Evidently the sky in its turn needed strengthening. The ancient Chinese, in the remote age when the gods still trod the land, discovered too late that this was indeed the case. The great sky arch—which was also a rainbow dragon, petrified and re-cast as it were[30]—was fractured and had to be repaired by the lady Nü Kua, the cosmic engineer and impulsive shaper of human beings. The story, as it was widely credited in Han times, went as follows:

Nü Kua smelted and refined the five colors of stones and used them to patch the gray-blue sky; she cut off the legs of the turtle-leviathan and stood them in the Four Extremities. But the sky was deficient in the northwest—and so the sun and moon are displaced thither; and the earth was deficient in the southeast—and so the hundred rivers pour out there.[31]

It is not clear whether the five standard colors fused to yield the white light of day, or—as seems more probable—blue stone dominated the melt. The former view is consonant with the opinion that the heavens are crystalline—and perhaps also permeable—a view that reappears from time to time in Chinese literature (sometimes metaphorically), as in this verse by Chang Chiu-ling: "The frost falls from the crystal of Heaven's eaves."[32] The latter agrees with the primitive description of the sky as *ch'ing*—a word that was also used nominally for the blue pigment which we aptly style "azurite," as in the phrase *shih ch'ing* "stony ching" ("ching" being an English color word, as the second edition of *Webster's New International Dictionary* testifies), the common Chinese expression for the natural blue crystals of the mineral. With this we can compare *ch'ing t'ien* "blue sky," or perhaps in early times "azurite sky." A lapis lazuli sky studded with pyrite stars would serve even better—as it has in many lands and cultures. But lacking sure knowledge of the Chinese word for lapis lazuli in many periods, and believing that mineral to have been rare in the Far East anyhow, we cannot affirm with certainty that the early Chinese entertained the conception of a sky composed of the best lazurite, studded with glints of fool's gold. From Sung times, at any rate, the Chinese knew a "gold star stone," deep indigo in color and brilliantly spotted, which they imported from Khotan.[33] This was certainly authentic lapis lazuli from the Kokcha valley of Badakhshan, where it is found in limestone, associated with balas rubies (red spinel), and served as a source of sky symbolism for the Sassanian Persians. (It is reported that the baldaquin over the throne of Khusrō II was constructed of lapis, with the stars and planets represented in pure gold.)[34] It is even possible that this pleasing sky-material was occasionally mined in China. In the eleventh century, at least, both "gold star stone" and "silver star stone" were produced there—in Anhwei (Hao-chou) and Shansi (Ping-chou).[35] During the same period the stars left their traces on plants as well as stones, and we may read in the Sung materia medica of a "gold star herb"—also styled "seven star (= planet) herb"—a kind of maidenhair fern (*Adianthum sp.*).[36]

This leads us to the suspicion that the five colors of Nü Kua's original mix were not fused, but segregated, and retained their separate identities manifest in the several planets. Were not the stars and planets stony? And did not each of the five visible planets (we must except the anomalous Rahu and Ketu) have its distinct color—white Venus, yellow Saturn, red Mars, and so on. These planetary colors were not merely symbolic; they were also true and natural. An example is Saturn, which manifested itself as a magical old man at the beginning of the Han dynasty. This personage was called "Lord of the Yellow Stone," and his name was explained quite simply: "Yellow is the color of Queller Star [Saturn]; stone is the stuff of stars."[37] If classical norms were strictly adhered to, we should expect the mineralogical composition of the planets to be as follows: Jupiter of azurite, Mars of cinnabar, Saturn of orpiment (or perhaps yellow ochre), Venus of ceruse, and Mercury (doubtless burned to a crisp by the proximity of the sun) of carbon black. I must confess I have never encountered such a tabulation—but the equivalences are normal, and must have been pondered by someone.

These arcane and almost alchemical correspondences apart, the stars and planets were commonly thought to be made of the same elemental substance as precious stones—or, to some writers, to seem to be. Next to "crystal" (with its strong linguistic associations with "germinal essence") perhaps, jade is most common—simply because it is most apt. Jade connotes "white, pure, translucent, mineral, noble, refined, magical" and so on. All of these flattering epithets belong equally to the stars. Accordingly Meng Chiao, who liked to write such lines as:

> Green water condenses as green jade,
> White waves give birth to white wands . . .

could also describe the spray thrown up by a boatman's pole as "white jade."[38] Here we have a clear case of metaphor. But when he and other poets wrote of the actual stars as if they were jade or some other shining mineral, it is not so easy to speak of "metaphor." What of Ku K'uang's verse: "The two phosphors like jade discs—the five stars like pearls"?[39] Here are clearcut similes: the "two phosphors" (*erh ching*) are the sun and the moon; the "five stars" are the naked-eye planets. But let us suppose that he had written "the two phosphors are jade discs, the five stars are pearls"—and indeed such statements are common enough in Chinese literature. Could we say with assurance that these are only pretty metaphors? Perhaps we should take the poets seriously. All men knew that transparent quartz was petrified ice. The moon was literally akin to the pearl. Where can the line be drawn between the jade deposited in veins on earth and the superior jade arrayed in the sky? The distinction between metaphor—that is, imaginative kinship—and true physical or metaphysical affinity is blurred. But let us accept a pusillanimous compromise, that the sky and its components and ornaments have *something* in common with precious stones and metals, with all that that implies. Accept that—for a literary man at any rate—they are more crystalline than plastic, that they differ from plants and animals in being inorganic, and from foods and drugs in being non-reactive, and so on. This much conceded, we can proceed to accept with some equanimity—coupled with understanding —that the watery Ho River below "connects once more with the silvery Ho above,"[40] and even that the "jade fairy" rises[41] in much the same way that Phoebus rises, while personifying the moon in quite a different way than that personage represents the sun.

"Doubt thou the stars are fire."
—Shakespeare, *Hamlet*, II, 2

5 The Stars

A star is a magic light in the sky. To the ancients a star, like a man, had both a spiritual and a physical nature, either of which might manifest itself at different times in different manners and degrees. Disoriented, dismayed, or disabled, it might change its color, its shape, or its brilliance. Fallen to earth—a cosmic disaster—it might take the form of a magical stone. To the men of ancient China none of these things were very surprising. But they found it troublesome to construct an acceptable definition of the true inwardness of the stellar substance. A kind of average view taken from random textual samples would suggest that the Primal Pneuma produced both the stars (*hsiang* "simulacra") and the creatures (*hsing* "forms") simultaneously. Each was the counterpart and intimate companion of the other, although in individual samples it sometimes appears that earthly creatures are emana-

42

tions from the stars.[1] In any case, they are interdependent. The stars are said to be *ching* (embryonic, germinal essences) as often as they are said to be *ch'i* (energizing breath). Consider the following: "The stars are the 'finest bloom' (*ying*) of the Primal Pneuma; they are the embryonic state (*ching*) of Water."[2] I take this to mean that they are at once an original product of the generative stuff of nature—the potentiality of which water is the concrete actualization. Similarly, it was said that "The stars are *formally* generated in the earth and *germinally* formed in the sky."[3] Wang Ch'ung put it that the origin of the sun and the moon was precisely the same as that of the stars: indeed that sun and moon were "akin to stars." However, they are at the same time the "embryonic essence" of the Myriad Creatures of the universe.[4]

Although some claimed that the stars were generated along with the sun and moon in the cosmic mists, in other texts they have a secondary nature. So it is in *Huai nan tzu*, where we read:

The hot pneuma of accumulated *yang* generates fire, and the embryonic essence of the fiery pneuma makes the sun; the cold pneuma of accumulated *yin* makes water, and the embryonic essence of the watery pneuma makes the moon; the embryonic essence left over from forming sun and moon makes the starry chronograms.[5]

Another Han source puts it this way: "When at its fullest growth, the essence of *yang* makes the sun, and the sun makes the stars by division. Therefore in forming the graph [representing *hsing* 'star'] one sets a 'sun' on 'give birth' to make 'star.' "[6] That is, graphically "star" represents "sun-generated."

The book of *Lieh tzu* raises the question, "Why do the stars not fall from the sky?" And the answer is, "They are merely shining concentrations of 'pneuma'," and so imponderable.[7] (Often the stars collectively are said to be "pneumatic dispersions" of "Metal," taking the latter to be one of the five energetic phases.[8] The same was said of the Sky River, which also has bilateral relations with both metal and water. The material evidence for this connection [other than the symbolic criterion of "whiteness"] is presumably the incontrovertible criterion of iron-nickel meteorites.)

None of these views, however much they differ in emphasis, is incompatible with my initial formulation: the celestial phosphors emerged from the primordial gas as brilliant condensations, while retaining intimate connections with the more solid and opaque emergents below on earth. This scheme underlies the significant doctrine of "counterparts" or "simulacra" (*hsiang*), which constitutes the theoretical foundation of astrology.[9]

The problem of the actual size of the stars exercised the imagination of the sages but little. We must assume that they generally took them to be little flickering points of light—fiery atoms. However, Wang Ch'ung, as contrary-minded as always, thought them to be more bulky than this: "If we figure the physical nature of the several stars, we get a hundred *li*."[10] A Chinese Buddhist source is rather more specific: "The largest stars are 700 *li* in circumference; the middle-sized stars are 480 *li*; the smallest stars are twenty *li*."[11] However, statements like these are exceptional.

The celestial simulacra were not all single stars. The majority were concatenations—quincunxes, girandoles, rosettes, and chains. We may think of each simulacrum as a distinct molecule in which the stars are atoms. (Some of these celestial molecules consist—like those of the diamond—of a single atom, in this being unlike our western constellations.) A very large number of these figurations are named for offices, ranks, incumbencies, and responsibilities of a high order, in particular those closely connected with the royal palace on earth. These asterisms, collectively, seem bureaucratic with a vengeance, and clearly do not carry the mythological burden that our own do. But in fact they are not so much a detached celestial hierarchy reigning above the world, or deviously directing the activities of their terrestrial counterparts, but a set of numinous but impersonal powers, the energetic equivalents on the fully numinous level of the political and social powers exercised by the earthly offices to which they correspond and with which they are in resonance. In popular lore, of course, they were degraded to imposing jacks-in-office puttering pompously about the fields of heaven. For the Taoists the vestments they assume are merely masks and disguises reserved for particular theophanies as token robes of office serving chiefly in lieu of identification badges, like St. Catherine identified by her wheel, or St. Mark in the form of a lion.

The most powerful constellations were concentrated in the region of the north pole, where they swing about their fixed focus—a grand procession visible at all seasons of the year. The focus itself is not plainly marked, but it is not far from our Polaris (α Ursae minoris). But during the first millenium of our era it was much closer to another star, which must be counted as the true pole star of T'ang times. This is a rather inconspicuous one at the northern tip of an asterism then styled "Northern Culmen" (*Pei chi*), which extends upward from the two stars marking the outer edge of the cup of our Little Dipper—that is, pole-wise from γ and β Ursae minoris. This star, which we know as Σ 1694 of Camelopardalis, or the invisible pole nearby, was called simply "Knot" or "Nexus" (*niu*), a very impersonal title for such a significant star, especially when compared with the grandiloquent labels attached to its more brilliant but less strategically placed neighbors. This was the "Pivot of Heaven" (*T'ien chih shu*) where the Twelve Mainstays of

44

the sky were tied together. (The name "Pivot of Heaven" was also affixed to Dubhe, the lucida of the Great Dipper, quite remote from the pole.)[12] Σ 1694 *was* the true pivot, but its *name* is "Knot" and we shall reserve the name "Sky Pivot" for Dubhe.

The brightest and most important star of the Northern Culmen was reddish Kochab (β Ursae minoris) which was the visible aspect of the Theocrat (*Ti*), or Grand Monad (*T'ai i*). He was one of the two greatest of the polar deities—the other being that represented by Polaris.[13] He "contains the Prime, emits the Pneuma, lets flow the Embryonic Essence, generates the Monad."[14] In a similar formulation he is said to "contain the Primal Pneuma and distribute by means of the Dipper."[15] The official astronomical books, kept free from the contamination of anthropoid deities, fail to give any descriptions of this mystic potentate, but a Taoist source, encumbered by no such ideological handicaps, informs us that "the Star of North Culmen" (which refers to the Theocrat rather than the Knot) has a circumference of 770 *li*: "In its midst is an Occult Terrace with a high building of jade." The presiding deity there is clad in purple damask.[16] The astrological significance of the star is plain: "If the Culmen star disappears, in the eighth year the Middle Kingdom will lack a lord and king, and all Under Heaven will be in disorder."[17] There is some evidence that the chief polar asterism played a significant role in ancient Chinese city planning. It seems probable that the palace of Ch'in at Khumdan (Hsien-yang) represented the true Pole and Knot of the universe, and was placed relative to the Wei River and the rest of the capital as that pivot was to the Star River and the other polar asterisms.[18] (See figure 1.)

Other stars in the Culmen include the Grand Heir (*T'ai tzu*; Pherkad; γ Ursae minoris), The Princes (*Shu tzu*; Flamsteed 4), and the Heirgiver's Palace (*Hou kung*; Flamsteed 5).

The most magnificent of northern star spirits, however, was our Polaris —although in T'ang times it was further from the Pole than now—which we also call Yilduz (Turkish *yıldız* "the Star"), Stella Maris, Lodestar, and which was the Scip-steorra of the Saxons. It is actually a double star whose major member is the color of yellow topaz.[19] The Chinese styled it "Heaven's Illustrious Great Theocrat" (*T'ien huang ta ti*) and "Radiant Moon-soul Gem" (*Yao p'o pao*).[20] Sometimes it is "Northern Chronogram" (*Pei ch'en*). Although it marks the northern tip of the asterism "Hooked Array" (*Kou ch'en*), it has close affinities with the neighboring asterism "Northern Culmen," of which it is an emanation[21]—a projection of the insignificant Knot which occupied the true mathematical pole. In the orthodox view this magnate among stars "governs the numerous spirits and holds the chart of

45

The Polar Asterisms

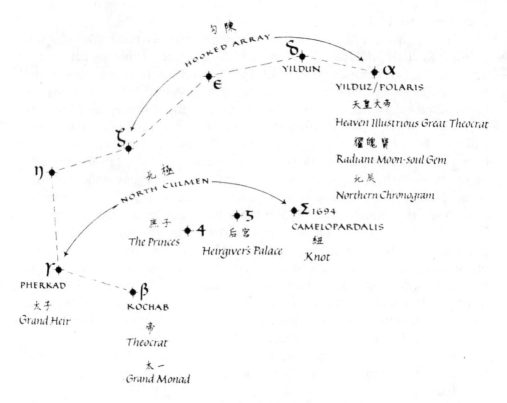

FIGURE 1.

the ten thousand deities."[22] The mighty monarch rides the carriage of the Great Dipper, which was also the Wain, in the shadow of the parasol of Cassiopeia, as in this Taoist description:

Deep under the lower tip of the Floriate Canopy, bright and glowing, is the immobile star—the Northern Chronogram. This is the separated germinal essence of the Lord of the Great Way, the Mystic Illustrious one of the Adytum of Jade, the true star of the Jade Theocrat. Shaded by the Floriate Canopy he rides in the Northern Wain as he journeys through the Nine Heavens.[23]

The chief stars of Hooked Array, which correspond to the stars of the handle of our Little Dipper below Yilduz, are greenish Yildun (δ Ursae minoris) and its two neighbors (ε and ζ Ursae minoris). They constitute the "regular residence" of the Great Theocrat, and also are the home of his True Consort.[24] This meaning may be the one alluded to in a poem by Li Shang-

46

yin on the women's palace of the ruler of Ch'en. It shows the sovereign drifting on the lake in his new pleasure park in his dragon boat:

> The lotus by the holm pays homage to the Orthodox Conveyance;
> The birds from the sand transgress on the Hooked Array.[25]

Here the asterism "Hooked Array"—parallel in the poem to the autocrat's vessel—plainly symbolizes the queen, the "True Consort," whose alter-ego hangs below the polestar.

These Dominations and Powers were domiciled within a great protective wall, whose penetration by some such monstrous apparition as a comet forboded ill both in Heaven and on Earth. Wang Ch'ung wrote of it soberly enough:

The location in the sky of a Sky Divinity is just like the residence of a king. The royal one resides within a double barrier, and so the divinity of the sky is suitably placed within a hidden and secret place. As the royal person resides within the buildings of a palace, so the sky too has its Grand Palace of Purple Tenuity.[26]

Wang Ch'ung made this statement not so much to glorify the gods of heaven as to assert that, secluded as they are in their remote enclosures high above the earth, it was unlikely, as commonly believed, that they could hear anything about the misdeeds of mortals, or even care about such distant and insignificant creatures. At any rate, the "Purple Palace," as it was sometimes called for short, was the residence of Heaven's Illustrious Great Theocrat, who reveals himself as Yilduz, which is itself, like him, the secret embryo and quintessence of the god Grand Monad, revealed as Kochab.[27] Surrounding both of them and their attendants, was the Wall of Purple Tenuity, a circumpolar constellation of about 15° radius, largely composed of the stars of Draco. The Chinese saw it divided into two sections. One of these was the Eastern Boundary (*Tung fan*), a crescent-shaped hedge which includes ι (Ed Asich), θ, η, ζ, φ Draconis; χ, γ Cephei; and 21 Cassiopeiae. The Western Boundary (*Hsi fan*) was another lunette comprising α (Thuban), χ, λ (Giansar) Draconis; minor stars of Camelopardalis; insignificant stars in Ursa Major; and others.

A ceremonial gateway, called Ch'ang-ho (*ch'ang-ghap*) provided orthodox admission to exalted spirits. This name has been etymologized thus: "*Ch'ang* is 'precentor'; *ghap* is 'hide away'. This means that the *yang*-pneuma directs the Myriad Creatures, but shuts out the Yellow Springs."[28] That is: living beings may be guided through but crasser entities from the stygian founts beneath the earth are barred. The great gate may be discovered north of Boötes, going towards Kochab and the pole beyond; its Left Pivot is orange Ed Asich, and its Right Pivot is pale yellow Thuban.

Unlike our Little Dipper, unobserved by the Chinese, who saw Hooked

Array and North Culmen instead, our Big Dipper maintained its integrity in the Far East. We know it also as Great Bear, whose Greek name *Arktos* is part of our vocabulary of science, the Ladle, the Plough, and (in southern France) the *Casserole* (saucepan). It is also the Carriage or Wain. The origins of all of these names are lost in the mists of time—doubtless all of them stem from folk nomenclature, transmitted from oral to written literature.[29] To the Chinese the constellation was primarily a kind of dipper (*tou*), but it was also a carriage, as it was for the Greeks. "The Dipper is the carriage of the Theocrat which courses round the very center, looking down upon and regulating the Four Countrysides."[30] The royal carriage in the sky is not unknown to the imaginative literature of the T'ang period. For instance, a whole rhapsodic *fu* entitled "The Dipper is the Theocrat's Carriage" is still extant—from the pen of Po Hsing-chien, an author better known as a writer of prose fiction, such as the "Story of Li Wa."[31] The same designation is given to the Dipper in an anecdote about I-hsing, the great Buddhist astronomer. The Astronomer Royal had reported to Li Lung-chi that the Dipper had vanished from the sky, and the alarmed monarch summoned I-hsing in the hope that he might devise some way of averting the omen. "In the time of Later Wei," said that great man, "the Dazzling Deluder [Mars] was lost. Now, however, it is the Theocratic Carriage that has disappeared. There has been no such occurrence in antiquity. Heaven is sending a great monition to your Enthroned Eminence!" The monk went on to advise the performance of a great act of charity to demonstrate the character of the regime—in particular, a general amnesty. And so it was decreed. "Later that evening the Grand Notary reported that one star of the Northern Dipper had appeared. In seven days all were restored." Tuan Ch'eng-shih, the author of our source, concludes with these words: "Ch'eng-shih takes this affair to be somewhat grotesque. However it has been widely disseminated by the mouths of the masses, and so I could not do other than set it down."[32]

But it was the dipper image, rather than the carriage, that held first place in poetry. A good example is the astrological comment on recent history written by Li Po soon before his death in 762:

Fate-shift is quick: Heaven and Earth are locked up;
Barbarian wind forms flying frost.
The hundred plants perish in the winter month;
The six dragons sprawl in the western waste.
Great White comes out of the eastern quarter;
Sweeping Star elevates its germinal light.
Mandarin ducks are not Yüeh birds;
Why are they set on flying south?
Only in antiquity did hawks lead hounds;
Now it is liegemen that take their lord.

Taking to water they become krakens and dragons;
Contending on pools they carry off phoenixes.
The Northern Dipper will not regale us with wine;
The Southern Winnower tosses up emptily.[33]

Clearly a commentary is called for:

1. The poem refers to the rebellion of An Lu-shan and its aftermath; the natural order is in disarray.
2. The wind of the power of An Lu-shan's army—with its nomadic horsemen and transport camels—sweeps through China, bringing the cold of northern steppes and forests.
3. So the flowers of T'ang, natural and spiritual, perish.
4. The dragon-horses that draw the carriage of the Son of Heaven languish in Szechwan, where Hsüan Tsung lives in ignominious exile.
5. "Great White" is the planet Venus (and Li Po's own name); from it emanate white comets.
6. The ominous comet rises high, a portent of war.
7. Mandarin ducks symbolize the good men of north China; the birds of Yüeh are the exotic wildfowl of the south.
8. They have been forced to flee into those strange and inhospitable lands.
9. In the good old days, true fighting leaders led the hunt.
10. Now the sovereign is at the mercy of his currish vassals.
11. The warlords show themselves in their true guise.
12. They carry off the wealth and beauty of T'ang as plunder.
13. The Dipper does not function to strengthen the royal party as it should.
14. The "Winnower" (asterism in Sagittarius) throws us no nourishing grain.

The comet referred to here is probably the one first reported on 16 May 760; it reappeared on 20 May, and gradually disappeared.[34] Abdicated Hsüan Tsung was still alive in that year, but subject to many indignities. Civil war still raged through the land. Tanguts and Tibetans harassed the frontier. Gloom hung over all—and the glare of the comet.

In a grand sense the Dipper presided over the welfare of the state and its sovereign. For instance, it showed forth symptoms of cosmic harmony when it was in the best resonance with the ruler: "When the filial conduct of the kingly one spills over, the Dipper lets its germinal essence fall: the seven stars of the Northern Dipper are bright and luminous as if about to fall."[35]

If anything, the asterism was even more significant to the Taoists than it was for conservative officialdom—for the Taoists it was the greatest of all celestial sources of power. Whole "sutras" in the Taoist canon are devoted to the study of methods of comprehending the esoteric meaning of the Dipper and its components, of learning to project one's secret self into it, of realizing it within one's innermost anatomical chambers, of conjuring it to inspire, to protect, to outlaw, to perform miracles. It is an active agent; its bowl will cover your head like an apotropaic shield, proof against the plague.[36] It is a

49

passive repository; if an adept's jade tablets of ordination are on file there, he belongs among the "Superior Transcendents."[37]

But it is possible that the power of each of the stars that make up the Dipper was as great as the composite power of the whole asterism. Certainly each had its separate name and individual identity, so that the Dipper can be regarded as much as a set of seven (or eight, or nine) independent luminous agencies as a single one. The commoner names for these divine apparitions are shown in table 1. There were still others.[38] The "Taoist" names there— one set among many provided in the canon—were regarded as appropriate to the noumenal entities whose "common" names referred properly to their epiphanies; that is, the supernatural power "*Yang* Luminosity" reveals itself visually as "Heaven's Pivot," and so on. Three of these major stars (numbers 1, 2, and 7) were classified as "cloud-soul/actualizing spirits" (*hun shen*); the remaining four (numbers 3, 4, 5, and 6) are styled "white-soul/embryonic essences" (*p'o ching*). "Grotto Luminosity," the little visual companion of Mizar, is the "cloud-soul embryonic-essence, *yang* luminosity" (*hun ching yang ming*) of the secret but mighty "Sustainer" (*fu*); the "Hidden Prime," a dark star or black hole, is the "cloud-soul luminous empty-space numen" (*hun ming k'ung ling*) of the disembodied power "Straightener" (*pi*).[39]

Each of the stars of the Dipper had, in the old oracle texts of Han—that is, the "weft" texts which were still read in T'ang times—a unique earthly manifestation under which it could be recognized. For example, "Jade Transverse" might show itself as a barnyard fowl, as a horned owl, as a hare, as a rat, as a plum-tree, as fagara, as a thorn-bush, as an elm, as sweet flag, and so on, showing a marked preference for the vegetable kingdom. "Gemmy Light" might reveal itself in an elephant, an elaphure, a crow, a small passerine, a swallow, a magpie, a goshawk—a very birdy star this one—and even as a turtle or a piece of ginseng.[40] This last-named star, the brightest in the Dipper, was honored more than its neighbors in Chinese mythology, in palace nomenclature, and in poetry generally. For example, the divine mother named Female Pivot (*nü shu*), as if she were a female version of the polestar, conceived the prehistoric king or demigod Chuan-hsü when she saw the star "Gemmy Light" piercing the moon like a rainbow.[41] Clearly she had been fertilized by the god of Alkaid, and her royal son was the imp and scion of that star. In T'ang times a basilica named "Gemmy Light," built under the rulers of Sui, still stood on an island in the Lake of the Nine Continents at the Lo-yang palace in the years of the Empress Wu,[42] and Hsüan Tsung's beautiful winter palace at the thermae of Mount Li was ornamented by a high building also called "Gemmy Light."[43] That bright shiner plainly had a great appeal, doubtless of astrological origin, for royalty.

50

Table 1
Stars of the Northern Dipper

Name	Ursae Majoris	Color	Magnitude	Standard Chinese Name	Taoist Name
1 Dubhe	α	yellow	2	Heaven's Pivot 天樞	Yang Luminosity 陽明
2 Merak	β	greenish white	2.5	Heaven's Jade-cog 天璿	Yin Embryo 陰精
3 Phecda	γ	topaz yellow	2.5	Heaven's Armil 天璣	Perfected Person 真人
4 Megrez	δ	pale yellow	3.6	Heaven's Beam 天權	Occult Tenebrity 玄冥
5 Alioth	ϵ	white	2.1	Jade Transverse 玉衡	Cinnabar Prime 丹元
6 Mizar	ζ	Brilliant white and pale emerald	2.1 and 4.2	Disclosed Yang 開陽	North Culmen 北極
7 Alkaid	η	brilliant white	1.9	Wavering Light or Gemmy Light 搖(瑤)光	Heaven's Bar 天關
8 Alcor	80		4.8	Grotto Luminosity 洞明	Sustainer 輔
9			0	Hidden Prime 隱元	Straightener 弼

NOTE: The Taoists regard the stars we see (and as named under *Standard Chinese Name*) as the epiphanous souls of the true spiritual beings (named under *Taoist Names*) which are invisible to laymen. Note that they give the name North Culmen, which in the standard nomenclature is a distinct asterism, to the god of the star Mizar. The secret stars "Sustainer" and "Straightener" will be discussed later.

REFERENCES: CCYTSC, 1a–1b; CS, 11, 1102c; *Huang lao ching* (YCCC), 24, 7a–7b; Needham, 1958, 232–233; Allen, 1963, 437–441; Schlegel, 1967, 502–503.

The star "Far-flight" (*chao-yao*, a third magnitude star in Boötes) was sometimes counted as one of the Nine stars of the Dipper, making a longer handle. "Far-flight" was master of the nomadic warriors of the north, and when it shot forth spears of light, men could count on imminent warfare.[44] We have a poem by the mid-eighth century poet Ch'u Kuang-hsi, one of a set of five about the new winter palace built by Li Lung-chi at the hotsprings on wooded Mount Li, in which this important star hangs ominously over the semi-divine court of the Taoist monarch, suggesting ambiguously both royal power and danger from the north:

> His Highness' Forest—and the palace of a divine lord;
> This place is surely a court of illuminati.
> The mountain opens up—colored by auroral light;
> But the sky turns round with the "Far-flight" Star.
> "Three snows are compensated by great possessions":
> Who will hold that this is not our Numen?[45]

The first three verses of this poem reflect the lovely distant days when Hsüan Tsung appeared before his worshipful court at the Floriate Clear Palace, like the epiphany of a Taoist deity among a host of demigods. The palace itself is wrapped in rose-colored cotton wool, at once signifying the divinity of its pneumas and shielding it from the horrors of the real world. But time passes; the imperial war-star swings high. Three snowy winters, according to an old tradition, bring a great crop yield: we assume that all will be well, and the star is our star, ruler of the sky. But the fate of the empire is poised at an apex—as the frontier peoples muster their strength.

Another imperial constellation was the Palace of Grand Tenuity (*T'ai wei kung*), whose name is similar to that of the potent Palace of Purple Tenuity at the north pole. This was the sovereign's southern palace, in the antipodes of the northern one—not factually but symbolically, since it was located on the ecliptic in Virgo and Leo, with the center of its great encompassing wall exactly at the autumnal equinox. This duplicate of the polar palace fell within the equatorial domain of the Red Bird of the south, and its long walls radiated their power among the generals and ministers of China.[46]

Comparable to these vast citadels was the Market of Heaven, surrounded by a wall composed of a long array of asterisms swinging down from Hercules through Serpens and curving back into Ophiucus. Each segment of this wall represented one of the feudal states of the classical age, and, when marked by a flareup—comet, nova, or whatever—had appropriate reference to those geographical regions. The area within had a special relationship with merchants and markets, and the ghastly appearance of a supernova there indicated conflict among tradesmen, and various kinds of mercantile chicanery.[47] Closely attached to this wealthy asterism, and immediately north of it, was a kind of necklace or chaplet of stars named "Threading Cord" (*kuan so*). It corresponds closely to our Corona Borealis, also called *Stephanos*, "Garland," or "Wreath," which Tintoretto painted over the head of Ariadne,[48] and whose lucida is brilliantly white Alphecca—also called Gemma and Margarita Coronae.[49] While it was no royal crown for the Chinese, it symbolized wealth, showing itself as a string of perforated coins. But it also had less pleasant associations: it represented bonds, fetters, and prison.[50]

There were other such great concatenations of stars, most notable among them being the "Forest of Feathers" (*yü lin*), the celestial counterpart of the imperial palace guards, the sometimes all-powerful pretorians of T'ang. But the enumeration of all the many scores of figures that ornamented the Chinese sky—a more richly furnished sky than our own—would not be appropriate here. There were lakes, and thunderbolts, and a "curling tongue"—the asterism of flatterers.[51] There were the Heavenly Barrier, the Heavenly Kiang (separate from the great galactic Sky River), the Heavenly Ship, the Heavenly Street, the Heavenly Cudgel, the Heavenly Court. There were individual stars so brilliant as to have a distinct identity, the most notable being our "Dog Star," Sirius, which for the Chinese was "Wolf Star," the governor of the Tibetan peoples.[52] But the list is almost endless.

Ya, Señor, sé lo que dicen las estrellas de tu
cielo; que sus puntas de diamante me lo
vienen escribiendo.[1]

Eugenio Florit, "La Noche"

6 Astrology

The Chinese arts of divination were manifold. Such techniques as scapulimancy and rhabdomancy, hallowed by blessed tradition, enjoyed, under a
great number of forms, the esteem of the elite. But they were only two
methods among many. When Li Lung-chi handed down the following decree
on 23 May 739 he was subsuming a great many popular arts of prognostication under a few rubrics: "All *yin-yang* extrapolative numerology; unless it be for choosing days for weddings and funerals by divination, is
forbidden."[2] (The elegant expression "extrapolative numerology" is an
attempt to capture the sense of the Chinese *shu shu*, "projective calculation" plus "computation; counting; number," which covers the several arts
of anticipating the future through the manipulation of symbols for the Five
Activities, *yin* and *yang*, and the like.[3]) Freely accessible to the lettered
classes were the so-called "weft books" (*wei shu*), chiefly written, it ap-

54

pears, in the Han dynasty. These passed under the deceptive guise of scholia on the old Chou dynasty classics (the *ching*, or canonical "warp threads"), and enjoyed a certain modest respectability without official approval—somewhat like the Sortes Vergilianae. Some of these books, now extant only in fragments, survived in the T'ang period, and were even consulted by the high court astrologers. "The Tokens Reactive to the Germinal: A Weft on the 'Springs and Autumns'" (*Ch'un ch'iu wei kan ching fu*) was quoted abundantly in an eighth-century treatise by the court astronomer Gautama Siddhartha, "The Divinatory Canon of [the reign] Opened Epoch" (*K'ai yüan chan ching*).

But whether one consulted a vulgar chooser of lucky days, or the quasi-sybilline writings of antiquity, or some other outwitter of destiny, one was inevitably faced with the stars. To put it another way, astrology was a component in virtually all of these arts since, after all, the stars were the great sources of supernatural power.[4]

In the golden ages of China's remotest past, astrology—at least, portent astrology—was unnecessary. In the language of the *Book of T'ang*, "In the Grand Tranquillity of antiquity, the sun was not eroded and the stars did not explode."[5] But after the rule of the godlike supermen came to an end, the skies over the Middle Kingdom were soon flashing with warnings from the All Highest. Accordingly the earliest Chinese astrology, like the earliest western astrology, was not a horoscope astrology. In Mesopotamia and China alike the astrologers were officers of the kingdom, devoted to the interpretation of strange lights and movements in the heavens, and the timely anticipation of disasters for the monarch or the nation.[6] In China, at any rate, the body of lore associated with such startling phenomena acquired a theoretical framework, apparently not long before the beginning of the Han dynasty. This was chiefly the cosmic dualism of *yin* and *yang*—along with the doctrine of the Five Activities, to which the five visible planets could readily be adjusted.[7]

Even more fundamental than these neat metaphysical packages was an assumption which is so obvious that it has often passed unnoticed by modern students. That is the "theory of correspondences," to which I have already alluded. The crucial word here is *hsiang*: (1) effigy, simulacrum (as a painted image); (2) analogue, counterpart, equivalent, other-identity.[8] Celestial events are the "counterparts" or "simulacra" of terrestrial events; sky things have doppelgängers below, with which they are closely attuned.

> In the sky are formed counterparts (*hsiang*);
> On the earth are formed contours (*hsing*).[9]

This poetic statement of the T'ang period was expressed more fully this way: "The germinal essences of the Myriad Creatures in every case have counterparts up in the sky, and form shapes (contours) under the sky."[10] "Correspondence"[11] has been defined as the relationship between the cosmic and political realms, or the natural and human worlds, or between macrocosm and microcosm, with the Son of Heaven as a critical nexus between them all, dedicated to maintaining the exactness of the correspondences by proper ritual observances. Wheatley, following Berthelot, has styled the study of these cosmic arrangements "astrobiology," representing the "parallelism" (perhaps too weak a word) between the astral and biological worlds. Against this background it was inevitable that early Chinese philosophers should ponder problems of *relationships*, little concerned with the matters which preoccupied the Eleatics, who were interested in *substance*. The divine mission of the Son of Heaven was reflected even in the roles of his officers, and accordingly of the titles they were given, which magically achieved the great ends of cosmic coordination. A significant example is the title given to the great army leaders of the eighth, ninth, and tenth centuries, who became virtually independent warlords, and many of them kings. They were called *chieh tu shih* "Commissioners for Order and Rule." The title quotes a common Han dynasty expression *chieh tu* "order and rule"—an expression with special reference to the orderly movements of the planets and the metaphysical processes behind them, which should have congruent orderliness among men. The charismatic captain, though he were a man of blood and terror, was, like an illiterate priest, invested with all the responsibility of an agent of the Supreme Being.[12]

It was within a universe of this sort that Chinese astrology made its investigations. This astrology, more particularly in its officially recognized form, has been styled both "portent astrology" and "judicial astrology"— both correctly. The former expression refers to the kinds of celestial events that needed understanding, and the latter to the mundane processes that they referred to. The spectacular asterisms that spread across the Chinese sky—corresponding with our own delineations only in a few cases—were the opposite numbers and harmonic equivalents of political incumbencies, geographical regions, and physical objects on earth.[13] Transient planets and comets, brief meteor showers, and the flareups of novae gave emphasis to the predetermined significance of the asterisms in which they appeared. It was these underlinings, and their significance for the represented natural realms or political offices that corresponded to the underlined asterisms, that constituted the subject matter of official astrology.[14] The enormous significance given to celestial portents by the government may be exemplified by the inauguration of a new era by Liu Ch'e (Han Wu Ti) in 134 B.C., after the appearance of a "long star"—doubtless a comet. The epoch was designated "Primal Light" (*Yüan kuang*).[15]

Astrology

There were always skeptics, however. An early one was Wang Ch'ung who, writing about the appearance of a comet and a baleful apparition of Mars reported in an ancient book, disagreed with the generally accepted interpretation, namely that "When an evil administration emerges, weirds and aberrances are seen." He argued that there can be no possible connection between celestial events and moral judgments.[16] P'ei Kuang-t'ing, a privy councillor of the T'ang period, objected to accepted dogma on different grounds: "A cognizor of stars said that the [stellar] analogues were altered, to the disadvantage of the great vassal [i.e. Kuang-t'ing] and asked him if he might exorcise this. Kuang-t'ing said, "If one could dispel misfortune by exorcism, one could bring fortune by invocation.' "[17] The magnate may well have accepted the notion of divine judgment, which Wang Ch'ung rejected, but if so, he regarded cosmic punishment as ineluctable.

The case of P'ei Kuang-t'ing can properly be treated as exemplifying an impersonal omen, directed more at the office than the man (except insofar as performance in office was concerned). But parallel to this kind of incumbency-oriented portent astrology, there was a variety—or rather a number of varieties—of popular astrology, poorly documented in the contemporary sources, in which knowledge of the personal destiny of the individual—especially with regard to wealth, position, and death—was the aim of the astrologer's diagnosis. This kind of belief was held even by scientifically sophisticated persons, as is shown by the following example concerning the Astronomer Royal Shang Hsien-fu. In A.D. 702 Mars transited the asterism "Five Barons" (*Wu chu hou*; stars in Gemini). Now the fifth star in this asterism (ϕ Geminorum) resonated with the office of Astronomer Royal ("Grand Notary"),[18] a fact well known to any official astrologer. But Hsien-fu was also privy to the fact that his personal destiny was bound to the Activity (*hsing*) of Metal. He brought this matter to the attention of his sovereign, the Empress Wu: "The 'inward tone'[19] of your vassal's fate is Metal. Fire [i.e., Mars] is the enemy of Metal: your vassal is going to die." Lady Wu thought to avoid this omen by transferring her astronomer to the bureau in charge of the public water supply—since Water is the enemy of Fire. But the unhappy Hsien-fu died the same year.[20] This form of personal portent astronomy goes back at least to the Han dynasty, when Wang Ch'ung inveighed against certain aspects of it. He held that death was predetermined by the vital energy (*ch'i*) with which each man is endowed at birth. During his lifetime the changes among the stars can effect only the vicissitudes of his career, they cannot alter the hour of his death.[21] His views did not prevail among most men however.

There are many references to "omen-taking from the stars" (*chan hsing*) in T'ang poetry, and normally the expression refers to the attempts of

57

private persons to read their future in the skies. Here is an example by Liu Yü-hsi entitled "Hearing a Zither on a Merchant's Boat at Night":

> A great flatboat, high masted—a hundred feet!
> New sounds press on the bridges—from thirteen strings.
> A merchant's daughter from the market of Yang-chou
> Comes to take omens from the moonlit sky west of the Kiang.[22]

(We must not take the "hundred feet" seriously. The merchant vessel looms over the poet's head. The bridges are the moveable bridges under the strings of the zither [cheng, the Japanese koto]).

Fate prognostication was called "Extrapolation of Destiny" (t'ui ming) and had been done at least since the end of the Han dynasty, using procedures in which the cycles of the Five Activities, the sexagenary cycle, and the zodiacal animals seem to have been more significant than events in the sky. The first book on this subject of which we have any knowledge was written by Kuan Lu in the Three Kingdoms period. He wrote: "My fate is with yin ["tiger"]); I was born at night during a lunar eclipse." He predicted that he would not live beyond forty-seven or forty-eight years. Note here the evil significance of an eclipse of the moon—the moon's condition was paramount in personal astrology. The expression "salary and fate" (lu ming), in the special sense of official emoluments and destined length of life, was almost synonymous with "extrapolation of destiny."[23] We might also render it "fortune and fate." As Nakayama has rightly emphasized, the "astrological" element in this sort of prognosis was not so much a matter of celestial events and movements, but of computational arts, employing tables of cyclical signs. He calls it a kind of "time-numbering."[24]

The great exponents of these practices in the T'ang period were the astronomer-monk I-hsing, a certain Sang Tao-mao, and, most celebrated of all, Li Hsü-chung who, in the early ninth century, wrote a book, the "Book of Destiny" (Ming shu), which is still extant.[25] We also have an inscription composed by Han Yü for his grave monument, in which he praises Li Hsü-chung for his great erudition, while remarking that his learning in "Five Activity" theory was especially profound, and that he was able to calculate the fortunes of a man's life as well as its length from plotting the day, month, and year of his birth against the units of the sexagenary cycle.[26]

Despite the fact that such systems of fate prediction had become very popular by T'ang times, not everyone accepted them. We have an example in Lü Ts'ai of T'ang, whose essay directed against "Salary and Fate" prognostics still survives.[27] In it he takes notable examples from antiquity to demonstrate the falseness of the doctrine which, as he describes it, is fundamentally based on the position of the moon at the time of a man's birth.

A closely related concept was that of "hemerology" (as Nakayama calls it) or "chronomancy" (as Needham styles it)—the art of predicting lucky and unlucky days as a basis for proper conduct of the various affairs of life. In China this art was practiced by persons called (roughly) "day persons" (*jih che*). I think it would be better to label their profession "hemeromancy" rather than either hemerology or chronomancy, since the one seems too broad in scope and the other inexact. The art itself goes back to Han times at least, and seems to have been best known on the lower levels of society. It is worth noting that no single technique was employed for the determination of the quality of the days: scapulimancy, aura-watching, astrological techniques—the findings of each or several might serve as a basis of judgment.[28] (Foreign, especially Buddhist influences, affected the development of these forms of calendar wisdom. They were particularly pronounced in T'ang times.)[29]

I have noted the importance to fate predictors of the moon and the lunar month—to which should be added the cycle of twenty-eight lunar positions. This gets us to the fringes of horoscope astrology. Now let us consider a more specific example of moon-oriented astrology.

The great writer Han Yü thought himself badly treated by detractors during his lifetime. He wrote the following verses as a mildly humorous complaint against this injustice, pretending to blame it all on the poor service provided by two of the asterisms that presided over his birth—the powerful pair "Dipper" and "Ox," the constellations which housed the famous magic sword of Feng-ch'eng, about which much more will be said later:

> For the chronogram of my birth,
> The moon lodged in Southern Dipper;
> The Ox stretched out its horns,
> Winnower gaped its mouth.
>
> But Ox was averse to submitting to my pannier,
> And Dipper has decanted neither treacle nor wine.
> Winnower alone had spirit power:
> At no season did it pause from raising its fan.
>
> When I failed in Goodness, my fame was much heard of;
> When I did no Evil, the only sound was abuse;
> So fame and sound banished one another:
> The gain was minor, the losses more than enough.
>
> My three asterisms are equally high in the sky;
> Their pentads and decads ranked to east and west:
> But alas! you Ox and you Dipper—
> You [two] alone have failed in strength of spirit![30]

To sum up: the winnowing fan has dominated his life—retaining little wheat while spreading much chaff. The astrology on which this sad tale is based is plainly lunar, in this differing from ours in which the solar path is central.

Pacing the Void

On the day of Han Yü's birth, the moon had been in Dipper (stars in Sagittarius); the previous day it had been in Winnower (further west in Sagittarius); the following day it was in Ox (in Capricornus). The entire triad should have determined the ups and down of his life; the two most promising members, including the central one, had conspicuously failed him, leaving his fate at the mercy of the whimsical winnower.

Evidently the moon had rivals in the Five Planets. It is clear by T'ang times at any rate that some astrologers took their positions and activities on the day of a man's birth into account—but apparently it was their relative position along the lunar orbit that was most significant, leaving the moon still in a paramount position. The personage most worthy of notice in this connection is Chang Kuo of the eighth century. He was a Taoist hermit of great repute, skilled in yoga breathing techniques. He was persuaded to come to the court of Li Lung-chi, and his influence over the mind of that monarch seems to have been largely instrumental in his conversion to Taoism. Chang Kuo has also the reputation of having been a great astrologer. This picture of him seems in part based on late accounts of his career or on books wrongly attributed to him.[31] But the evidence of both history and tradition makes it highly probable that Chang Kuo actually predicted life histories and ultimate fate on the basis of the positions of the five planets on the day of one's birth. However, we are not provided with a clear procedure for plotting the astrological data. The astrological writings plausibly ascribed to Chang Kuo introduce traits quite unlike those known in the west. For instance, a person born during the light of day rejoices in the solar energy, and so benefits from the first six (upward) phases of the zodiac ("the six yang"), whereas a person born during the night profits from the last six (downward) phases of the zodiac.[32]

At any rate the importance of the twelve stages of the Jupiter cycle—i.e., the Chinese "zodiac"—and the effects on human destiny of the several planets as they passed through these "palaces" was well established in T'ang times. Jupiter himself is still a maleficent planet in popular belief, notably on Taiwan. In Han times, as we know from Wang Ch'ung, his power extended into the zone of geomancy, and when he tarried in a sky palace whose relative position offered a threat to an earthly house, the latter had to be protected by powerful cantrips.[33] Possibly his personal malignancy goes back as far as T'ang times. Certainly by the ninth century at least the twelve phases of his twelve-year circuit of the heavens were important in calculating the fate of individuals. Consider the case of the famous statesman and poet Tu Mu. This talented magnate wrote his own obituary, which fortunately remains accessible to us. The first part of it is given over to a conventional statement of his successes in the world of government. Most of the rest is dedicated to alarming premonitions of his end. Among these is an astrological record which ties both the fortunes of his life and the time of his ultimate demise to the positions of the planets on the Jupiter cycle:

My birth was in the Horn star [Spica, and other stars in Virgo]. Mane [Pleiades] and Net [Hyades] are the eighth palace from Horn, which is called the Palace of Sickness and Frustration. It is also called the Palace of the Eight Killings. Earth Star [Saturn] was stationed there. Fire Star [Mars] followed after Wood Star [Jupiter]. The artist Yang Hsi said, 'Wood was in Spread [stars in Hydra], the eleventh from Horn—a Palace of Fortune and Virtue. When Wood is in Fortune and Virtue, a great and princely man is saved at its side. There is nothing to be concerned about.' I said, 'Since I was transferred from Protector of Hu to Chamberlain within the course of a year, Wood was then fortunate enough for Horn. That Earth and Fire should be deadly to Horn is quite suitable.' "[34]

Taking Horn as position one, and counting back along the animal cycle through "Rat" and "Pig," the eighth position corresponds to Hyades and Pleiades in Taurus ("Cock"), and the eleventh to part of Hydra ("Horse"). I am unable to deduce a system from this instance, but some of its major assumptions are obvious enough.

The twelve positions of the Jupiter cycle also figure in a method of fate computation known at least by the end of T'ang, in which a person's destiny is bound to one of the seven stars of the Great Dipper, in accordance with its special relationship to the year of his birth as placed in the cycle of twelve.[35]

We have another remnant of ninth-century horoscope astrology from the pen of Tuan Ch'eng-shih. This fragment shows the use of the twenty-eight lunar lodgings but makes no reference to the twelve Jovian palaces. Presumably, then, it emphasizes the positions of the moon, an emphasis characteristic of the astrology of this age:

A person born in the lodging "Horn" [Spica et al.] is fond of mockery and jesting. For a person born in the lodge of "Woman" [in Aquarius] no affair he undertakes will come off on days [when the moon is] in "Gullet" [in Virgo, next to Horn], in "Triaster" [in Orion], or in "Roof" [in Aquarius and Pegasus]. But in "Barrens" [in Aquarius and Equuleus] and Horn he will prevail.[36]

This specimen is unique in attributing traits of personality and character to the moon-sign of a man's birth.

Horoscope astrology of the kind familiar to us apparently began in the western world in about the fifth century B.C., supplementing if not superseding the judicial astrology of Mesopotamia. It seems to have been developed chiefly in Hellenistic times, and gradually found its way to other parts of the world, arriving in China during the latter part of the T'ang period. There it did not supersede the native varieties, which were based on the locations of the moon and Jupiter, but it did enjoy a modest and specialized career in the underworld of popular Buddhism and mingled with the nameless brands of vernacular astrology.[37]

STELLAR CHROMATICS

. . . The colored stars
Sparkle like gems—capricious Antares
Flushing and paling in the Southern arch,
And azure Lyra, like a woman's eye,
Burning with soft blue lustre, and away
Over the desert the bright Polar-star,
White as a flashing icicle, and here,
Hung like a lamp above th'Arabian sea,
Mars with his dusky glow, and, fairer yet,
Mild Sirius, tinct with dewy violet.

Nathaniel Parker Willis, "The Scholar of Thebet
Ben Khorat," in *Melanie and Other Poems*

Despite the allegations of my epigraph, Antares, although a binary which looks fiery red to the naked eye, is not a variable star; Vega (α Lyrae) does indeed look like a "pale sapphire," but Sirius seems brilliantly white to our eyes.[38] Nonetheless, Willis' verses contain, in miniature, a number of matters that must now be pondered: the colors of individual stars and planets, or variable or apparently variable stars, and differing reports on star color. It is fitting that star color be treated as a special subject since, although most men agree and have agreed in the past that considered collectively the stars are white—candent diamond points against a velvet sky—there are certain remarkable exceptions to this general rule. It is these exceptions that have attracted the attention of astrologers and other professional worry-warts— and also that of poets—in most times and places, to say nothing of modern persons equipped with optical lenses, who can see the night sky as a veritable jewel box.

If not white, stars tend to appear red—or occasionally yellow or orange. Those that seem pale blue or green to the naked eye are very rare, and often unnoticed at a casual glance. Most of us recognize red Antares, red Betelgeuse, and red Aldebaran quickly, along with a few other reddish stars. It is easy to suppose that the Chaldean mage and the Chinese sage saw these spectacular sky-sparks just as we do—but apparently we should be wrong in so doing. Along with reds just mentioned, classical writers listed, among other surprises, Sirius as "fiery red." This apparently represents an actual change that has taken place during the short span of human history. "It is thought that the present white lustre of Sirius does not date back more than a thousand or 1200 years."[39] There are other similar cases, and even in recent times such changes have been verified by the use of a colorimeter.[40]

When Bryant wrote of "The great Orion, with his jewelled belt"[41] he probably had a diamond-studded cincture in mind. In the light of modern astronomical observations, he wrote better than he knew. Among the prominent stars in the belt, a good telescope reveals Almitak to be a triple star colored topaz yellow, light purple, and gray; Alnilam is brilliantly white; Mintaka is a double, shining white paired with pale violet.[42] Such wonderful sights are not accessible to the unaided eye because the delicate tints are generally seen only in stars that burn with low intensity.[43] These undistinguished lights, which we overlook entirely when we survey the clear sky, appear to the astronomer quite differently: "They not only vividly sparkle in green and gold, azure and crimson, but shine in the sober radiance of fawn and olive, lilac, deep purple, and ashen grey. Chalcedony, aquamarine, chrysolite, agate, and onyx have counterparts in the heavens as well as rubies and emeralds, sards, sapphires, and topazes."[44] Even the red stars we know so well—Antares and the rest—seem washed out when compared with the red stars of low magnitude, which are true rubies or drops of blood.[45]

However, although early Chinese records comment frequently on the sidereal palette as if it displayed the complete range of the spectrum, they are not always referring to "objective" colors determined by the actual radiation received from outer space. Some of the reported colors derive from subjective states in the observer—as in day-dreams and visions; some are due to personal aberrations of vision—as under the influence of drugs; some supervene because of factors in the earth's atmosphere—such as diffraction near the horizon; some are purely symbolic—as in the writings of Chinese cosmographers for whom color may be latent, and visible only to the inner eye; and some are literary—as in the verses of Chinese poets for whom color may be metaphorical, magical, or melodramatic. It might be averred that, however limited the stellar spectrum implanted on the retina of the sober citizen, the Chinese had correct instincts about the true variety of stellar chromatics.

Since long before T'ang times the doctrine of the "Five Activities," each with its appropriate symbolic color, had become part of the standard mental equipment of every Chinese. Even among skeptics, who seem to have been few, these arbitrary associations were so natural and obvious that, at the very least, they were inevitably employed in instantly recognizable figures of speech. In antiquity, however, and for the intellects that revelled in uncomplicated metaphysics, these associations were not conventional tropes but powerful models of changes in living reality, and not only symbolized but in some sense "explained" both cosmic revolutions and political upheavals: "When the Mandate of Heaven operated with Black, in Hsia there was a Smoky [as in "smoky quartz"] Baton; when the Mandate of Heaven operated with White, in Yin there was a white wolf biting on a hook; when the Mandate of Heaven operated with Red, in Chou there was a red bird biting on a message."[46] In T'ang times this historical formula was as familiar as the story of the sky-arc that God displayed to Noah, and its implications even more extensive. Naturally, this cosmic symbolism was revealed among the planets. There were exactly five visible planets. The color of only two of them is conspicuous, but Mars was "Red" to the Chinese in more than a merely visual way, and Venus was not just phenomenally "White." Indeed, the "classic" name of Venus was "Grand White," and many ambiguous references to a "white star" must refer to that planet, just as Mars, the "Dazzling Deluder," must frequently lie behind such expressions as "pink star" or "red star," as it commonly does "fire star." The remaining three colors could be fixed in Mercury, Jupiter, and Saturn by the imagination. Even the sun and the moon could partake transiently in the great pattern. For example, when the sovereign reigned by virtue of the power of Wood (implying necessarily Spring, Dragon, Blue, etc.), "the sun was yellow with a blue nimbus," while if he reigned under the power of Earth the sun displayed all five colors, with none of them dominant. Not surprisingly, the five basic musical modes and the notes which generated them partook in this scheme, so that each planet had its appropriate pitch, almost as if Pythagoras had had a hand in the arrangment.[47]

It is true that planetary position and planetary movement were taken into account in astrological reckoning, each in differing degree. The old tradition was that "motion [i.e., speed] prevails over color, color prevails over position; holding position prevails over lack of position; having color prevails over lack of color."[48]

The mystic phenomenology of color was applicable to all parts of the sky, from the lowest to the highest. Indeed some of its manifestations in the lower regions, which we should classify as "meteorological," were the most striking and often the most significant. We are familiar with these displays either in or conditioned by the contents of our atmosphere. We do not regard them as substantial changes but as "discolorations" whose explanations are not to be sought in high celestial purposes. Such are the appearances of sun or moon or stars altered by dust storms, or high clouds of ice crystals. The polar auroras and the flashes of meteors fall into the same class as bloody suns and haloed moons. To the medieval Chinese, however, all of these events were as portentous as any other unpredictable celestial event, and they had their equivalences in human affairs which could sometimes be quickly identified or otherwise patiently explained in retrospect. The inter-

64

pretations were usually simple equations, with parallels in other levels of omen lore. Accordingly, when the sun was "red as an ember," it forboded war,[49] while a purple sun was a portent of barbarian invasions.[50]

But these were transient colors which stood for transient events. The inherent or native colors of the planets were constantly effective, not only in mundane affairs but in the celestial order. The most obvious planetary influence was exerted upon the aspect of comets, those frightening apparitions that came and went without regard for reason and system. It was commonly believed that comets were temporary emanations of planets, bearing messages decipherable by the wise.

> ... To-night
> The ghost of a dead planet shall walk through,
> And shake the pillars of this dukedom down.[51]

Their substantial place of origin was signified by their show of color. Hence, in accordance with "Five Activity" doctrine, Mercury generates black comets, Venus generates white comets, Saturn generates yellow comets, Mars generates red comets, and Jupiter (anomalously) generates watchet comets.[52] Jupiter's color should have been designated by a simple traditional word, like those given to the colors of the other planets and their comet children—that is, *ch'ing* "blue." This is a common enough word in astronomical descriptions. But I have not observed a single report of a straightforward "blue" comet. Instead, there are regular reports of comets whose color is said to be *ts'ang*, a color regularly attributed to lichens, that is, "gray; greenish gray." The word is cognate to *ts'ang*, the color of the iron-gray or blue-gray sea, for which I like the old word "watchet," appropriate here for the color of the off-blue comets. In any event they were an evil sight, responding to "baleful and disastrous" conditions on earth, as the T'ang scholiast Chang Shou-chieh informs us.[53] Although reports of colored comets are fairly abundant in older books—notably in the histories of the Han and Chin periods—they are seldom alluded to in the T'ang histories, possibly because the grip of Five Activity theory on the imaginations of the men of T'ang was less strong than it had been on the color vision of the ancients. An exception to the general rule is the comet of the summer of 770. This white comet—not a color likely to cause much comment—appeared in the northern skies on 15 June. It moved eastward until, on 19 June, it was close to Lynx and Camelopardalis. By 9 July it was approaching Virgo, and on the 25th of the same month it disappeared. The report concludes: "As its color was white, it had been given birth by Grand White," i.e., by Venus.[54] But there is no further comment.

Meteors came as unexpectedly as comets, and were just as ominous.

Reports of them frequently tell of their color. Not surprisingly they are most often red, yellow, and white.[55] At least one old source, still extant in T'ang times, had a simple scheme for the classification of meteoric color portents: red was "men-at-arms," yellow was "earth," white was "civic responsibility; dutiful activity," blue was "anxiety," and black was "death."[56] Not a subtle system, but one that allows considerable latitude for reasonable comment. An example: "If a yellow star falls into the sea, its waters will overflow."[57] This is plausible under the scheme.

Although many T'ang poets show a vivid awareness of the solemn significance of a meteor of any color, some were able to treat them in untypical ways. Unalarmed, they might transmute them into flowers instead of hot ferrous lumps on their contact with the earth. Beginning with the ninth century when the native red azaleas (rightly called *ignescens*[58]), of the balmier regions of China, were first made famous in the celebrated gardens of Po Chü-i and Li Te-yü, many writers composed verses about these rose-red or crimson blooms that cloaked the hillsides, associating them with blood, with woman, and with fire.[59] Finding a fine description of the metamorphosis of red star into red flower in a double quatrain of Meng Chiao, who lived only into the second decade of the ninth century, confirms the suspicion that the new appreciation for this handsome plant had probably developed already during the closing years of the eighth. That poet expresses wonder at the fallen stars, their redness retained, glowing like hot coals in a green garden.

> It seems they were not placed here by men's hands
> Nor, surely, has it to do with the odd set of the land;
> Lonely lights—threading through a surplus of halcyon blue,
> Or only reflections—dancing full of artful beauty.
>
> Shattered fire—burning through the leisurely land,
> Ruddy stars—fallen out of the blue sky.
> Struck by astonishment at how one species is displayed in another,
> The favored visitor is held spellbound by them.[60]

These fallen stars remained immobilized as flowers, but the potential mobility of transformed meteors was more characteristic of old Chinese imagery. To take another example from the red part of the spectrum, this quality made a fallen star a better metaphor for a spark from a magical hearth than for a gem fixed in a setting of gold. Li Po provides us with a typical example in a poem about the fire in a brazier—apparently an alchemist's furnace—which lights up heaven and earth with "red stars turbulent in the purple mist."[61] These pseudo-stars are ominous, ruddy meteors disturbing the glowing vapors so characteristic of Taoist milieus.

Astrology

Leaving such undependable stars as meteors and comets, let us look at some examples of the significance of the hues of fixed stars—if any stars can be said to be fixed. Commonly a brilliantly displayed color, or the assumption of color accompanied by some such phenomenon as unusual movement or extravagant sparkle, attracted attention and comment. Thus Lo Pin-wang, early in T'ang times, wrote of the relation of a ruler to the celestial power that "if his Virtue is derived from Heaven, a white star is allowed to descend; it is the germinal essence of the Royal Sustainer."[62] Here the approaching star is an emanation of another star, just as comets are emanations of the planets. (The star called Sustainer [*fu*], an important body despite its apparent insignificance, was an adjunct to the Northern Dipper. Specifically it is the "pale emerald" companion of bright white Mizar [ζ Ursae majoris], a visual double and a prominent member of the asterism next to the tip of the handle. Mizar was called "Disclosed *Yang*" [*k'ai yang*] in Chinese astronomy [—one Taoist name is *wu ch'ü*]. We for our part give the name Alcor to the Sustainer.[63] That star, whose Chinese name might just as well be rendered "Steady Prop," was the reliable support of the great star-kings seated in the Dipper, and was therefore a symbol of dedicated service to the throne, and the asterism of minister and viziers.[64] It was very important in Taoist practice, as will appear later.)

While dramatic change or motion was necessary to call attention to a white star—it was the common color, after all—and served even to emphasize the color of other stars, it appears that color in itself, especially if not characteristic of a known star or exemplified in a new star, was sufficient to signalize a message to earthlings. Yellow or yellowed stars, for instance, seem to have been—all else being equal—favorable omens. A *locus classicus* is the old tale of Huang Ti—the Divine King of Yellow—who ruled by Virtue of Earth, and at whose birth the splendid omen of a yellow star appeared.[65] A similar token of good—in this case of military victory (the defeat by Ts'ao Ts'ao of his rival Yüan Shao)—was the apparition of a yellow star during the reign of Han Huan Ti, fifty years before the event.[66] Such notorious events as these may underlie a trivial fashion of T'ang ladies—that of dimpling their pretty faces with a painted yellow star.[67]

The most celebrated of all lucky yellow stars was one of the fixed stars, Canopus—not a familiar star to most northerners. Canopus (α Carinae in Argo Navis) is the second brightest star in the sky, yielding first place only to Sirius, which is almost due north of it. This handsome object was named after the unfortunate pilot of Menelaus, the Canopus who died far from home off the coast of Egypt. (The star itself has other Egyptian connections—for one, it is the star of Osiris.[68]) Canopus is not visible above 37° N. latitude. Nowadays it reaches its northernmost culmination close to that parallel early in February, a few minutes ahead of Sirius. Ch'ang-an, the T'ang capital, lay between 34° and 35°, and so the great star would have been just visible there in winter months, peeping over the southern horizon—weather and the Chung-shan Mountains permitting.

Canopus is, in fact, a brilliantly white star. Indeed it is said to appear steely blue-white over the deserts of Egypt and Arabia. It is a little surprising to find it described in Chinese sources as yellow—a diamond transmuted into a topaz. The explanation seems to lie in the regular presence of a haze of yellow dust—the loess blown in over north China from Central Asia by the winds of the winter monsoon—which would seriously alter the color of a source of light close to the horizon, if indeed it did not obscure it altogether. The traditional name of Canopus was "Old Man Star," and sometimes "Longevity Star." In the old Taoist hierarchy it was styled, more grandiloquently, "Supreme Verity of Cinnabar Tumulus, Old Man of the Southern Culmen."[69]

Although it appeared regularly in the south, Canopus was not so readily or steadily visible in the north as other stars. In short, it fell into the class of variable phenomena, with something of the character of a repeating supernova on the murky rim of the world. In the years when it was clearly visible, it was welcomed with a rash of congratulatory "manifestos" (piao) addressed to the throne, since its luteous rays were thought to beam with special benevolence on the royal house. These complimentary memorials, which would also appear on such auspicious occasions as the discovery of an undoubted dragon or a red-maned white horse, normally began with descriptions—sometimes rather gaudy ones—of the wonderful star. A typical one, written apparently in an early October late in the seventh century and telling of the happy apparition, took note of the tradition that "when it shines with a yellow color, then shall the Lord of Men be glorious in his longevity."[70] In the ninth century, on a similar occasion, Li Shang-yin wrote of it in somewhat more glowing language: "The Old Man Star was seen at the Southern Culmen. Its color was luminous yellow, and it was large and lustrous. The Peerless One [i.e., the sovereign] is in accord with Virtue, and the divine outcome is demonstrably auspicious. Hence a coruscating figure is let down to demonstrate the limitless blessings."[71] Such glorious occasions as these were also celebrated and certified by "Rhapsodies" (fu), often composed as command performances. One of these departs somewhat from the yellow stereotype and tells of "the downward darting of its sorrel rays . . . the peripheral melting of its purple flames."[72] A comparable effusion, rich with planetary and stellar images, and comparing the event to the apparition of a Phoenix, or of springs of wine, or of the South Sea without waves, and calling down heaven's blessings on the imperial household, emphasizes the lambent scintillation of the star:

> In the court of the Southern Culmen—the Star of the Old Man;
> Blazing, flashing, resplendent, sparkling!

Elsewhere in this same confection we find the following couplet:

> For Yin K'uei a yellow star appeared in Ch'u;
> For Lei Huan a purple breath approached in Wu.[73]

The "purple breath" was the mana of the famous star swords connected with the names of Chang Hua and Lei Huan in the Chin period. We shall encounter it elsewhere in this study. Yin K'uei, on the other hand, was the Han astrologer who explained the purport of the yellow star which appeared during the reign of Huan Ti, commented on above. Our present writer, then, takes that unnamed yellow star of the distant past to have been Canopus, and indeed it appears that this is the usual identification made of yellow stars known only from legend and classical literature, even in the absence of information placing the apparition with certainty on the southern horizon.

Usually apparitions of the Old Man Star were classified under the highest categories of good omens. Canopus was a *jui hsing*, that is, "a star as auspicious token or talisman."[74] We have just taken note of its special relationship with the Chinese sovereign. Its benefits extended to the realm generally, as personified or represented by the ruler: "When the Old Man Star appears, the ruler is secure; when it does not appear, armed men rise up."[75] This follows from the divine affinities of the sovereign: "When the royal one receives the pattern of heaven's virtue, the Old Man star approaches."[76] (Favorable stars were often described as "approaching," "drawing near," or "descending.")

Canopus also had special relations with the Son of Heaven as a longevity star:

> The mystic signs—to what do they now refer?
> The seasons are harmonious—the government too is tranquil;
> The auspice is for the One Man's longevity,
> Its color radiates—the Nine Empyreans are illuminated . . .

And so on. These lines are from a ninth-century poem celebrating the sight of Canopus.[77] But the One Man, empowered by his star, is capable of bestowing more specific and concrete favors than general prosperity and abstract peace:

> The misted moon in the west lights up the pool-side kiosk;
> In the fagara room at night's start, covered by painted screens,

The palace women call to each other—what is the matter?
From the top of the storied hall they watch together—the Old Man Star.[78]

In this characteristic "palace poem" written in the tenth century by Ho Ning
we must expect a somewhat frivolous note—hardly a mere pious sentiment
of awe and good wishes to an aging monarch. No—plainly the young ladies
are thinking that the celestial sign portends years of manly vigor ahead, and
(spoken or unspoken) the question of the moment is, "Which of us shall
play first fiddle in the seraglio?"

The great yellow star had special significance for the unfortunate exiles
who were condemned to spend years or even decades in the steaming jungles
of Nam-Viet,[79] in expiation of serious political malfeasances. To them Canopus was a beacon of hope, hung above the haunted, hated forests of the
south, suggesting a pious and fervent regard for the welfare of the ruling
dynasty, and offering the expectation that the benighted viewer might out-
live ("longevity" would be putting it too strongly) his period of banishment
in such a baleful environment.

South of Canton the citrine star—presumably less golden than from the
viewpoint of Ch'ang-an—became a prominent sky object from about the
eighth lunar month, that is, from about the middle of September.[80] Here is
Chang Chi sending off a high official to Canton. His first quatrain is courtly
and somewhat pompous (what a marvelous choice our virtuous sovereign
has made!); the second is devoted to the wonders of the south:

Our Peerless Levee has chosen a Leader—he takes up his tally and token.
Now the messenger from Within proclaims it—the Hundred Barons take heed.
From north of the seas—pagans and savages come dancing and leaping.
From south of the passes—governors of the marches send us their charts
    and handbooks.
White silver-pheasants fly in circles, welcoming the official barge.
Red hibiscus open up right by the kiosk where the visitor is feted.
Let no one speak of too much malarial pestilence in this place
Where, at the sky's edge, well within sight, is the Old Man Star![81]

Malaria, the bane of Lingnan, should have no terrors for the eminent
traveler who basks in the royal favor and can expect the prophylaxis promised by the royal star of longevity. The celebrated poet Shen Ch'üan-ch'i
was certainly out of royal favor when he arrived at the port of Hanoi and saw
the spectacle of red Antares high in the sky, and Canopus floating serenely
with our constellation the ship Argo, but the view cheered his heart greatly,
even on this pestilential coast.[82]

The Taoist initiate could also read the messages of star color—without
even looking at the night sky. Accessible to him was the secret art of cranial
astrology: stationed in a dark place, and pressing upon his eyeballs, the adept

70

could see the powerful asterisms as points of colored light. Prominent among them was the Stabilizer Star, supporter of the northern Dipper. In general, if it shone red, it meant "virtue," yellow meant "joy," white meant "armed men," blue meant "affliction," and black meant "malignancy."[83] These internal stars were called "spirit lights" (shen kuang).

Few if any stars were regarded as idiochromatic. Changes from one color to another—especially away from a normal color—were highly significant. Some planets seem to have been prone to this sort of alteration, particularly when infected by the proximity of influential constellations. Thus Venus, as everyone knows, is properly white. But let that planet come close to the "Wolf" (Sirius) and it will appear to be red. Similarly it becomes yellow when near "Heart" (Antares), watchet by Triaster's (Orion's) left shoulder, and black by Triaster's right shoulder.[84] While we moderns are seldom conscious of such a spectral range as this, color variations in the inhabitants of the sky are not unknown to us. Many are caused by phenomena we should style "meteorological," but some—such as the "copper penny" moon we see in some lunar eclipses—take place in outer space.

Generally speaking, color changes were disastrous. For example, if the Ox-hauler (Altair)[85] "loses its usual color, most cattle will die and the cereals will not take shape."[86] Similarly, if there were an alteration of color in "The Seats of the Five Divine Kings" (wu ti tso, in Leo), it would portend dire cataclysms.[87] The color of a planet might be contaminated by that of another, and this improper hue was a sign of dreadful things, too. Thus, "When Dazzling Deluder [Mars] loses its color, it has been smothered by Grand White [Venus]. This means that Fire is not able to fuse Metal. Since Metal is at maximum hardness, Fire is going into decline. There will be great uprisings of armed men in the western quarter; men-at-arms will be weakened in southern countries, and useless for doing battle . . ."[88]

The assumption of a red hue by a star was frequently reported, and appropriately dramatized by the astrologers. This change might betoken another celestial change, such as an eclipse: "When the sun is on the point of eclipse, the color of the second star of the Dipper will be altered: it will be faintly red and non-luminous, and on the seventh day there will be an eclipse."[89] Or the reddening might have a desirable import. For instance, when the Weaving Woman (Vega) shines red, "women's exploits will be excellent."[90] Or again, the lunar lodges "Straddler" and "Harvester" (stars in Andromeda, Pisces, and Aries) "govern matters of weaponry. When their color is red, it is excellent: a great leader will have meritorious achievements."[91] But the military outlook was not necessarily benign, as when Venus was seen near the Pleiades on the evening of 16 March 904, "colored red, ablaze and flaming like fire." Then, the following night, "it had three horns, like a flower, while it trembled and shook. The prognosis is: 'There will be a walled city in revolt; there will be a fiery disaster; foreign men-at-arms will rise up.' "[92]

Other color variations, of course, were also significant. Here is an example. When the sun, moon, or one of the planets intrudes on the Hooked Lock (*kou ch'ien*, a pair of stars in Scorpio which lock up the "heart of Heaven"), "if their color is blue there is affliction in the nation and losses to its arms; if their color is white, the great weapon-bearers kill each other and corpses are heaped like hills."[93] (It will be noticed that these color auguries are the same as those given above for the hallucinatory stars seen behind the eyelids.)

Uniformity of color among the planets or the stars of a constellation was prognostic of peace: "When the Five Stars [i.e., planets] are alike in color, all under Heaven will lay down arms, and the Hundred Clans will have security and peace."[94] In these excellent circumstances the planets appeared uniformly yellow.[95] Such occurrences, which must have been rather rare, since it was required that all of the planets be visible at once, were occasions of jubilation, and of felicitation due to the Son of Heaven, whose merit was the immediate cause of the wonderful display. At least three "rhapsodies" on the theme "The Five Stars are Alike in Color" by writers of the T'ang period still survive.[96] All tell of divine approbation of the ruling dynasty, plainly and gloriously certified in the sky. As for the starry figures, not only was agreement in color favorable, but disagreement was unfavorable. For instance: "When the color of the Three Eminences (*san t'ai*, in Ursa Major) is uniform, there is harmony between liege lord and liegeman; if it is not uniform, there is great discord."[97] The meaning was much the same when the color of a planet was considered in comparison with that of the components of an asterism through which it was passing: if alike, good; if not alike, bad.[98]

STELLAR VARIATION

. . . the fiery Sirius alters hue
And bickers into red and emerald.
Alfred Tennyson, "The Princess"

It is thought that Tennyson, in his lines about the color changes of Sirius, was referring to the phenomenon known as "scintillation," which is the rapid change of brightness or color in a star due to refraction of its light through air whose density is changing quickly. In short, he was describing "twinkling," which tends to be more pronounced when the star is near the horizon.[99] For the Chinese, the pronounced shaking of a star, beyond mere twinkling, portended evil. For instance, when there is "motion and shaking"

72

of the Heart (Antares) "the Son of Heaven is afflicted."[100] Or again, "when Great Horn (Arcturus) shakes and disperses in five colors, splendent and fiery, the interpretation is 'the kingly one will find it offensive.'"[101] Or yet again, when the four stars of the Heavenly Kiang (*t'ien chiang*, near Galactic Center) "shine as if moving and shaking, there will be a great emission of water and great uprising of men-at-arms."[102]

"Variable stars" are those which show a regular alternation in the quantity of light they emit:

> Pellucid Forms! whose crystal bosoms show
> The shine of welfare, or the shade of war.[103]

Such stars are familiar to us as "Cepheids." The most famous of these is the "Demon Star," Algol (Arabic *al-ghūl*[104]), an eclipsing variable in Perseus —actually a binary whose bright component is regularly eclipsed by its dark companion. To the Chinese, this star was "Heaped Corpses" (*chi shih*). Following the usual rule that exceptional brilliance emphasized the basic meaning of a star, Algol betokened evil when it was most luminous. For the Chinese, Algol controlled "the rites and ceremonies of death and mourning," and when it glared evilly, "dead men [are piled] like hills."[105] But naked eye variables are rare, and although they sky is spangled with variable stars, especially in globular clusters and along portions of the Milky Way, their variation is apparent only in a telescope. Probably most slow variations in brightness recorded by the Chinese were caused by atmospheric changes. That this is the case is suggested by meanings assigned to the total disappearance of fixed, familiar stars—not variables in our usage. For example, consider the Minister (*hsiang*), a star just below the Great Dipper in Canes Venatici: "when the star is luminous, it is benign; when dark, malign. If it disappears the minister will die, but otherwise he will be exiled."[106] On the other hand, as already illustrated in the case of Heaped Corpses, unusual brightness enhances the symbolism of the asterism, as was the case—in a different manner—with Cavalry Officer (*ch'i kuan*) in the southern constellation Lupus: "when its stars are numerous, there is security under Heaven; when its stars are few, armed men rise up."[107]

While the wary reader of medieval Chinese poetry must always be watchful for rubescent stars, tremulous stars, and brightening or fading stars, whose literary appearances are seldom untainted with spectral overtones, he must heed above all the import of stars that approach the earth with leisurely and solemn dignity. Such grave processions must never, of course, be confused with the violent and hasty actions of meteors. Indeed the two kinds of activity are usually separable by simple lexical distinction. Meteors are "floating" or "drifting" stars (*liu hsing*, and a few rare variants), carried, as it were, involuntarily across the dark sky. Those which plunge to earth to be magically transformed into stones are said to "fall down" (*to* or *yün*). But stars which approach with majestic mien are said to "drop" like the petals of flowers (*lo*), a word most often used of sun, moon, or stars approaching the western horizon; to "be suspended" or "hung" (*hsüan*) like a picture in the sky; "to be lowered or let fall" (*ch'ui*) like a beaded curtain; or simply "to come down, be sent down" (*chiang*). A sampling of T'ang poems yields: "Night on the shoals—the stars are lowered toward earth";[108] "Quiet frontier walls—an uncanny star drops down";[109] "Suspended from heaven—the stars and moon: and still no clouds."[110] All of these verses describe the apparent closeness of the stars, as they stand forth vividly on clear nights. The first two of the three are from scenes on the northern desert frontier. Of these the first is set by a sandy desert river bed—a dry wadi. The second is symbolic of peace with the nomads, telling of the disappearance of a threatening star (most likely a comet); it occurs in a dirge for the late Wu Tsung, to whom the credit is given. The third is a simple scene on a brilliant night, as in any of a hundred other poems. Less commonly, but with no sense of strangeness, the stars "sink" into the depths of a great inland lake in the west. Here is the view from a divine balcony:

> Stars sink to the sea's bottom—visibly at my window;
> Rain passes by the source of the Ho—to be watched behind
>   my chair.[111]

But more typical than these peaceful scenes is the vision of the following quatrain, by the Taoist initiate Tu Kuang-t'ing. It is pregnant with awe:

> Soul quiet, thoughts in focus—looking up into the blue abyss;
> This very evening the enduring sky let down an auspicious star.
> Last night, out on the sea, I heard the plumed wings of the *p'eng* bird,
> Just now, here among men, I see the ceremonious shape of a crane.[112]

The adept sees, in the hidden landscape of his own mind, the divine birds, potential vehicles for a longed-for journey to paradise. Their advent is betokened by a symbolic star, possibly Canopus, sent for his eyes alone.

Similar verses, intended to give form to a poet's appreciation of the hospitality shown him at a Taoist friary or Buddhist monastery, are common. The writer flatters his hosts by describing the approving descent of the stars towards their obviously blessed establishments. Frequently it is the entire Milky Way that descends towards the holy precincts, which are described as if they were themselves crystalline sky palaces, attracted up above the clouds to the banks of the great sky river. Or the stars may be mystically reflected in the monastery's lotus pool, duplicating on earth the idealized setting above: monastery buildings become the habitations of star

74

spirits. Even the royal palace could be figured as a home of the gods by the simple device of showing how "the Starry Ho lowers to brush its trees."[113]

DISASTROUS GEOGRAPHY

A star for every State, and a State for every
star.

Robert Charles Winthrop, "Address on
Boston Common"

By "disastrous geography" I mean that important aspect of Chinese portent astrology which they themselves call *fen yeh*, which would be rendered more literally by some such phrase as "apportioned champaign" or "allotted countryside."[114] Although "disastrous geography" is no translation, it has the advantage of incorporating the word "disaster," which connotes "astrally unfortunate." The doctrine is yet another version of the idea of sky-earth correspondences and other-identities. It holds that the map of the night sky may be profitably projected onto a map of China. The overlay instantly reveals what constellations are in a state of assonance with the several eternally established segments of the civilized world, so that with the appearance of a prodigy or dislocation in one of the regions of heaven, one can point unerringly to the spot on earth where danger is imminent.[115] (See table 2.)

The most generalized form of disastrous geography divides the sky into four great precincts, sometimes called "palaces," which were named according to the symbolic animals supposed to preside over them in the scheme of cardinal directions. Accordingly each of these regimes subsumed seven of the twenty-eight lunar lodgings, and their individual significance could be taken to be a kind of abstraction of the least common multiple of the atomic meanings of all seven. The palaces, then, were named Blue Dragon, for the east, Dusky Warrior, for the north, White Tiger, for the west, and Red Bird, for the south, as on the accompanying chart. But these areas of responsibility were so large and complex that little opportunity was afforded for precision in astrological predictions. It was the smaller divisions that made reliable forecasting possible. There were three sets of these sky realms: one of nine, one of twelve, and one of twenty-eight.

The system of nine domains in the sky corresponding to the same number of fields on earth is the truly "classic" one, known at least by the end of the Chou dynasty,[116] and the one best fitted to the ancient and hallowed plan whereby the sage engineer Yü divided the land into nine "islands" raised from the flood for human habitation. These drained provinces corresponded ideally to the eight directions plus the center.[117] The correspondence was only approximate in late, degenerate times, and it was a perpetual

Table 2
Important Starry Chronograms

Palace	Lunar Lodging		Animal Cycle	Jupiter Station	Classical State	
玄武 Murky Warrior	斗	Dipper	丑 OX	星紀 Star Chronicle	吳越	Wu and Yueh
	牛	Ox				
	女	Woman	子 RAT	玄枵 Murky Hollow	齊	Ch'i
	虛	Barrens				
	危	Roof				
	室	House	亥 PIG	娵訾﹕陬訾 娵訾 Loggerhead Turtle'	衛	Wei
	壁	Wall				
白虎 White Tiger	奎	Straddler	戌 DOG	降婁 Descending Harvester (?)	魯	Lu
	婁	Harvester				
	胃	Stomach				
	昴	Mane[2]	酉 COCK	大梁 Great Plank-bridge	晉	Chin
	畢	Net				
	觜	Beak	申 MONKEY	實沈 Shih Ch'en (name of deity in Orion)	魏	Wei
	參	Triaster				
赤鳥 Red Bird	井	Well	未 GOAT	鶉首 Quail Head	秦	Ch'in
	鬼	Ghost				
	柳	Willow	午 HORSE	鶉火 Quail Fire	周	Chou
	星	Star				
	張	Spread				
	翼	Wing	巳 SNAKE	鶉尾 Quail Tail	楚	Ch'u
	軫	Axletree				
青龍 Blue Dragon	角	Horn	辰 DRAGON	壽星 Longevity Star (same as name of Canopus)	鄭	Cheng
	亢	Gullet				
	氐	Base	卯 HARE	大火 Great Fire (Antares)	宋	Sung
	房	Chamber				
	心	Heart				
	尾	Tail	寅 TIGER	析木 Split Wood (said to represent the ford of the Sky River)	燕	Yen
	箕	Winnower				

T'ang "Allotment"[3]	Common Monitions[3]	Approximate position in western constellations[5]
Coast south of the Yangtze	High officials; rank and status	$\phi, \lambda, \mu^1, \sigma, \tau, \zeta$ Sagittarii
		$\beta, \alpha^2, \xi^2, \pi, o, \rho$ Capricorni
Shantung	The north; wasteland	$\epsilon^2, \mu, 3, 4$ Aquarii
		β Aquarii, α Equulei
		α Aquarii, θ, ϵ Pegasi
Lower Plain of Yellow River	Royal ancestral temples[4]	α, β Pegasi
		γ Pegasi, α Andromedae
Southern Shantung to Huai region	Armory	$\eta, \zeta, \iota, \epsilon, \delta, \pi$ and other stars in Andromeda and Pisces
		β, γ, α Arietis
		34, 39, 41 Arietis
Southern Hopei	Northern frontier, soldiers	17, 16, 19, 20, 23, 7, 28, 27 Tauri, The Pleiades
		$\epsilon, \delta^3, \delta^1, \gamma, \alpha$ (Aldebaran), θ^1 (Alcyone)[6] and others in Taurus, Hyades
Shansi	Weapons[4]	λ, ϕ^1, ϕ^2 Orionis
		$\zeta, \epsilon, \delta, \alpha$ (Betelgeuse), γ, κ, β (Rigel) Orionis
Shensi, parts of Kansu and Szechwan (especially Ch'ang-an)	Drought (cf. "Well"), Death (cf. "Ghost").	$\mu, \nu, \gamma, \xi^2, \epsilon, \zeta, \lambda$ Geminorum
		$\theta, \eta, \gamma, \delta$ Cancri
most of Honan	Lo-yang	$\delta, \sigma, \eta, \rho, \epsilon, \zeta$ and others in Hydra
		$\alpha, \tau^1, \tau^2, \iota$ Hydrae
		$\upsilon^1, \lambda, \mu, \kappa$ and others in Hydra
Hunan and southward	Southern barbarians; horses and carriages (cf. "Axletree")	$\alpha, \gamma, \zeta, \lambda, \nu, \eta$ and others in Crater
		$\gamma, \epsilon, \delta, \beta$ Corvi
Huai region	Nature[4]	α (Spica) ζ Virginis
		$\kappa, \iota, \phi, \lambda$ Virginis
Eastern Honan	Residence of the Son of Heaven; great officials	$\alpha^2, \iota^1, \gamma$ Librae
		π, ρ, δ, β Scorpii
		σ, α (Antares) τ Scorpii
Northern Hopei	Queen's headquarters; ladies' part of palace	$\mu, \epsilon, \zeta, \eta, \phi, \iota, \kappa, \lambda$ and others in Scorpius
		γ, δ, ϵ Sagittarii

1. *Ch'u hsüeh chi*, 1, 11. The meanings of the second and third pairs of graphs are uncertain. Some scholiasts say "lips of a beautiful woman."
2. Cognate to 失 (*Shih chi*).
3. *T'ang shu*, 32–33; *Chiu T'ang shu*, 36; *T'ang liu tien*, 10, 30a–31b; Schlegel, *Uranographie*, 557–558.
4. From sources other than the dynastic histories.
5. From A. Wylie, "Catalogue of Fixed Stars," 110–138.
6. Following Schlegel, p. 341.

problem for the Chinese to fit the recalcitrant facts of physiography to the ideal Ennead of traditional cosmography. This scheme may also be regarded as classical in the sense that the ideal set of nine had many obvious metaphysical supports, and indeed it can be found in a number of other cultures, where it fits the notion of the domains of the Seven Planets along with the two "pseudo-planets" at the lunar nodes.[118] It would also have occurred naturally to the Chinese to think, for instance, of the powers of the nine stars of the Northern Dipper—the bright seven plus the two secretive ones.

Although in existence long before the T'ang,[119] it was the set of twelve "champaigns" that counted for the most in that era—or at least it is the one emphasized in the dynastic histories.[120] The twelve celestial zones correspond to the twelve lodges of the Jupiter cycle, and have rather fantastic names, such as "Quail Fire" and "Split Wood"—the translations in some cases are quite tentative—which are listed in the accompanying table. The corresponding terrestrial "countrysides" represented approximately the locations of twelve of the more prominent feudal states of the late Chou period. Thus, "Murky Hollow" on the ecliptic corresponded to the old state of Ch'i in Shantung, while "Quail Tail" matched the ancient state of Ch'u in the central Yangtze region, and so on. Allusions to these occur constantly in poetry, and contribute greatly to making many T'ang verses unintelligible to the reader who is not as familiar with the system as were the educated men of T'ang. Here is a sample:

This night, in the allotted champaign of Yen,
They shall see the passage of a messenger star.[121]

This couplet is taken from a martial poem written on the occasion of the departure of a military commander for his post on the northeastern frontier. "Yen" is the ancient name for the Peking region, whither he is destined, and he himself is the messenger star—that is, an envoy of the Son of Heaven. The Jupiter station (tz'u) poised over Yen, its "allotted champaign," is "Split Wood," said to mark a ford in the Sky River, and also corresponding to the solar or zodiac sign "Tiger" (yin), as our table shows. This falls within the larger realm of the Blue Dragon of the east, who reigns in Scorpio and part of Sagittarius.

In T'ang times a liberal spirit which infected the office of the Grand Notary for a while led to attempts to reform the traditional system in order to make it agree better with recent astronomical observations and with the fact of a greatly expanded empire. The innovations began in the age of T'ai Tsung, when Li Ch'un-feng, in his "Treatise on Normalized [Star]-counterparts" (Fa hsiang chih), reworked the boundaries of the Twelve Stations to conform to the actual divisions of T'ang administrative geography. This program was supplemented and corrected by the monk I-hsing in the following century. In this new scheme, greater attention was paid to actual topography than ever before, and the boundaries of classical provinces and ancient states were ignored, although their nomenclature was retained by conservative astrologers. This revisionism led on to a novel system—which seems not to have been generally adopted—in which the twelve champaigns, corresponding to the Jupiter stations as newly conceived, were grouped into five pragmatic sets, and these in turn were made to match the Demesnes of the five sacred mountains—the Marchmounts—the authority of the Five Planets, and the new regional deities of the Four Seas, and the Central Province, all very neatly indeed.[122]

Even more refined and meticulous predictions were made possible by adapting the twenty-eight lunar lodgings (hsiu) to the ancient system. Their twenty-eight monosyllable names, meaning "Dipper," "Ox," "Woman," and so on, were considered as fractions of single Jupiter stations, with two or three lodgings assigned to each station. For instance, the lodgings "Dipper" and "Ox" represent approximately halves of the station "Star Chronicle," and the lodgings "Woman," "Barrens," and "Roof" together make up the sky realm called "Murky Hollow."[123]

Longfellow's vision of the character of the moon and its daily changes of lodging could have been written by a Chinese: "So glides the moon along the damp/Mysterious chambers of the air."[124] The restless satellite had twenty-eight resthouses available for the nights of its sidereal round—a course that repeated itself endlessly. Each was an overnight hostel, a selenian caravanserai—a khān al-qamar in the desert of the air. Although some scholars have traced this post-house system—a reasonable and natural one, after all—back as early as Shang times,[125] it is generally supposed that it was introduced to the Far East from outside at a considerably later time, possibly from India (the Chinese hsiu "lodgings" corresponding to the Indic nakshatra).[126] The case for an original homeland in Mesopotamia or even Egypt remains unproven.[127] Whatever their antiquity, they formed an important part of astronomical nomenclature and of astrological practice in the Far East from the time when the earliest known books were written in China—that is, the late Chou. The system flourished still in T'ang times. We are fortunate in having a chart of the whole sequence of twenty-eight asterisms as they were envisaged by the men of mid-T'ang preserved on the ceiling and upper walls of a burial chamber recently excavated at Turfan (ancient Qočo) in Chinese Central Asia. The lodgings are shown as geometrical patterns in white, with the component stars represented by dots connected by (usually) straight lines. There is also a red image of the sun on the northeast wall, blazoned with a golden crow. On the southwest wall there is a white representation of the moon, containing the figure of a cinnamon tree, with the celebrated alchemical hare next to it: evidently they represent the point of the beginning and end of the cycle. There is also a set of parallel white streaks on the ceiling which may represent the Star River.[128] The four corners of the ceiling represent the solstices and equinoxes. This unique sky diagram is reproduced as figure 2.

The position of the moon with respect to her assigned nightly lodgings meant nothing astrologically, but an eclipse, a comet, or a nova in one was—like a burglary, a fire, or an explosion in an earthly hotel—an event of great moment.[129] Such dire omens are discussed in many studies of Chinese astrology and I shall not try to tabulate them. Rather I shall note a few points of special interest about some of the lodgings—mere snippets of lore—to serve as a sampling of the rich store of folkloristic, linguistic, and literary riches that lies virtually unexplored behind their names and traditions.

Dipper: A Taoist book provides the secret names of its six stars. I list them beginning at the tip of the handle and working toward the lip of the bowl: "Vermilion" (*chu*), "Light" (*kuang*), "God-king" (*ti*), "Yao," "Glorious" (*ch'ang*), "Luminous" (*ming*).[130] This sequence may be read as a kind of invocation or mantra: "God-king Yao of Vermilion Light, Glorious and Luminous." (This is the Southern Dipper, which will receive greater attention under "Star Swords").

Woman: Imbalance in this asterism forbodes the ultimate unpleasantness: a solar eclipse, for instance, signified that "the realm shall have a woman for its master, afflicting all under Heaven; the womanly arts will not be practised."[131] A comet boded worse, if that is possible: "There shall be a great armed rising in that country, and a female lord shall devise disorder."[132] The lodging also had its regional connotation: its champaign lay along the southeast coast of China:

> Beside the Minx Woman star the joyous vapors are incessant:
> On top of the Viet King's platform they have seated a true poet.[133]

The referent of the couplet is a man of literary talent given symbolic sway over the vanished Kingdom of Nam-Viet, under the rays of Woman. She is one of the most common of the lunar asterisms in poetry, and often appears in a purely fairy guise, as when Ch'en T'ao imagined that he might meet her cruising in her boat of stars,[134] very unlike the demoniac being suggested by the omens.

Mane: One of the cognates of this member of a family of hairy words is *mao* "yaktail," an object often used as a military pennon. Doubtless the fibrous image refers to the cosmic nebulosity in which the Pleiades are enmeshed, described most elegantly by Tennyson in *Locksley Hall*, where

80

Astrology

The Twenty-eight Lunar Lodgings
(FROM A TURFAN TOMB ROOF)

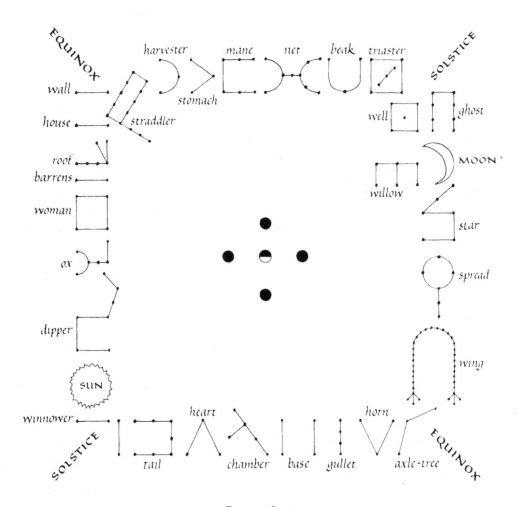

FIGURE 2.

they "glitter like a swarm of fire-flies tangled in a silver braid."[135] In many non-Chinese traditions this lovely cluster has been visualized in female form—for instance as a company of virgins by the Lapps, and a group of girls playing at corroboree by some native Australians.[136] No such pleasant connotations adhere to the Chinese asterism "Mane"; it is above all the sign of nomadic warriors, the Tibetans in particular, but it is also associated with other barbarians, with war, and with executions.[137] In T'ang times an astrologer named Wei Chien-su foretold the death of An Lu-shan, most de-

81

structive of all men of alien origin to the house of T'ang, when a planet intruded upon the constellation of Mane.[138]

Net: For us, since classical antiquity, the Hyades have been associated with rain, and they had a similar meaning in ancient China, where the new moon in that asterism was the presage of rain.[139] This is one of several lunar asterisms with specific meteorological associations; compare Well, in which imbalances forbode drought.

Triaster: This is not an invented word, although I have transferred it from its usual biological context. The old (Greater Seal) form of the graph shows a figure surmounted by three stars, suns, or phosphors (i.e., the modern form of *ching* "crystal").[140] Since there is no firm tradition which explains this configuration, I have simply named the lodging "three-stars." It seems reasonable to assume that the three are those which for us constitute the belt of the giant Orion.

Well and *Ghost*: Together these make up the Jupiter station named Quail Head, which rules the champaign of Shensi. They have particular reference to the capital city. "When Well-star is faintly luminous, wine and food diffuse their aromas. When Well-star is dark red, food and drink conceal poison. When Ghost-star is faintly luminous, libertinage (?) is on the rise. When Ghost-star is rayed or horned, the dead and gone will fill our eyes."[141] In modern belief, the Ghost asterism is made up of five stars disposed in a quincunx. These are the terrible "Five Ghosts," among the most destructive powers in the astrological hierarchy.[142] This evil reputation was already established in T'ang horoscope astrology; a person's health was affected by the relative position of the planets in his personal asterism and among the Five Ghosts.[143]

Axletree: This lodging is interesting in that it, like Well and Ghost, had a specific urban reference. The constellation, which lies south of the great gate of Heaven at the Autumnal equinox, represents the far south generally, but its central star stands for the city of Ch'ang-sha.[144]

Chamber: In judicial astrology this lodging represents the Hall of Light, the sanctum of the Son of Heaven: "When Chamber-star is luminous, the royal one will be illuminated."[145] But it was also a dragon horse, that is, one of the divine horses in the stable of the Son of Heaven.[146] The poets certainly preferred this version. So Li Ho, in one of twenty-three poems simply entitled "Horse," wrote:

This horse is no common horse:
Chamber Star is first and foremost his star.
Knock on the lean bones of his foreparts—
He still carries with him the sounds of bronze![147]

82

The quatrain is certainly ironic: it describes an emaciated nag whose prominent bones, smartly rapped, suggest the bronze fittings of a great warhorse—a noble courser of Heaven. Possibly the reference is to the poet's own jade.

The twenty-eight lunar lodgings had special meanings for the Buddhists and the Taoists. For the latter each ruled one of the many holy hills of China and a corresonding spirit realm above. For instance, Horn (Spica) presided over the spirit world of Mount Yang-p'ing as well as over unformed and endless reaches of the outer empyrean, while Ox had Mount Ch'ang-li under its control as well as the Nine Heavens above.[148]

Probably the *nakshatra* were even more important to those Chinese whose beliefs and habits, under the aegis of Buddhism, were impregnated with Indian thought and tradition—a very large body of men in T'ang times. For them, as for the Taoists, the stellar panorama showed a fantastic theocracy whose deities presided over the separate aspects of the physical world and human beings. The lunar lodgings constituted a very important sub-pantheon.[149] The natural taxonomy of living creatures apportioned among the individual *nakshatra* provides a glimpse of Indian culture, but at the same time yields a viewpoint which was evidently also entirely congenial to the men of T'ang China, as a celebrated Chinese Buddhist compendium of that era testifies. For instance, the deity in the lodging Horn rules over birds; the god of Gullet governs the destiny of seekers of the Holy Way; Chamber presides over travelling salesmen; Heart over women; Well over goldsmiths; Ghost over kings and magnates; Willow over snow, mountains, and dragons; Spread over bandits; Straddler over seafarers; Mane over water buffaloes; Triaster over the Kshatriya caste; and so on. Many governed regions of India: House, for example, is the lodging of the lord of Gandhara.[150]

In one account at least the presiding spiritual power was not always one of the recognized devas, such as Brahma in Ox, whose asterism is said to show the shape of an ox's head and to whom clarified butter should be offered, or Vishnu in Bird, whose stars show the form of a heart and to whom the flesh of birds is a suitable gift. Sometimes it was an abstraction, such as "Wood" for Triaster and "Fire" for Star; sometimes it was a deity conceived in moral terms, such as the "Compassionate Deva" for Chamber and the "Deva of Joy and Happiness" for Horn; and sometimes whole classes of spirits, such as the divine musicians, the Gandharvas, whose starry emblem is a horse's head, and who ruled from Harvester, and "the children of the sun and moon," whose blazon is a deer's head and who ruled from Beak. Particularly fascinating is the sun deity in the form of "a woman's dimple" in Triaster. This scheme, in its ninth-century Chinese form—the one upon

83

which I am drawing—does not agree with the usual Indian arrangements of the lunar lodgings or system of deities that are revealed in them, but it does agree with very ancient Indian belief in an important essential: it begins with the Chinese sign Mane, in the Pleiades, which was the Indian *Krittikā*, and marked the vernal equinox in remote antiquity before the equinox moved into Aries. As in ancient India, this medieval Chinese text associates this critical asterism with Fire (and so with Agni and the planet Mars), and describes its shape as like a razor.[151]

ATMOSPHERICS

Or going up with music,
   On cold starry nights,
To sup with the Queen
   Of the gay Northern Lights.
William Allingham, "The Fairy Folk"

"Meteorological" and "astronomical" phenomena were not categorically distinct for the early Chinese. Some descriptions of them treat them as exactly the same in kind, while differing in strength, persistence, or location. In the dynastic histories thunder, lightning, rainbows, hail, and the like are conjoined with earthquakes and storms in a "Monograph on the Five Activities." However, like stellar events, these were regarded as generated by the workings of pneumas and the interactions of *yin* and *yang*—but we rarely read of them as "embryonic quintessences" (*ching*) as we do in texts that tell the nature of stars and their high-born cousins. (There are occasional allusions to frost and hail as "quintessence of *yin*" in the "weft" interpretations of the classics of the Han period.)[152] These subtle events or creatures in the lower air, then, are not so subtle as those above them, but neither are they so gross as animals, vegetables, and minerals. Moreover, meteorological shapings differ among themselves in the degree of their supernatural power: rainbows are classed with clouds and other vaporous entities, but are more mana-filled—more truly numinous. The old book of *Huai nan tzu* told that rain was caused by a fusion of the pneumas of earth and sky, that a thunderclap marked a clash of *yin* and *yang*, that mists were confused minglings of the two, that rain and dew evolved when the *yang* pneuma was dominant, and that frost and snow congealed when the *yin* pneuma was paramount.[153] One of the "weft" books attributes atmospheric phenomena to the mode of union of *yin* and *yang* rather than the predominance of one over the other:

When *yin* and *yang* are amassed they make clouds;
When *yin* and *yang* are harmonized they make rain;
When *yin* and *yang* are congealed they make snow;

84

Astrology

When *yin* and *yang* are mated they make thunder;
When *yin* and *yang* are compressed they make lightning;
When *yin* and *yang* are intermingled they make male and female rainbows.[154]

We might wish for a little more precision of definition here, and more subtle explanations of the processes involved, but in the end we must be satisfied with the intuition that degrees of conjunction are somehow related to the showiness of the product.

Our sources are unclear about the spatial zones in which these clashes and fusions take place: are they up close against the blue dome along with the comets and novae, and with the sun and moon, or are they much lower, as they seem to be to the untutored eye? Wang Ch'ung, at least, had been definite on this point. For him, rain does not fall from the sky—that is, from *t'ien*, the celestial vault. Rather rain, snow, frost and dew are generated by the pneumas of the earth, and fall or condense just above its surface, all manifestations of the same natural process.[155] Indeed, the notion that the clouds—somehow closer to the invisible pneumas than their other manifestations—were generated by mountains, where they so often seem to huddle, was widely held. A tenth-century source expresses the belief in these words: "Clouds are the pneumas of mountains and rivers: the village people of modern Ch'in and Lung [Shensi and Kansu] call them 'floss of Heaven's lord'."[156]

Clouds were pneumas made visible, as ghosts are casual apparitions of spiritual beings. So also were other misty visions, such as fogs, aurorae, and lurking miasmas. All were classified and interpreted according to their shapes and colors, which could sometimes dazzle the eyes. Already in Han times—and probably much before—the forms assumed by plastic vapors, auras, and clouds were given supernatural significance: if they resembled flocks of sheep or colored tents of felt, they referred to the nomads of the northern steppes; if they could be identified as replicas of ships, and watery wastes, they represented the lands of the southern barbarians.[157] From such signs as these the probable course of significant future events could be anticipated—and such was the mode of the courtly aura-watchers. Wang Ch'ung again had taken a skeptical view: ominous vapors in the sky were comparable only to the signs of physical deterioration on the face of a dying man. They were the symptoms of disease, but held no political significance.[158] If he had disciples in T'ang they were silent on this point.

Of the lower atmospheric phenomena the most important were the rainbows, which the Chinese separated into two sorts. The more conspicuous one—the rainbow everyone sees—was called **ghung*, a word cognate with Chinese words for "dragon," "arch," "bow," and the like;[159] we know this as a primary rainbow. Its twin, the secondary rainbow, a paler counterpart outside the primary, whose inner rather than outer edge is red, was named *ngei—a word whose etymology is uncertain, but could have meant "child," that is, the progeny of the primary rainbow. The Chinese lexicographers like to think of it as a "female" rainbow. The two were respectively the embryonic essences of yang and yin—and their matings produced rain, snow, frost, thunder, and lightning. Or, to be more precise, yin and yang display themselves separately to our physical eyes as a pair of colorful arches, but show the effects of their cosmic copulations in flashes of unearthly light and crystalline florescences.[160] It was known in pre-T'ang times that rainbows had a special connection with the sun, but some texts also suggest that they were a "confused essence" of the Southern Dipper.[161] "Confused" appears to have something to do with irregular matings and fantastic couplings—the shadowy displays of excited dragon spirits, perfusing their sweaty hazes and drizzles with magical pigments. In astrology, they were—like so many other ambiguous nebulosities—ominous. Frequently they were tokens of divine disapproval,[162] especially of lechery in high places—debauchery with political implications.[163]

Dangerous white rainbows turn up from time to time in medieval texts. They had the alarming and ambiguous power of their multicolored fellows, but whiteness suggests ghosts and death, or, at best, direct supernatural intervention. White rainbows meant especially the unnatural domination of the female essence, and hence every sort of perversion of civil life.[164] Physically, white rainbows, or "fog-bows," are rare, but by no means nonexistent: they happen occasionally when the rainfall is extremely filmy and fine. Our eyes—and the eyes of the medieval Chinese—have also been bewildered by nocturnal rainbows. Properly these are "moonbows," formed when the moon replaces the sun against the dwindling drizzle. Red rainbows are equally prodigious, and equally normal, even though rare; they always flame out just before sunset.[165]

Next in potency to the rainbow was the aurora borealis:

Goblin Clouds, uncanny perfusions,
Winter thunders, summer snows,
Solar nimbi, drawn-out rainbows,
Stream of stars and crouching turtle;

With a surplus of yin the earth is moved;
With a shortage of yang the sky is split.[166]

Writing of fearful vapors in the T'ang dynasty, Yang Chiung concludes his list with a rift in the sky through which pours an unearthly light. This seems to be a description of at least some manifestations of the aurora.[167] These wonderful displays are caused by the interaction between blasts of ionized particles ejected from the sun and the upper atmosphere of the earth, modified by the earth's magnetic field. They are usually seen in high latitudes, but during periods of maximum solar activity they are even visible in the tropics. A notable instance was a display over Thailand when "sky and sea flamed throughout three nights."[168] These colorful shows often assume fantastic shapes, which are readily interpreted as the forms of supernatural entities. The great aurora of 1716 was described by Edmond Halley in these words: "There arose a very great *Pyramidal* figure like a *Spear*, sharp at the *Top*."[169] It would have been hard to resist the inference that a weapon of divine vengeance was hovering over the world—and such interpretations were the norm in China as well as in Europe.[170] The Chinese had many names for different kinds of aurorae, just as they did for variations among comets. They were not only rifts in the sky, but flags, streamers, curtains, and malignant glares. From Han times we have a report of an emission from the planet Venus "like a strip of linen stuck to the sky; when such is seen armed men will rise up."[171] This is typical of Chinese descriptions of the aurora, although it also shows some similarity with accounts of spectacular comets equipped with wavy tails. In the following lines I recapitulate some T'ang reports of lights in the sky that seem likely to have been auroral:

18 October 707: "There was a red pneuma extending across the sky; its light illuminated the ground; it ceased after three days."[172]

24 August 708: "A red pneuma to the limit of the sky; its light illuminated the ground; it ceased after three days; red vapors are omens of blood."[173] (This report from the *Book of T'ang* may be the same as the previous, from a Sung source, the date having been garbled in transmission.)

20 October 796: "There was a red pneuma in the night, like fire; it appeared in the northern quarter, mounting up to the Northern Dipper."[174]

[20 July–17 August] 825: A brilliant red display in the northwest cutting across the polar Palace of Purple Tenuity.[175]

7 May 828: "There was a red pneuma like blood in the northern quarter."[176]

25 November 868: This apparition is described in terms very like those applied to the Han aurora—if such it was—referred to above. It is given the name *ch'ang keng*, "prolonger of the *keng*-hour," ordinarily used for Venus as evening star: "There was a 'star' which came forth like a strip of shimmering satin; it extended across the void, was transformed into clouds, and then disappeared. This was in the 'allotment' of Ch'u, and refers to an appearance of 'prolonger of *keng*,' at which armed men will rise up."[177] It is conceivable that this and the Han example refer to unusual meteor trails, but the effect of a tremulous sheen seems more appropriate to an aurora of the curtain type. The connection with Venus is obscure: it seems possible that the name "prolonger of *keng*" may, despite our dictionaries, have been applied to more than one kind of fire in the sky.

A class of phenomena whose nature must be similar to those just discussed but is not readily identifiable was called "phosphor clouds" (*ching yün*), i.e., luminescent vapors signifying great good fortune and more particularly universal peace. What are these bright nebulosities? The Chinese omen books say "neither pneuma nor haze," and "like haze but not haze—like cloud but not cloud."[178] The classical descriptions suggest a source of light diffused, as it were, through a frosted glass, with transient flashes of haloes and glinting circles. A notable apparition is recorded for 25 July 710. Li Chung-mao, the unhappy son of the murdered Li Hsien (Chung Tsung), had been compelled to abdicate after a hardly noticeable reign, to be succeeded by his uncle Li Tan (Jui Tsung). The happy sign in the sky seemed to bless the advent of the new sovereign, and accordingly on 19 August of the same year he proclaimed the inauguration of a new era, named "Phosphor Cloud."[179] It seems that among the celebrations that marked this notable event was the performance of a dance called "Dance of the Phosphor Cloud," in whose honor we have a short ritualistic poem by that name penned by Chang Yüeh, who held a high ministerial post under the new sovereign.[180] This begins with the verse "Snow-flurry scintillation of the phosphor cloud."[181] The representation of this staggering sight must have given pause to the royal choreographers—but we can readily imagine that discrete pyrotechnics played a role in the masque. The notice does little, however, to help us identify the phenomenon that occasioned the courtly rejoicing. None of the official records suggests that these were particularly nocturnal phenomena, which might point in the direction of lunar haloes and noctilucent clouds. Possibly some of these reports told of unusual daytime apparitions of bright stars—let us say Sirius, Alpha Centauri, or Arcturus—at dawn or dusk, their light refracted through the icy traceries of high cirrus clouds. More often they must have referred to "iridescent clouds"—diffraction phenomena, whose colors are related to the size of component water droplets. These have nothing to do with sunset or sunrise coloring, but yield lovely hues of green, purple-red, and blue at any time of the day. They are especially attractive in flaky alto-stratus clouds. Some "phosphor clouds" may have been the faintly luminous band of interplanetary dust spread along the ecliptic which we call the "zodiacal light." This celestial glow, extending up from the horizon, is best seen on early spring evenings or early autumn mornings. Possibly, however, some apparitions of this class were indeed "noctilucent clouds," which are also fine particles of interplanetary dust about fifty miles above the surface of the earth, sometimes lit up by the sun when that star is already far below the horizon.[182] In fact, I have detected in a Taoist source an exact Chinese equivalent of the expression "noctilucent clouds"—*yeh kuang yün*. This occurs in a description of the costumes of cosmic deities dating from pre-T'ang times, and preserved in T'ang and Sung anthologies: it describes one divinity distinguished by a "damask skirt of seven-colored noctilucent clouds."[183] The Taoists were far better acquainted with the upper world than were ordinary mortals.

Another class of phenomena in the high atmosphere which struck the

Astrology

Chinese as ominous was solar and lunar haloes. These are concentric rings, analogous to rainbows, caused by the refraction or reflection of light through or from clouds of minute ice-crystals. Those produced by reflection are colorless, while those caused by prismatic refraction can be vividly colored, displaying a range from inner red to outer blue. They are both normal in the daylight sky.[184] The Chinese name for these attractive circlets is *yün*, and they were minutely classified according to their thickness and the combinations of colors that were most conspicuous in them. In general solar and lunar haloes betokened disastrous struggles for power, but if all five standard colors were plainly shown together, this was taken as a benign token for the empire.[185] A typical example is this one: "When the moon enters Heavenly Market and Ho [River] with a triplicate halo, men-at-arms shall arise, the roads of the subcelestial Realm shall be cut off, and the commanders of armies shall lose the advantage."[186] A spectacular event reported for 12 January 776, although it does not seem to describe a circular aureole, probably falls into this class:

The moon emerged in the eastern quarter with more than ten paths of white vapor, like strips of shimmering stain, penetrating Five Carriages, Net, Beaked Turtle, Triaster, Eastern Well, Carriage Ghost, Willow, and Hsüan-yüan [i.e., through Taurus, Orion, Auriga, Gemini, Cancer, and Leo]. In the middle of the night they dispersed and were gone. The omen reading is: "a female ruler is lethal; a white vapor is losses at arms;" Five Carriages governs the storage of weapons; Hsüan-yüan is the Rear Palace; the Lodgings are the allotments of Chin and of the Capital City.[187]

This purely atmospheric event, then, was for the official astrologers in no wise different from occurrences of comets and novae in the depths of outer space. All were baleful and unusual lights, showing the uncanny and unwished-for attention of mysterious powers, whose intentions they dearly hoped to fathom.

METEORS AND METEORITES

I know the woe the meteor brings
When it on high doth flaming range;
Storm, famine, plague, and death
   to Kings
War, earthquake, flood and direful
   change.

      Guy Gilpatric, "Star Dust and Corn,"
       in *Three Sheets in the Wind*

The clouds of miscellaneous debris which appear to us periodically as meteor showers are probably cometary waste. These heavy, ragged clouds may even follow the orbits of comets, as the Aquarids of the beginning of

May trail along the path of Halley's comet.[188] They have always been portents to mankind everywhere, and have regularly been confused with comets and even occasionally with aurorae.[189] Like comets, meteors are often described as swords or banners in the sky, and are intimately related with warfare on earth. They may even be seen as actual combat above the clouds. These commonplaces are as true of China as of other places. In the case of China, however, we have fortunately inherited seemingly reliable and detailed records both of regular meteor showers and of the spectacular plunges of individual bolides. But despite all of these carefully kept data I have seen no evidence that the Chinese took note of the periodicity of the great showers.[190]

For the Chinese, meteors were stars like any other stars, and they were called "stars," with some epithet appropriate to their unusual mobility prefixed—such as "flying," "streaming," or "running." They were also said to have "tails" (wei), just as we speak of the "tails" of comets, which the Chinese called "awns" (mang).[191] Meteors were further subdivided according to brightness, apparent size, and the nature of their movements. Thus relatively dull ones symbolized the common herd of men, but brilliant ones represented the gentry; random showers stood for the dispersion of peoples; undulant or sinuous motion showed forth plots and conspiracies, and so on.[192] Certain set phrases recur over and over again in reports of meteors and meteor showers. Very common is "the trace of its tail adhered to the sky," as if some celestial painter had drawn a brush loaded with phosphorescent paint across the astrodome. Often linked with this is the statement that the surface of the earth was lit up. Such high illumination, coupled with persistence, was particularly ominous. Great numbers of meteors skittering this way and that stood for sheer numbers of individual people without regard to identity, while great fireballs that plunged towards the horizon with a thunderous roar told of high political dramas and their actors, and the tactics of great warlords.

More particularly, in astrology, the baleful intent of meteors appears in their visible lunge into the ground, and the court notaries seem to have taken particular note of such events, expecially when they occurred during political and military crises. They represented the direct retribution of Heaven in the form of the physical attack of a star—a divine missile. For our period, the years which saw the fall of Sui and the slow struggle of T'ang for ultimate primacy against many powerful rivals were marked by several notable examples, as we shall see. But there are lesser instances in abundance. A typical one tells of the magnate Ling-hu Ch'u, who, after taking command of the T'ang armies "south of the mountains," fell ill. A great star fell over his bed and lit up the whole courtyard outside. The warlord took leave of the members of his household and then passed away.[193] The actual fall of a burning meteorite in the vicinity of an army, and above all directly into a camp, was an infallible sign that the army was doomed.[194] To a lesser extent, the portent was adjudged from the apparent point of origin of the shower among the asterisms, or its destination. In such cases it was usual that the inherent significance of the constellation counted for much less than the overwhelming sense of death and destruction brought by the meteors themselves—as compared with the entry of planets into an asterism, for example, when they were often no more than pointers.

Although different types of meteors were often given appropriately fierce and threatening names, by far the most common label affixed to them was "streaming stars" or "flowing stars" (liu hsing), using a word that connotes "following a current in a set direction."[195] The books of omen lore are replete with their malignancies: "If a broom star or streaming star enters it [the Literary Glory asterism], a great commander will rebel, with revolutionary disorders;"[196] "if a streaming star violates Attendant Woman, none of the people will preserve his house and family;"[197] "if a streaming star violates and holds to Southern Dipper, the Prime Minister will be ostracized and stationed on a remote frontier."[198] And so on—a repertory of disasters. When the meteor seemed to be aimed at a planet or a lunar lodging, like a dagger thrown at a throat, special meanings could be assigned. For instance, if Mars were the object of the attack, there would be devious kinds of subversion among the magnates, and cares for the ministers of state.[199] But if the meteor shower should flood the asterism Tail (in Scorpio)—a lunar lodging that was attuned to the royal seraglio—it could be inferred that members of the queen's family would be put to death.[200] Rarely, a meteor brought good tidings: "When the color of a streaming star as it breaks up and enters the east is like the rosy red clouds of dawn, there will be joy in the eastern quarter."[201] And so it was for the other directions: color was determinative. But usually a meteor was fraught with dire consequence. Consider now a "streamer" in a poem by P'ei Yüeh, written early in the tenth century, a time of doom and dismay. The setting is a gaming board: the progress of the military game which we know by the Japanese name of go, a game with ancient connections with both strategy and divination, has been halted for a while. The forces of good—the government—and the forces of evil—the rebels—are at an impasse. The nineteen "roads" are the lines ruled on the gaming board, marking the various directions of play:

> Nineteen strips of level road—
> "Level" we say, but perilously steep.
> Human hearts have no vantage-point for contriving,
> Though the nation's hand holds a season for winning.

91

The trend has shifted—the streaming star is far off;
Humors dry up—the falling hail is slower.
HE approaches the balcony for one more game—
The cold sun hangs once more in the west.[202]

This is a mysterious, scary, doom-laden poem. It ends on a note of inde-cision, but with some prospect of success offered by a new game: the nation has a secret advantage. The faceless "He" of the seventh verse appears to be the Son of Heaven, the crucial player.

Much less common is the expression "flying star," a term which sug-gests somewhat more vigor and independence than "streaming" or "flowing" star. We have a standard distinction in the following pair of definitions: "When the traces of light are linked with each other, it is called 'flowing star'; when the traces lead off in fragments, it is called 'flying star'."[203] The aptness of the definitions seems arguable, and the reputation of the scholiast is doubtful. I prefer to think of "flying stars" as characterized by more whiz and vim than "flowing stars." On the other hand, we may be deceived by taking a rigid view of "flying." Tu Fu writes, in a midnight poem set far from home:

Flying stars cross the water's whiteness;
Dropping moon moves the sandy waste.[204]

It would seem gratuitously slighting to assume that Tu Fu was using a trite image for meteors and that by "move" he means only "move across." If Tu Fu was a master of writing, as is generally assumed, it seems only fair to take him at his word: the dropping moon is the moon descending from the zenith; the flying stars, then, are the fixed asterisms descending as rapidly as the night's passage—but reflected in the water. (The word "moves," of course, as so frequently in Chinese poetry, represents the seeming motion of a fixed background or foreground against a truly moving but apparently stationary object—as skylines against shifting clouds. Here, the apparent shifting of the mobile sands is an illusion caused by the shadows cast by the setting moon.)

"Running stars" (pen hsing) is a rather rare and bookish expression,[205] as is its handsome homonym "spurting stars" (pen hsing), but even the lexicographers are somewhat at a loss; might they not also be novae or comets?[206]

"Crooked arrows" (wang shih) are hairy (in the good western tradition of cometae) but distinguished by their wavy path and dark color. This is rather troublesome. It is easy to suspect that the early Chinese were often at a loss how to apply the venerable term to the proper apparition, just as we are. Some such shining bolts seem to have been meteors in irregular flight, but perhaps some were comets, and the appearance of others—like an unfurled roll of cloth—strongly suggests the aurora borealis.[207] Whatever the proper form—if indeed it was not believed to assume many guises, the serpentine figure being determinative—"crooked arrow" betokened evil on the highest level of government, the union of seditious armies, for instance,[208] or regicide by a great lord,[209] and comparable deeds of horror.

In poetry meteors had yet another name—a very apt one, but one that suffers from ambiguity: "fire star." Vivid as the connotation of this expression was, it was readily adaptable to a variety of referents other than sparkling meteors. It was regularly used of the planet Mars, of the red star Antares, and even of the puny sparks shed by mortal fires. Indeed it is not easy always to decide with certainty whether a given "fire star" is indeed a star, a planet, or a bit of incandescent debris, whether celestial or terrestrial:

> Night deepens on the purple paths—dew drips from the sophoras;
> Snow is vanished from the cyan void—star(s) of fire
> stream(s) by.[210]

This couplet is from a rather feverishly lit night scene in Ch'ang-an, and either the baleful light of Mars or a swoop of fiery meteors would fit the language well.

Ghastly in the extreme were the Heavenly Dogs which leaped from the sky to terrorize humanity: "a sky dog is like a great running star, but makes a noise . . . it shatters armies and kills commanders."[211] Evidence like this makes it certain that these canine stars were explosive meteors which struck the earth or rushed through the lower atmosphere as fire-balls.[212] But there was also a constellation called "Dog of Heaven," in Cancer, southwest of the lunar lodging Ghost, and the lodging Tail in Scorpio was also sometimes given the same name.[213] The demoniac aspect of the deity was well known in antiquity. Possibly the designation of a brontide as a sky dog depended on what part of the sky it emerged from. At any rate, human sacrifices were made in T'ang times to appease the glowing monster—although evidently the government took some pains to cover up these needful offerings, lest everyman assume that he was a potential victim. The agents of the (supposedly official) priests who obtained the needful materials for the holy rites were described in the following lurid terms: "When they come, their bodies are clothed in the skins of dogs, with iron claws. They always take the hearts and livers of men, departing under cover of darkness." The citizenry, shaken and wary when they heard rumors that these official werewolves were abroad, ventured from their homes only if armed with bows or swords.[214] The official histories take an attitude of incredulous contempt towards what was obviously commonly accepted belief. But cases of these alarming if pious activities were still lodged in the imperial archives: possibly the error of the man in the street lay in putting the responsibility on great ministers of state—exalted theorists—when it belonged in fact to minor and more practical ritual functionaries.

While I have not tabulated here the T'ang reports of comets and novae, already well studied by other scholars, a table of recorded appearances of meteors—or what seem to have been meteors—follows. These are, of course, of less interest than comets and supernovae to historians of astronomy, but I have chosen to set them down, since I know of no comparable list elsewhere. Perhaps some of the famous showers, such as the Aquilids (e.g., the red meteor of 11 July 834) or the Pegasids (e.g., the brilliant meteor of 14 June 644) may be identified and peculiarities noted. Moreover, the student of astrological omens and divination may wish to consider the meteors from that point of view. I have, in a few instances, suggested the probable event to which the court astrologers, in their retrospective wisdom, thought the apparition applied, but for most of these reports I have left it to the interested reader to investigate, for fun or profit, the likely referents.

14 January 616: A "flowing star" fell on the encampment of the outlaw Lu Ming-yüeh destroying some of his military engines and killing a dozen or so men. The "standard prognostication" here was, "Where a running star falls, it will smash the army and kill its leader." That year Lu Ming-yüeh was destroyed. The same source gives two further examples of similar phenomena in 617, both bringing about—immediately or remotely—the loss of a nation, the death of a king, a great battle, the destruction of an army, and the killing of its general—all as predictable.[215]

29 November 620: A meteor fell inside the Eastern Capital with thunderous roar. A tactful courtier explained to T'ang Kao Tsu that precedent showed this must signify the death of a great villain at the place where the "star" fell. So it turned out: Loyang was the scene of the overthrow of Wang Shih-ch'ung, a powerful rival of the rising Li family, and his abortive dynasty.[216]

22 July 642: A great meteor "like the moon" appeared in the west. Unriddled as, "When the star is exceedingly large it is the Lord of Men." (The same thing is said of the moon itself, but the riddle remains.)[217]

14 June 644: A brilliant meteor fell thunderously from the Eastern Wall (in Pegasus, especially white Algenib on the equinoctial colure). Unriddled as, "With a sound like thunder, it is a sign of anger."[218]

[27 October–25 November] 653: A woman named Ch'en Shih-chen led a rebellion from Mu-chou in Chekiang, claiming supernatural powers. A star fell on her camp. On 27 November she was finally defeated and her head was cut off.[219]

17 March 708: A meteor fell with a thunderous roar "in the southwest." "All the wild pheasants screamed."[220] The meaning is obscure; pheasants were often associated with fire, the planet Mars, and martial authority generally, and with the south. But sometimes they were evil omens—examples of wild creatures found inappropriately in civilized places, foretelling their wasting.[221] (Compare the similar event of 31 March 811.)

Astrology

3 *October 711:* A meteor passed from Auriga—curiously in large part coincident with the Chinese "Five Carriages"—to Ursa Major.[222]

28 *October 711:* A meteor in Ursa Major. The history relates this to the dismissal of four great ministers on 17 November.[223]

4 *March 712:* A meteor in the polar region, which disappeared in Ursa Major.[224]

[9 *July–6 August*] 712: On the evening when the great marshal Sun Ch'üan set out on a punitive expedition against the Manchurian tribes a "great star" fell into his camp.[225] On 27 July his army was utterly crushed.[226]

15 *July 714:* A great display of meteors crossed the pole, and "the stars in the sky all shook." By morning the sparkling host was gone. The diviners explained that the mass of stars represented masses of people, and that they flowed across the sky symbolized folk who had nowhere to go. The *Han shu* was cited to the effect that "When the stars are shaken, the people are shaken."[227]

27 *October 724:* A meteor "like a peach, its color red-yellow," which lit up the ground. Interpreted as "When the color is red it is a messenger of the Army Leader."[228]

4 *April 744:* A meteor "like a moon" fell in the southeast, followed by a noise. It was falsely rumored in Ch'ang-an that the magistrates had sent out agents to obtain human livers to sacrifice to the Heavenly Dog. There was so much alarm that messengers had to be delegated to pacify the citizenry.[229]

19 *May 757:* A rebel general, Wu Ling-hsün, had laid seige to Nan-yang; a great star, colored reddish-yellow, with a train several hundred feet long, fell into his camp.[230]

? 757: A star with a long serpentine tail exploded into fragments of light. This was called a "crooked arrow."[231]

15 *October 767:* A large meteor with a yellow tail came out of the south and disappeared in the northeast. One source notes a rebellion of the Lao tribesmen in Kuei-chou that autumn—this would explain the southern source. Another connects the apparition with the northeast part of China, but with no specific event.[232] In any case, there were many other evil omens in the latter part of this year.

19 *October 768:* A brilliant north-bound meteor was interpreted as a "noble messenger."[233]

20 *June 798:* A star fell in the northwest.[234]

17 *January 808:* A meteor crossed the sky in the north; its tail broke up into scattered beads. It was taken to be "a noble messenger."[235]

31 *March 811:* A great meteor fell out of the sky with a thunderous roar. The wild pheasants cried out. A red vapor like a snake arose from the spot where it fell. The sun was in the zodiacal sign *hsü*, associated with the destiny of Shantung. The history notes that "within ten years the wild lord of that place was killed and his land divided."[236] (Note: "wild pheasant" equals "wild lord"; the symbolic color of pheasants is red, that is, fire.)

27 *April 814:* A great meteor in the north, whose long tail lit up the earth. It vanished west of Boötes.[237]

27 *February 816:* A long meteor appeared from west of Gemini.[238]

13 *August 818:* A white star with a long tail plunged into the southeastern murk.[239]

11 *September 818:* A great red meteor passed from east to west, where it disappeared in Aquarius.[240]

18 June 819: A great meteor moved from the Dipper to Leo and there disappeared. This was taken to be a sign of an amnesty![241] (Cf. 2 February 825 below.)

4 September 820: A great meteor ran from the vicinity of the Lesser Bear to the north of Aries.[242]

25 February 821: A great red star with a tail-trace thirty feet long appeared in the south, originating north of Sirius (the Wolf). It went to the northeast and disappeared near Hydra.[243]

? 821: A great yellow star with a tail sixty to seventy feet long appeared in the northwest. Illuminating the ground, it disappeared in Aquarius.[244]

21 May 822: A brilliant meteor exploded noisily from the "Wall of the Heavenly Market" (Ophiuchus, Serpens, and Hercules) and disappeared in Coma Berenices. The text hints, but does not clarify, a connection with the palace guard.[245]

7 September 822: A great star flowing from the east to west vanished in the Pleiades with a thunderous sound.[246]

27 August 824: "A great star emanated from northeast of Heaven's Great Army Leader [an important asterism in Andromeda, especially the orange-emerald-blue triple Almach] and vanished in the murk."[247]

29 October 824: A large meteor went from the Sky Ship (in Perseus) to the polar region and disappeared there. The reading: "There will be some affair concerning boats and oars."[248] (This is a good example of the common type of stellar omen where the meteor—or comet or whatever—serves only as a neutral marker, drawing attention to the significance of an asterism.)

12 January 825: A meteor emerged from the "Double-floored Way" (in Cassiopeia) in the northwest and disappeared over the pole.[249]

2 February 825: A brilliant meteor fell from the north pole into the horizon murk. Divined as meaning "there will be an amnesty."[250] (Compare the meteor of 18 June 819; it seems probable that a meteor originating at the North Pole represents a messenger from the imperial court bearing an amnesty).

15 September 825: A meteor went from the north pole to the handle of the Southern Dipper and disappeared.[251]

5 July 826: A meteor with a long and brilliant tail came out of the northwest and vanished into the Heavenly Market. Interpreted to mean "There will be an execution" (because public executions were held in the market-place).[252]

27 August 826: Meteor heading south just at sunset.[253]

6 September 826: A great meteor emerged from *wang liang* and vanished at the handle of the northern dipper. The text notes that *wang liang* designates an officer in charge of carriages.[254]

22 July 830: A great shower of meteors of all sizes, lasting through the night. Comments in text: the people lose their place; the king loses the Tao; the cords of Heaven are done away with.[255]

11 July 834: A red meteor with a long tail which lit up the ground and disintegrated into beads went north from "Drum of the Ho" (in Aquila) to "Heavenly Cudgel" (in Draco). It vanished with a thunderous crash. Omen: the Ho Drum is the Leader of the Army; the Heavenly Cudgel is the military preparations of the sovereign.[256]

22 July 835: More than twenty meteors, great and small, crisscrossed the sky in the vicinity of the sky river.[257]

21 December 836: A large meteor with a bright persistent tail "congealed and adhering to the sky" in Ursa Major.[258]

18 December 837: A star fell at Hsing-yüan (in Shensi).[259]

Astrology

18 May 838: A brilliant display of long duration in the east in Ophiuchus, vanishing in the southeast.[260]

30 July 838: A brilliant meteor with tail "more than thirty feet long" came out of the east and disappeared with a sound like thunder.[261]

27 September 838: A meteor in the north: "traces of its tail congealed against the sky, and the radiance of its light illuminated the ground."[262] (The common formula of *T'ang hui yao* descriptions.)

13 April 839: More than two hundred meteors with persistent tails passed westward from the zenith.[263]

30 April 839: More than a hundred meteors of all sizes, everywhere in the sky.[264]

3 October 839: A meteor with a tail over eighty feet long emerged from the Forest of Feathers (in Aquarius). It vanished with a thunderous sound. "The Forest of Feathers is the Army of Heaven."[265]

5 December 839: A brilliant meteor came from the right leg of Orion and vanished in the south.[266]

23 June 841: Fifty small meteors seen about midnight.[267]

21 July 841: Dozens of small stars crisscrossing the sky through the night. "The omen is: Small stars are symbols of the people."[268]

23 July 841: A meteor, illuminating the ground, went from north to east. There was a noise like thunder.[269]

22 December 841: A great meteor from the northeast, lighting the ground, with a noise like thunder. "A mountain collapsed" (great avalanche). A comet appeared in the lunar lodging House (in Pegasus) and lasted for fifty-six days.[270] (Evidently all of these evil omens portended something dreadful. Nothing is immediately obvious, although serious ravages were committed by the Uighurs in the north.)

22 July 842: A meteor crossed the sky in the north; there was a plague of locusts; the comet reappeared.[271]

30 March 846: A reddish meteor, like a peach, whose tail illuminated the ground, appeared in the northeast and moved west to penetrate the wall of Purple Tenuity.[272]

1 August 865: A great meteor with a long tail as brilliant as a flash of lightning appeared, followed by a host of small stars, going from south to north. Its meaning is that a rebellious host will march from north to south.[273]

[13 August–10 September] 877: A large meteor entered the Forest of Feathers (in Aquarius). Read as "Weapons without," presumably meaning war with foreigners.[274]

21 November 886: A white star with a wavy tail fell out of the west. Interpreted to mean the spilling of blood.[275]

[27 May–24 June] 887: A great star fell out of the daylight sky and fell into the camp of Ch'in Tsung-ch'üan at Pien-chou (K'ai-feng). This symbolizes the death of the commander. (Ch'in Tsung-ch'üan was crushed by Chu Ch'üan-chung later this year.)[276]

[Summer] 894: A star fell in Yüeh-chou (in Chekiang); then a light more than ten feet long, shaped like a serpent. This was called a "crooked arrow."[277]

[14 July–12 August] 896: A star like a teacup rose in the southwest and fell in the northeast with a sound like a flock of ducks. Interpretation: traitorous plots.[278]

[25 November–24 December] 900: A great star drifted slowly across the zenith from the east, like a wavy light with a persistent trail. It was a crooked arrow.[279]

30 June 904: On a dark rainy night there was a star trailing two hundred feet from east to southwest. It had a black head and red tail, with white in the center. This was a crooked arrow; it is also called "long star."[280]

13 April 905: A great star, with a tail like a red and yellow flame, moved like a serpent from the northwest, accompanied by a rain of small meteors. Later a greenish-white vapor "like a bamboo thicket" rose to the sky: a crooked arrow.[281]

[*18 December 906-16 January 907*]: A star like Venus rose slowly in the east until it reached the zenith, where it resembled a half moon. Then it moved crookedly and divided into two. The omen: "There will be a great scourge."[282]

In the late classical Mediterranean world some authorities thought that the angels of heaven were made of "immaterial fire."[283] The description is not alien to the Chinese notion of messengers (*angeloi*) sent by Heaven—or by Heaven's surrogate on earth. They were fast, fiery beings, members of the host of stars, void of crass earthly matter, with missions important both for gods and mortals—in short, meteors. The analogy was not lost on western poets. Milton described the angel Lucifer in terms which would have been entirely comprehensible to the men of T'ang:

> Sheer o'er the Crystal Battlements: from Morn
> To Noon he fell, from Noon to dewy Eve,
> A Summer's day; and with the setting Sun
> Dropt from the Zenith, like a falling Star.[284]

Granted we have to do with a fallen messenger, lapsed from divine service —the point remains the same.

Just as all the stars had their appointed role and meaning, so the subclass of mobile planetoids and starlets had errands to run. Among them was the rapid moon: "The moon, now—it is the chronicler of the company of *yin*, the envoy of Highest Heaven."[285] But among the noble servants of Heaven who carried information or admonitions to mankind, the meteors were preeminent, as testified by a multitude of texts. "A flowing star enters the Southern Dipper: there is a messenger who will come to the nation."[286] Here the grammar suggests that the flowing star is symbolic; other texts state bluntly that it is the actual messenger descending from the sky. Neither interpretation is accurate: according to the doctrine of correspondences the vision in the sky and the agent on horseback—if he can immediately be identified—are ultimately the same, two faces of a single being.[287] The Chinese expression is *t'ien shih*, translatable as "envoy of Heaven," "messenger of Heaven," or "agent of Heaven." But sometimes meteors were called "mandated stars" (*ming hsing*)—that is, they carried the commands of Heaven for implementation below.[288] The terminology, like so much else of the vocabulary of celestial beings, was readily adapted to eulogy, and even to flattery. The poet Chang Chi, praising his late friend the eccentric genius Han Yü, wrote of him as a condescending star:

In life he was the very figure of a great statesman:
A heavenly messenger—he gave light to our T'ang![289]

"Star Messenger"—this shall be our standard name for these celestial couriers—were sometimes styled "messenger stars" (*shih hsing*), as in an ancient divination text which says "When a streaming star is like a pot or basin in front, glistening white behind, and just before its entry moves haltingly like a crossbow trigger(?), we shall call that a 'messenger star'. The Lodging into which it enters will have good fortune in a year's time, or two years at the latest."[290] Since meteors were normally ill-omened things, this class contravened the general rule, making it an important matter to recognize its representatives. The incandescent pot referred to in the text just quoted could not have been a common apparition—its peculiarities seem to disqualify it as the "messenger star" so often mentioned by T'ang poets, which was regularly a bearer of good news, or else simply caused rejoicing by the fact of its presence. Ch'en Tzu-ang, lying ill, was informed that "two stars had entered Well"—that is, had floated into the Lodging that represented the region of the capital. Presently he heard a bullock-cart arrive, and it was indeed two expected messengers—an event he describes as a "happy meeting."[291] Or sometimes it was a distant and inferior prince who had cause to rejoice in such a visit. In a poem written on the occasion of seeing off a friend as appointed ambassador to the Nan-chao nation, Ch'üan Te-yü wrote:

Off towards the southwest goes the messenger star—
Far into the distance, carrying through the goodwill offer
    of our levee.[292]

In this couplet the messenger star is an envoy to a nation conceived as vassal to the Son of Heaven, who carried the shining light of civilization in his own person. The symbolism is obvious—but we may be certain that, even if obscured by clouds, his high counterpart and simulacrum was making the transit of the evening sky.

Statistically, the construction "star messenger" was far more common than "messenger star," but practically, its referents were the same: courtly imperial agents, thought of flatteringly as irradiated with the divine light which brought fortune to their destination. Repeatedly these lucifers were described as descending "from Heaven"—implying of course that their terrestrial version was a representative of the Son of Heaven. So a ninth-century writer and high official, Lu Wo, in a setting of splendor and excitement at "the Post-station of Welcome Fortune" describes the approach of a bearer of glad tidings from the divine court:

The star messenger descends from Heaven with the vermilion decree.[293]

Or again, in such a suitably holy environment as the guest's quarters of a Taoist friary, the envoy might be described as descending from the "Transcendent Capital,"[294] or born by angel's wings (as it were) out of the sky city.[295] Indeed, the close analogy of a star messenger to a divine being of Taoism was obvious and frequently exploited in verse, as in the example of a parallel drawn between "star messenger" and "cloud transcendent" by the poet Sun T'i.[296]

A special variety of star messenger in T'ang times was the "star esquire" (*hsing lang*), an honorific epithet bestowed on palace administrators holding the title of *lang chung*. The latter officers were, in effect, the executive secretaries of the Six Departments (Rites, Arms, Works, *et alia*) which dealt with the practical problems of the government. They were mandarins of the fifth degree, and did not enjoy the prestige of their department heads (*shang shu*), nobles and members of the Privy Council whose functions seem to have been in considerable degree restricted to the making of policy and to participation in state ritual. One gets the impression that the literary politicians of T'ang were more often the boon-companions of the real businessmen of the imperial administration than of the noblemen close to the throne.

Similarly "star carriages" were the official vehicles that carried "star messengers" to their appointed destinations. They appear in verse in moderate abundance.

A meteor might have a more subtle and fruitful errand: it is reported that the Jade Woman of the Occult Miracles dreamed that she was visited by a "streaming star" which she swallowed. Eighty-one years later she gave birth to Lao-tzu.[297] This generative power was often attributed to rainbow dragons,[298] but as the divine stellar energy was itself germinal and embryonic, we must grant that a falling star was easily the equal of any misty sky-reptile. (The miraculous birth of Lao-tzu has been described in many ways: there is the alternate opinion, for instance, that the great lady swallowed a five-colored magic pill which fell from the sky—but no doubt this account is reconcilable with meteor doctrine.)

"The flowing stars fall, ah! making a rain!" These are the words of an ancient text,[299] and the Han scholiast Wang I comments: "Germinal elements of *yin* came down together like falling rain." This sounds straightforward and conclusive enough about the nature of meteorites. But the matter was never really settled. The planets remained firmly in place, and it could well be said that they shared the qualities of subcelestial rocks: indeed the philosophers' opinions about the subtle constituents of the "fixed" stars could easily be confuted by the evidence of human eyes: some "flowing stars"—that is, the ones we style meteors—actually reach the surface of the earth and are palpably stony. These might then be called "fallen stars" (*lo hsing*) although this phrase refers more frequently to the nightly descent of the stars from the zenith to the western horizon. Other synonymous names also occur, especially *yün shih* "fallen star-stones." The problem was how to account for them. The ideal would be a reconciliation of the "essence" theory with the "stone" doctrine. One solution is offered in a T'ang resumé of Buddhist wisdom, although the charming conception it offers is not especially Buddhist:

Some say that these are not stars, but are stones fallen from the Sky Ho. Accordingly, in vulgar books, the Sky Ho and the Earth Ho are linked together. Hence in Ho-nei [the province "within the Ho"] there are sometimes falls of stones.[300]

Other authorities, however, tend to take no account of the possibility that all meteorites are pebbles displaced from the bed of the Milky Way. The prevailing belief was that "falling stars" were transformed into "fallen stones," by some sort of terrestrial contagion that robbed them of their ethereal nature. So, in Han times, Wang Ch'ung, referring to the famous rock-fall on the ancient state of Sung, put it that "The five stones were stars . . . When stars leave the sky they are no longer spirits."[301] A source more or less contemporary with Wang Ch'ung—that is, one of the "apocryphal" commentaries on the classics—states it less crudely, adding the symmetry of cosmic cleavage:

The pneumas of *yin* and *yang*, in dense agglomerations, form germinal elements (*ching*): if these rise above as condensations they are stars; if they clot below as condensations they are stones.[302]

This solution was also suggested by the Buddhist compendium we have just cited with respect to the erosion of the bed of the Sky River. The passage begins with the establishment's view:

The vulgar say that the sky is germinal essence or pneuma: the sun is *yang* germ [*ching*] while the stars are the germs of the Myriad Creatures. The teaching of the *ju* [conservative pedagogues] is firmly based on this![303]

This popular opinion is contradicted by experience, the book observes. Stars fall to earth as stones; it follows that since they shine in the sky like the sun and moon, they must be pneuma like them, or else the sun and moon must also be stony. But if the stars are pneuma, and they float upwards against the sky whose ethereal essence they share, how could they move against the background of the sky? Or if, contrariwise, the sun and moon are stony, how could they remain suspended in the midst of pneuma? The solution that the "essence" of stars may change to stone is a possibility, even if an unsatisfactory one.

A writer of the early eighth century avoids this baffling dilemma by distinguishing the density—or perhaps even the substance—of stars and meteorites:

> When the Primal Breath was first altered,
> After the Form-possessors had erupted,
> Those endowed with a clear and shining exterior leaped up to
>   become starry chronograms;
> Those which received a heavy and turbid reserve fell down to
>   become the earth and stones.[304]

The lines just translated occur in a poetic effusion on "fallen star-stones" and must plainly be interpreted to mean that the difference between stars and meteorites was a matter of secondary quality rather than of primary substance. Accordingly, the transition from one state to another no longer seems such a puzzle. (The passage reminds us strongly of the description of the separation of heaven and earth in *Huai nan tzu*, in which the clear and sunny [*yang*] part of the primal substance thins out to become the sky, while the "heavy and turbid" part clots up to make the earth.[305]) By the same token, the boundary between sober fact and lively simile becomes difficult to discern—as in a poem by Li Po:

> Blue sky—how vivid and clear!
> Bright stars—like white stones![306]

Nowadays it is usual to classify meteorites roughly according to their composition. Siderites[307] are composed of iron and nickel; siderolites are mixtures of metals and other minerals; while aerolites are made up of non-metallic minerals—silicates and oxides—in short, stony meteorites.[308] Meteoric iron—that is, iron found in siderites—was used in many parts of the world before the development of an iron technology that permitted extraction of the metal from its natural compounds, such as hematite. Sometimes it was hammered into artifacts even in the presence of such a technology, possibly because of spiritual as well as superior physical qualities. Recently Chinese swords beaten from meteoric iron have been discovered. They seem to be of late Chou date.[309] Since the Chinese believed that meteors were star messengers, each with its unique message and that usually of war, the natural destiny of a meteor was to take the shape of a sword of power, a sword for kings and conquerors. Such may have been the original meaning of the new Chinese finds. In any case, there is no doubt that meteors were regarded as weapons. Some were "bolides" (fire-balls) that careened through the upper air.[310] If a meteor fragmented with a loud explosion (as a "brontide")[311] or preserved its unity until it plunged into the earth it was either a threatening or a dangerous thing—a prefiguring of the cannon-ball or explosive shell. Moreover, meteorites were regularly confused with polished stone tools surviving in the soil from the Neolithic era, which, to the Chinese, were numinous. Stone-age axes were regarded as the former weapons of the thunder-gods, hurled maliciously at the earth. Accordingly they were called "thunder axes" or "thunderclap axes" (*p'i-li fu*).[312] One such weapon, "presumably dropped by mistake from the grasp of the thunder spirit," is described somewhat ambiguously in a T'ang account. It has the attributes both of a manmade axehead and of a natural siderolite. But there may be fictional elements in the source—perhaps this accounts for the fusion.[313] At any rate, the reporter tells of a perforated axehead three inches long, "neither iron, nor stone." Could it have been hammered out of meteoric iron and provided with a hole by some magical smith in remote antiquity? That the porcine bat-winged Thunder Lords (whose cult was particularly well-developed on the Lei-chou Peninsula—which was in fact named for them) wielded swords as well as axes is demonstrated by a line by the poet Wei Chuang, written in A.D. 900: "Temper a sword for a thunder lord—its shadow will move the Ford of the Stars."[314] Although such a blade would appear most commonly as the flash of a meteor high in the sky, it might also melt and fall to the earth as a siderite—sky-metal to be reborn as the sword of a mighty warlord.

In recent times considerable attention has been given, both by geologists and jewelers, to an enigmatic kind of stony object which has been discovered in considerable concentrations in widely separate parts of the world. These are small glassy objects, shaped variously like rods, discs, buttons and teardrops—all in all, like splash forms. They tend to show a blackish, brownish, yellowish, or dark green color, with pitted surfaces. To some extent, they resemble drops of obsidian. They are called tektites. Chemical analysis reveals that they consist chiefly of silica and alumina, with small amounts of metallic oxides—especially those of iron, sodium, and the like, all common in ordinary earth rocks. But although some investigators assign a terrestrial origin to them, others have asserted that they are the debris of volcanic eruptions on the moon which has escaped the surface gravity of that satellite only to be recaptured by the earth and left in the soil of Australia, Indochina, and other places. Recently quantities have been found in southernmost China—in beach sands on the east coast of Hainan Island, and in Tertiary and Pleistocene deposits on the Lei-chou Peninsula, the domain of the Thunder Lords. In the last-named region they are well known as "stones of the Thunder Lords," and are so named because they are often exposed by erosion in the wake of passing thunderstorms.[315] It is reported that, although western peoples give these glass nodules such commonplace local names as "australites" and "moldavites" (for those found in

Czechoslovakia), the peoples of Indochina have named them rather more imaginatively as "devil's droppings," "droppings of the stars," and even "moon stones," all testifying to a belief in their superterrestrial origin.[316]

Tektites have long been known to the Chinese from the very region where they are found today. Two good accounts survive from the T'ang period, in both of which they are reasonably named "Ink of the Thunder Lords"—a name which aptly describes the appearance of those of them that resemble sticks of Chinese ink. I have translated one of these sources elsewhere.[317] Suffice it to note here that there they were said to emit a "bell-like clang" when tapped, and were "as bright and lustrous as lacquer." My other T'ang authority uses just the same language, but adds that plaintiffs and petitioners before a magistrate, bribing him with a gift of ordinary inksticks, must include an admixture of the precious "thunder inksticks."[318]

Whether or not tektites originated on the moon, it suffices that the medieval Chinese thought them to be natural objects from the sky. However, they were not the only species of star or sky detritus that was treasured by the gentry of T'ang. In late T'ang times, when rock gardens were in high fashion, meteoritic fragments became most suitable ornaments for a gentleman's establishment—much in the same way that they would have been cherished in the cabinet of curiosities of an eighteenth-century English squire. Doubtless their attraction was enhanced by a numinous aura lent by their former habitat among the sky spirits. This fact, too, was sufficient to make a meteorite a good subject for a poetic conceit. Po Chü-i, a notable rock-fancier,[319] referred to fallen stars more than once in his verses. An example is a double quatrain written to congratulate a friend on his reappointment to a prestigious office in the imperial library. The new archivist will be missed from his old neighborhood, where:

> On top of the fallen-star stone the gray lichens are old,
> In front of the painted crane parlor the white dew is cold.[320]

(In this period, it must be remembered, a fine and fantastic stone was thought by connoisseurs to be just as valuable as a fine painting.) The thought that such pleasing objects might be actual fragments of the sky is proved by the following couplet, by Han Tsung, in which it is suggested that the meteorite might choose to reassume its former identity and float up towards its former habitat. The poet is writing about a "fallen star stone" which was found in a field and transferred to his study in town:

> At what time will you assume your five colors,
> And rise away up to Nü Kua's sky?[321]

This conceit is more than matched by one employed by Po Chü-i, who

104

surmised that a stone he found and turned into a zither prop might have had a divine origin. Appropriately the poem, translated here in its entirety, is titled "I Ask the Stone that Props my Zither":

> Doubtless you fell out of space—bound to the lapse of a star.
> I sighed that you underwent muddy burial, plunged to the bottom
>   of a creek.
> Firmly fixed up in the sky must be superior to set on the earth,
> But propping a loom surely does not equal propping a zither.
> I carried you away and wiped you off—were you aware of my
>   kindness or not?
> Even if you cannot speak, it is fitting that you should have a heart.[322]

The loom alluded to is, of course, the starry loom of the Weaver Woman in Vega, and this stone, it appears, might well be a pebble she found in the bed of the Starry Han, lately and ignominiously removed to the sordid bottom of a stream on earth.

COMETS

> When beggars die there are no comets
>   seen;
> The heavens themselves blaze forth the
>   death of princes.
>
> Shakespeare, *Julius Caesar*, II, 2

When, on 16 April 1066, Harold Godwinsson came to Westminster to celebrate Easter he was greeted by a terrible sight: "Then over all England there was seen a sign in the skies such as had never been seen before. Some said it was the star 'comet' which some called the long-haired star."[323] The king was only three months crowned and his doom was already upon him. This great prototype of the monitory sense of a fantastic star in our own culture had close parallels in the Far East, where the lords of the land trembled for their thrones on the appearance of hairy stars, i.e., of comets.

For the Chinese, comets were messengers of doom, but they made up only one among many classes of ominous stars, and indeed were sometimes classed in one group, sometimes in another, depending on factors that we would consider superficial. Most prominent among non-specific terms for starlike apparitions—including comets—conceived as intrinsically ominous rather than as mildly numinous and taking on particular significance from the asterisms through which they pass, was the class of "uncanny stars" or "weird stars" (*yao hsing*). The various species of this genus had many senses and many referents, as did their analogues in medieval Europe.[324] A good general description of the whole group occurs in an esoteric Han text:

Pacing the Void

All alterations and aberrations of the effigies in the sky are based on induction from human affairs. Therefore, when effigies are formed by perverse pneumas, uncanny stars may be among them.[325]

Uncanny stars do not, as a class, correspond to any exact scientific rubric: they dazzle, they astonish, they terrify, but their mystery does not guarantee a constant genesis acceptable to the twentieth-century mind. Planets and fixed stars, dulled, enlarged, discolored, or deformed by atmospheric dust close to the horizon, probably account for some of them.[326] Even the medieval Chinese were sometimes hard put to classify them:

An uncanny star was seen: neither "broom" nor "explosive"—we cannot tell its name. Men of our time call it "uncanny star"—although some call it "star of evil."[327]

This statement, whose note of bafflement is unusual in the dynastic histories, evidently expects that classification as either "broom" or "explosive"—that is, as one of the two kinds of comet—would account adequately for most uncanny stars, but the specimen of late summer, 894, here reported, did not seem to fit either specification. The phenomenal aspect of a hasty, unsteady, unreliable star might support the view that comets were ruled out in certain cases. Comets are slow movers with a definite conatus: "There was an uncanny star that fell to earth in Fen-chou."[328] The language of this passage clearly points to a meteorite that survived the incendiary atmosphere of earth—but we are provided no clue as to the special quality that separated it from the usual "streaming star." Consider, on the other hand, the following anomaly, reported for 18 November 868:

An uncanny star emerged from "Woman"; it extended over the void as a strip of white satin, was transformed into a cloud and so vanished. This was in the allotment of Ch'u.[329]

There can be little doubt that this fine language describes the aurora borealis. It is our own concepts that need readjusting, however, not the Chinese language: *hsing* "star" means "shining heavenly apparition (not necessarily a glittering pinpoint)."

But in the overall the reported "uncanny stars" of the Chinese histories fit our definitions of comets, and these were almost exclusively calamitous visions.[330] Indeed we are warned of the intimate association of "uncanny" (*yao*) and "broom[-star]" (*hui*) in T'ang literature in the ominous couplet:

The Dipper's handle invades the uncanny broom;
The fount of Heaven transmutes to adverse scales.[331]

(The "adverse scales" on the throat of the dragon are as nasty a sign as the uncanny broom.)

Unlike meteors, comets move with great deliberation, and give no

106

impression that they are falling stars. Their slow progression through the constellations is comparable to that of the planets, and indeed the Chinese thought they saw an intimate relationship between planets and comets. Generally speaking, however, comets did not appear as distinct spots of light, but as blurred ones, often with a softly luminous tail pointing away from the sun. Tailless comets, which we call "aphelial comets," were plainly distinguished from tailed comets—"perihelial comets"—by the early Chinese although the former were sometimes confounded with supernovae ("stranger stars"). The former they styled *po* (M.C. *bwĕt), and the latter were *hui*.[332] *Po* suggests "bursting out, exploding," and the astronomical records frequently use the term as a "verb," having "a star opened out (in such-and-such a constellation)," for the appearance of an aphelial comet. *Hui*, on the other hand, connotes "sweep, broom," and was regularly used as an attributive to "star"—so a perihelial comet was a broom star or sweeping star. Naturally the aphelial comet normally developed a tail as it approached the sun closely, and it was then that it was most menacing. (Sometimes, however, the head never becomes more than a fuzzy ball.[333]) As the *T'ang shu* tells us, "An exploder and a sweeper are both extraordinary engenderings of evil vapors, though the disaster will be extreme in the case of a sweeper."[334] Exploders and sweepers (or brooms) differed in that the evil effects of their foul breath was more intense in the case of the sweepers.[335] They were the perfect and final tools for the cleansing of a corrupt civilization, and the preparation of the land for a new and purer order.[336]

A broom[-star] is for the removal of filth and the distribution of what is new. This is the means that Heaven uses to establish whatever adheres to virtue after the removal of whatever lacks the Tao.[337]

"Malignant pneumas" is not, for us, a very satisfactory description of the nature of comets, any more than "emanations from planets." In any case, these two conceptions of the genesis of comets ultimately mean the same thing. Other attempts to define their nature can be found in astronomical texts. In one, comets are essentially watery, and are intimately connected with the creatures of the deep sea—that is, with *yin* creatures. Accordingly, "When the whale-fish dies, a broom star comes forth. The whale-fish is a *yin*-creature and is born from the water."[338] In this at least—aside from the particular relationship with whales (which remainds us of the moon's special affinity with pearls)—comets had something in common with the Sky River, the moon, and even the stars generally. But they seem to have been less attended to as physical entities than these other inhabitants of the sky, and we see little if anything in the old books about a crystalline structure, like that which is constant in accounts of the moon and stars. Surprisingly too, comets may be threatening weapons, but the notion that they might be vehicles or messengers (as the meteors are) seems never to have occurred to the Chinese.

The official court astronomers, at least, knew of the relation between tailed comets and the sun before T'ang times. We have this excellent (from our point of view) account of them in the *Book of Sui*:

The body of a broom[-star] has no light, but makes a light which is transferred from the sun. Therefore when we see it in the evening it points to the east, and when we see it in the morning it points to the west.[339]

In short—the filmy tail is always driven away from the sun, and is made visible by it. (It should be noted that this statement of the derivative nature of the comet's light differs from medieval Chinese accounts of the moon's light, which was not a transference, but a reaction to a stimulus.)

Tailed and tailless were only the two most common modes of classifying comets. There were many others. Here I mention only a couple of examples from a list of subjective and casual names that seem even to have puzzled some of the scribes who recorded their advents. The uncertainty is understandable if we consider the following account of a "heavenly lancehead" (*t'ien feng*):

The "heavenly lance broom[-star]" has the figure of the point of a spear; when there are criss-crossed alignments Under Heaven the heavenly lance star is seen.[340]

The only clue provided here is the bald statement that this sky-darter is a species of broom-star—but the "criss-crossed" suggests the language of meteors. Editorial tampering with an ancient definition can be expected. There are also dubious features in the criteria for determining a "gouge of heaven" (*t'ien ch'an*), which is a "Broom star" with a hook shape, like a latheman's gouge—but the overall effect suits a crescent-headed comet with a curved biting edge.[341]

Aside from nomenclature, classification was sometimes based on comparable behavior. For instance, the meaning of the appearance of comets and novae in an asterism was often identical. The reason is plain: both were temporary stellar visitors among the fixed stars, and it would be surprising if their omens were different:

When a Visitor, or Sweeper, or Exploder enters into the Dipper, the barons will contend for authority and oppress the Son of Heaven.[342]

When a Visitor, or Sweeper, or Exploder enters Horn and has a white color, men-at-arms will rise in the nation, and there will be a Great Loss; the armies will also be defeated, and cities will fall.[343]

But let us leave these refinements and pass on to the standard cometary omens. First, in astrology the aphelial explosives were in all respects like the perihelial sweepers except in degree or amplitude. The exploders stood for armed uprisings, plots, and invasions—but for minor, insignificant ones, which could often be ignored. It is therefore possible to go on to the Brooms without further ado.

Tailed comets swept the sky clean, but the cleansing meant death. The omens are armed conflict, royal claimants and pretenders, avenging swords, spears, the execution of great men, widespread slaughter, insurrections in the marches (pointed to by the comet's tail), treason, female usurpers, conflict between suzerain and vassal.[344] In all of this the character of the asterism in which the dreadful creature showed itself was a factor—military or civil official, male or female, northern state or southern state. The message was always the same—disaster.

The possibilities for the poets were enormous, and they took good advantage of them. A favorite treatment was to represent the comets as ghostly, ectoplasmic emanations trailing across the sky. One anonymous T'ang poet, for instance, represents them as seeping through fissures in the dome of heaven, like the luminous curtains of the aurora borealis, to shed their spectral light both on its inner surface and on the frightened earth below:

> Flying light of Sweeper or Exploder illuminates Heaven and Earth—
> The Nine Heavens have cracked like tiles—to splash out
>   vengeful auras;
> The wail of ghosts sounds and resounds . . .[345]

The Buddhist poet Kuan-hsiu made a similar linkage in a poem full of melancholy and desperation, yellow leaves and blinding dust; battle and death are in the offing:

> The sharp awn of the uncanny star stabs at Yüeh—
> The set of the ghosts' keening is linked with Ch'in.[346]

In particular the T'ang poets liken the dreadful emanation from the sky to the stench of barbarian horsemen menacing the northern frontier:

> High in the sky a hunnish star explodes:
> Here among men the breath of revolt is ranging.[347]

The year was A.D. 758: Su Tsung was striving to bring stability to the shaken T'ang empire, eroded by internal and external attacks. The comet's glare lights up sweating chargers, humping leviathans, and all the dreaded manifestations of divine hatred. Its "horn" (possibly the "spike" showing ahead of some comets—but more likely another word for its tail, like "awn") reveals its malignance. Thus Chia Tao:

> Many years have fended off buckler and axe;
> But our present levee encourages wine and song.
> A covetous lord—with no white hair—

> When racing horses are crossing the Yellow Ho!
> Old houses destroyed by warriors' firing—
> In the new palace—memorials in abundance each day;
> While the uncanny star still has a horn—
> Many feet of steel are honed—and honed again![348]

This is a song of the sloth and effeminancy brought on by years of peace; a young self-indulgent prince who ignores the reports of invasion while the sky-sword reproduces deadly blades below.

Poetically, a comet—a hideous foreigner at the palace gates—could be treated as an emanation of the yaktail banner of the Pleiades, instead of (as the official astrologers had it) of a planet: "Polehead of the Yaktail sweeps the Purple Tenuity."[349] (The word rendered "sweeps" here is normally a noun, a perihelial comet.) But so great was the numinous power of Heaven, when transmitted through a good ruler, that even this ghastly vapor gushes in vain over the imperial palace of "Purple Tenuity":

> Uncanny star is sunk in his rain and dew;
> Harmonizing vapors fill the capital's gateways:
> The Supreme Commander's patriotic service is established—
> Our Illuminated Suzerain's lawful equipage has returned.[350]

(The "rain and dew" of this fragment of a longer poem represent—as frequently in conventional imagery—the rich liquor of the royal grace poured forth on the Chinese land: here it absorbs the poisonous gas from the corrupted sky.)

Such verses as these are commonplaces of the literary heritage of T'ang, when events in the sky filled the thoughts of men. The accounts of them that are sprinkled through the officially sanctioned annals may provide less pleasing metaphors, but some knowledge of the interpretations of court astrologers may help to illuminate our understanding of the hundreds of yet untranslated—and often misunderstood—poems in which comets figure prominently. We return then to the prosaic for pedagogic purposes.

The long record of comets observed by the Chinese has been well studied by competent authorities. In what follows I have essayed a short chronicle of those comets seen during the T'ang era that have some special interest. Among these are reappearances of Halley's comet, comets taken by the T'ang astrologers to have unusual significance, and comets which received marked attention in contemporary European documents.[351] As for the rest, the "uncommented-on comets of T'ang," I suppress them here to a footnote.[352]

22 September 634: An aphelial comet appeared in Aquarius, and moved on into Libra.[353] By early November it had developed a tail, and was considered ominous enough to excite a colloquy between Li Shih-min (T'ai Tsung) and members of his

110

court. The sovereign inquired of Yü Shih-nan: "This Sweeping Star that has appeared in the sky—what is its weird?" In reply, the minister cited classical examples to show that no personal blame attached to the ruler. But Li Shih-min was unwilling to accept an easy evasion: "I have had some self-glorifying ambitions, tending to diminish the gentlemen of my subcelestial realm." Heaven, he maintained modestly, had marked the hubris implicit in his meteoric rise to a supreme power and had sent this apparition as a warning.[354] The comet vanished some time during the ten days after 26 November.[355]

1 August 641: An aphelial comet was seen in a sector of the Palace of Grand Tenuity (*T'ai wei kung*, corresponding to some stars in Coma Berenices). Obviously the omen was aimed directly against the T'ang throne. At this time Li Shih-min was making an eastward progress towards T'ai Shan to perform the unique sacrifice to Heaven called *feng shan*. His advisers immediately told him that the omen was unpropitious and that the project should be discontinued. It was not until 8 August, however, when the cortege was in Lo-yang, that this was done. At the same time austerity measures—such as the simplification of the imperial diet—were temporarily introduced by way of penance. The comet disappeared on 26 August.[356]

18 May 668: A perihelial comet appeared in Taurus in the vicinity of the Pleiades and Hyades—asterisms which controlled the destiny of the northern barbarians and military affairs on the frontier.[357] On 27 May Li Chih (Kao Tsung) took the ritual measures of austerity, such as staying away from the chief basilica, and reducing his diet. His courtiers, especially Hsü Ching-tsung, a poet whose verses we have translated elsewhere in this volume, pointed out that although the comet's tail was bright, it was short, and so could hardly symbolize divine disapproval of the Son of Heaven. They urged, rather, that it presaged the imminent destruction of the kingdom of Koryŏ. Li Chih retorted that as master of all under Heaven, he was personally responsible for the Koreans, his own subjects, and that divine punishment would not be visited upon them, since he, their lord, was the guilty one. The comet vanished on 6 June.[358]

4 September 676: A perihelial comet appeared in the lunar station Well (in Gemini), which has power over the imperial capital. It had a brilliant tail which appeared to be thirty feet long—though some sources say only three feet—that is to say, stretching across 30° of the celestial sphere. This tail swept through the asterism *Wen ch'ang* "Literary Glory," which was linked with the destinies of accomplished men of letters. A commentary notes that the Censor Yüeh Yen-wei died soon after—obviously intending to let us know that this had been foretold by the comet. That month the Tibetans raided a frontier country, and the realm was troubled by droughts and flooding rains. On 17 September the usual austerities, including the abolition of musical performances, were decreed in the palace, to be followed by others on 11 October. The comet did not disappear until 1 November.[359]

17 October 681: "Liu Ching-ning, a woman of Wan-nien Township [in Ch'ang-an], entered the office of the Grand Notariate. She was riding a white horse and was dressed in white clothing, with a suite of eighty or ninety men. She came up to the couch in the Director's audience hall, sat down, and asked if there had been any disaster or strange event recently. The Director of the Grand Notariate, Yao Yüan-pien, had her held for questioning. That night a broom-comet was seen within the Heavenly Market."[360] This asterism is a complex arrangement of stars in Hercules, Ophiuchus, and Serpens, and had authority over the basic commodities of mankind, notably food and clothing. (Some said, however, that this event foreshad-

111

owed a change of capital city.) The comet's tail was fifty feet long—again, some texts say only five feet—and became shorter as it moved eastward across the sky to River Drum, a military constellation in Aquila. On 3 November it vanished.[361] (The woman was clearly a sybil, dressed in the spirit color white, with her entourage of dedicated believers. The Grand Notary's office made the official astrological evaluations on behalf of the government. Such manifestations of popular meddling in the royal monopoly of divine matters were frowned on.)

11 November 684: A star "like a half moon" appeared in the west.[362] There is surprisingly little to-do in the Chinese sources about this apparition of what must have been Halley's comet, which reached perihelion on 6 November of that year.[363] The demilunar shape apparently describes a broad comet flaring at both sides. The dynastic history only remarks laconically that this, like the moon, was an extreme manifestation of yin.[364] (There is doubt about the perihelial comet reported to have appeared on 6 September and disappeared on 9 October of this year. Was it an early view of Halley's or another one altogether?)[365]

28 July 708: An aphelial comet appeared near the Pleiades and northern Aries. Comment: "Allotment of the hu," i.e., this region of the sky is significant for the foreigners north of China.[366] There were troubles on the Turkish frontier later that year, after the relaxation of T'ang vigilance, and the portent probably referred to these.[367]

10 August 712: A perihelial comet which had been observed earlier in Hsüan-yüan (part of Leo) now entered the wall of the Palace of Grand Tenuity (mostly in Virgo and Leo)—in particular, a region marked by the asterisms representing the imperial throne and the Grand Heir, both in Leo, inside the wall. It finally disappeared in "Great Horn"—that is, Arcturus in Boötes. The T'ai-p'ing Princess and her clique, who favored the accession of Li Lung-chi, employed a soothsayer to persuade Li Tan (Jui Tsung) that this sequence of events signified the passing of divine favor from him to the heir designate. The maneuver was successful. Li Tan abdicated verbally on 31 August and Li Lung-chi acceded formally on 8 September.[368]

30 June 730: A perihelial comet in "Five Carriages" (Auriga). An aphelial comet was also reported on 9 July in Taurus. Possibly this was the same comet retreating from the sun.[369] This may be the comet reported by Bede for the year 729, although it is not easy to explain the discrepancy.[370] It is noticeable, however, that Bede also reports two comets, one going ahead of the sun, and the other following it—"multum intuentibus terrorem incutientes." Osric, King of Northumbria, died that year, and the Saracens were ravaging France.

16 May 760 (some sources have 20 May): A perihelial comet appeared in Aries. It was white, and had a four-foot tail (some sources have "several tens of feet"). It moved quickly through Taurus, Orion, Gemini, and Cancer to Zavijava (β Virginis). This "uncanny star" vanished during the first week of July.[371] This was a regular visit of Halley's comet, also recorded in Syriac texts. It reached perihelion on 11 June.[372]

26 May 770: A perihelial comet, "more than fifty feet long" according to one report, appeared in Auriga. The same comet was noted again on 15 June in the north. It was described as a white emanation of Venus. It vanished on 25 July. In gratitude a great amnesty was declared, including the reduction of sentences for capital crimes.[373] This is the great comet also reported by the Bishop of Hirta and recorded in Syriac.[374]

17 January 773: It rained mud. "That night a Long Star came forth in Triaster."

112

It stretched across the sky. "A long star is a kind of broom-star. Triaster is the asterism of T'ang." (Compare the comet of 852.)[375]

9 July 821: Tibetans raided the frontier, but were repulsed. Meanwhile a perihelial comet had appeared in the Pleiades. A characterstic Aldebaran omen: fighting on the northern or western marches.[376]

9 October 834: A perihelial comet was seen in the wall of the palace of Grand Subtlety. A tail "more than ten feet long" pointed to the west. It moved northwestward.[377] It was seen in the east on the night of 18 October, with its tail glowing fiercely.[378] This was one of many terrible portents of the disastrous "Sweet Dew" affair of 835 in the reign of Li Ang (Wen Tsung), when the young sovereign and some of his advisers tried to break the power of the eunuch-commanders of the imperial guard. The result was the brutal execution of some of the most eminent writers and statesmen of the time, and the virtual isolation of Li Ang behind the palace walls by the victorious eunuchs. The remainder of 834 and most of 835 were marked by other disasters such as a great earthquake in the capital, a storm that uprooted trees on the palace grounds, the occultation of Jupiter by the moon, the apparition of a great meteor, and, finally, near the end of 835, the abortive plot, leading to the death of the poet Wang Yai and many others.[379]

22 March 837: A perihelial comet appeared in the east near Sadalmalik (α Aquarii) with a tail more than seven feet long, pointing west,[380] towards the Southern Dipper in Sagittarius. On 24 March the tail had become strikingly brilliant. Li Wei, Prince of Chün, died on 4 April. By 6 April the tail was more than ten feet long, and the comet was moving west, with its tail gradually pointing more to the south. The tail was more than twenty feet long and three feet wide by 7 April, with the comet still in Aquarius. The tail was still growing on the next day. On 9 April, the comet had reached the Southern Dipper. Forty-eight female musicians and dancers were expelled from the palace. On 10 April the tail was fifty feet long, and had divided into two branches, one pointing towards Libra, and the other in Scorpius. On 11 April it was in Virgo, pointing north, with a tail sixty feet long and undivided. On the advice of the Astronomer Royal Chu Tzu-jung, the daily allotment of food in the palace was reduced to one-tenth of normal, and all construction on the palace grounds was halted "in response to Heaven's denunciation." Court picnics at the Ch'ü Chiang Park were discontinued. On 12 April the comet was moving northwest, and on 14 April the tail was more than eighty feet long. On 16 April, classical musicians were dismissed from the palace, and on the following day the young monarch issued a decree outlining the grave implications of the dreadful apparition, and announcing the need for further austerities, soul-searching, and penance.[381] An amnesty was declared. Wen Tsung wore only white clothing: the royal hawks and falcons were released. On 19 April the monasteries of the capital city were commanded to hold public readings from the "Sutra of the Humane King" (jen wang ching).[382] By 28 April the tail was reduced to three feet long, and the comet vanished in Leo. This spectacular comet was, of course, Halley's, which came very close to the earth on this visit.[383] It had reached perihelion in Aquarius on 1 March;[384] accordingly the Chinese records of its passage by our planet refer only to the first part of its outbound journey after rounding the sun. For the constantly harassed Li Ang this must have been a season of unbearable dread. (There is one anomalous report of a perihelial comet in Aquarius on 9 September.[385] Perhaps an error in the court records?)

11 November 838: A perihelial comet appeared in Corvus; it was thirty feet

long, extending east and west. This was five days after the death of the Heir Designate—Li Ang's troubles continued unabated. By 21 November it stretched clear across the sky. On 28 November the sovereign issued the now familiar kind of decree, blaming himself for Heaven's anger, and, to temper it, ordering the reduction of criminal sentences by one degree, e.g., exile instead of death—with the exception of such unforgivable offenses as treason, murder, and robbery. The Japanese monk Ennin observed the apparition on the nights of 31 October and 1 November. He describes the comet as "the light of a sword," and reports that the T'ang government ordered the great monasteries to have readings of the Nirvana and Prajñaparamita Sutras.[386] A comet reported in Aquarius for February of 839 is the same comet.[387] The comet was also mentioned in the "Life of the Emperor Louis," which tells of the Carolingian Louis the Pious, and also in other European sources.[388]

20 March 840: A perihelial comet appeared in Pegasus. This was just after the accession of Li Yen (Wu Tsung). Later that year the troubled Li Ang was buried.[389] Then, on 3 December, just after the state funeral, another tailed comet appeared in the east.[390] The difference in time makes it unlikely that these were the inbound and outbound journeys of the same comet. In any case, it seems probable—although the sources fail to mention it—that the two apparitions were regarded as the last of the string of evil apparitions to haunt the reign of the late Wen Tsung.

22 December 841: A perihelial comet appeared near the south gate of the Army of the Forest of Feathers—that is, ruddy Fomalhaut in Piscis Australis. This is on the same meridian as the lunar lodging House (in Pegasus, especially the stars Markab and Menkib) and so the significance of the comet was indicated by this asterism. It also "entered the Purple Palace" of the north polar region. (The Book of T'ang notes a comet in Aquarius, most of which corresponds to the Forest of Feathers Army, in the late summer of this year—but a dating error is possible.) Ennin had already observed the fearful apparition on 17 December; he reports that the officials ordered readings of scriptures in the monasteries.[391] On 31 December Wu Tsung abandoned his chief basilica, had his diet reduced, and extended his mercy to prisoners. The "Virtuous Intonation" (a name for special imperial announcements of a beneficent nature) he issued on this occasion still survives.[392] The comet vanished on 9 February according to one source—on 16 February according to another.[393] Li Shang-yin wrote a memorial of congratulation to the throne on behalf of a noble client. This too is still extant.[394]

[25 March–22 April] 852: A perihelial comet was seen in Orion. "Triaster is the asterism of T'ang."[395] (But, despite some civil disorders and Tangut border raids, this year seems not to have been more unhappy for T'ang than most.[396])

20 April 864: A yellow-white perihelial comet appeared in Aries; it was three feet long. Two days later the Astronomer Royal, claiming the authority of the "Star Canon" (Hsing ching), persuaded Li Ts'ui (I Tsung) that this was an auspicious star, much to that monarch's delight. The commentator on the Tzu chih t'ung chien, one of our sources for this event, disparages the opinions of later T'ang astronomical officials, "who take the uncanny to be auspicious!"[397]

[29 January–26 February] 868: A perihelial comet was seen in Aries and Musca Borealis. This comet was also referred to in English and Byzantine records.[398]

[10 September–8 October] 869: A perihelial comet appeared in Great Tumulus (in our Perseus, including the Demon Star, Algol). "The prognosis was 'men-at-arms of the external barbarians and also a watery disaster.'"[399] The court astronomical

114

bureau recommended the strengthening of frontier defenses, including drilling the soldiers and augmenting the fortifications.[400]

13 June 886: An aphelial comet was observed in Scorpius and Sagittarius: it traversed the northern dipper and an asterism in Boötes. "The prognosis was 'a noble vassal shall be executed.' "[401]

12 May 891: A perihelial comet appeared in Three Eminences (*san t'ai*, in Ursa Major).[402] It moved into the Palace of Grand Tenuity (in Virgo) close by the star Great Horn (Spica) and the Market of Heaven (in Serpens and Hercules). It was more than a hundred feet long. On 5 July it disappeared. The eunuch Ch'en K'uang, who was knowledgeable in the ways of the stars, reported the following opinion to the throne: "There will be disorder: a vassal will enter the palace. The Three Eminences are the three steps to the Grand Unity. The Grand Tenuity and Great Horn are the court of the divine king. The Market of Heaven is the markets of the metropolis."[403] This is the comet that is known as the "Great Comet of King Alfred." "And the same year after Easter, at the Rogation days or before, there appeared the star which is called in Latin *cometa*. Some say that it is in English the long-haired star, for there shines a long ray from it, sometimes on one side, sometimes on every side."[404]

[*23 November–22 December*] *892:* An aphelial comet was observed in the lunar lodgings Dipper and Ox—that is, in the Jupiter station "Star Chronicle," in our Sagittarius and Capricornus. "The prognosis was 'Someone will set himself up [as a usurper] in Yüeh.' "[405] (The depredations of Chu Ch'üan-chung were flagrant by this time.)

(*896:* For a possible split comet in this year, see under "Novae.")

[*21 May–18 June*] *901:* "There were three red stars with sharp awns in the southern quarter. Then it was the same in the western quarter, the northern quarter, and the eastern quarter. Time passed, then a single star was added to each making sixteen stars altogether. In a little while they were snuffed out, beginning in the north. The prognosis is 'They are drizzle stars (*meng hsing*); when they appear the barons and men-at-arms will assault one another.' "[406] This passage occurs in the *Book of T'ang*, among the "visitor stars" (novae) at the end of the section on comets. I record the apparition here for lack of a more suitable place: the Chinese classed it among the "uncanny stars," which (like comets and novae) appear where no star has been before.

18 May 905: "At evening in the northwest corner [of the sky] there was a star of a kind with Grand White [Venus] but with light over it. It resembled a broom[-star], and it was thirty or forty feet long, and its color was brownish. On 19 May it was as white as white silk and some said that it was a 'water star' from Auriga."[407] This may be an earlier view of the following apparition, perhaps obscured by dust in the upper air.

22 May 905: A perihelial comet appeared in Northern Ho [River] (Castor and Pollux in Gemini), penetrating Literary Glory (*wen ch'ang*) in Ursa Major. It was more than thirty feet long and brushed the middle and lower Eminences (minor stars in Ursa Major).[408] This was during the fateful reign of the last T'ang ruler, Li Chu—posthumously Ai Ti, Chao Hsüan Ti, and Ching Tsung. The apparition was treated as an admonition from Heaven, and by way of penance there was a release of prisoners held in the vicinity of Ch'ang-an, and a general reduction of criminal sentences.[409] The text of this decree—or of one related to it (the date is uncertain, but

115

could well be 26 May)—in which Li Chu announces the usual austerities in the palace, still survives. On 6 June the office of the Astronomer Royal announced the reappearance of the comet (after perihelion?) and the construction of a "Tract for the *Tao* of the Yellow Tablets of the Arcanum" in the "Palace of Grand Clarity" was ordered. In short, the Taoist adepts, as masters of star lore, were called in to assist in the crisis.[410] On 12 June the comet extended from Leo to Hercules, shining with a "fierce and angry" light across the sky. The next day it was hidden by clouds, and so it continued until 18 June when the sky cleared slightly—but the comet could not be seen. It had left, and the doomed monarch issued another public statement on its departure, of which the following language is characteristic:

> Up in the sky our sentence is seen;
> Down on earth we quake in alarm.
> It brings deep affliction from daybreak to nightfall;
> We fear much hardship for live and numinous creatures.[411]

The astrologers took the episode to be menacing to the heart of the Chinese nation and its capital.[412] This comet was widely observed around the world—as in the "Mercian Register" contained in the Anglo-Saxon Chronicle.[413] There was a great famine in France, and King Alfred died in the following year.

NOVAE

> A Star, not seen before in Heaven
>   appearing
> Guided the Wise Men thither from
>   the East,
> To honor thee with Incense, Myrrh,
>   and Gold,
> By whose bright course led on they
>   found the place,
> Affirming it thy Star new-grav'n
>   in Heaven . . .
>
> John Milton, *Paradise Regained*, I, 249-254

Watch with unmoving eye, where Cepheus bends
His triple crown, his sceptered hand extends;
Where studs Cassiope with stars unknown
Her Golden chair, and gems her sapphire zone.

These elegant verses are from the pen of Dr. Erasmus Darwin, that excellent exemplar of the eighteenth-century ideal, combining the amateur talents of both scientist and poet. Fortunately he has informed us exactly what he meant by the "stars unknown" in the throne of Cassiopeia: "Alluding to the star which appeared in the chair of Cassiopeia in the year 1572, which at first surpassed Jupiter in magnitude and brightness, diminished by degrees and disappeared in eighteen months; it alarmed all the astronomers of the age, and was esteemed a comet by some."[414] This anonymous star-gem was none other than the "new star" observed by Tycho Brahe.[415] We style it and others like it "supernovae" (the less spectacular ones being merely "novae").[416] They have sometimes been called "blaze-stars" because they flare up unpredictably where no star had been seen before.[417] The standard Chinese name for them has long been "stranger star" or "visitor star" (k'o hsing). However, in very ancient times the preferred expression was "new star" (hsin hsing). This term has been identified at least twice on the Shang oracle bones. One of these archaic texts reads "A new great star, alongside of Fire." The celestial "fire" is Antares (anti-Ares, peer of Mars), and the appearance of the nova can be dated to approximately 1300 B.C. It is thought that it corresponds to a strong radio "star" detected in the vicinity of Antares in modern times.[418]

The suggestion has been made that reports of "visitor stars" may sometimes—like other omens—have been fabricated to promote political ambitions,[419] but a study of the distribution of reported "visitor stars" has revealed that they show a symmetrical distribution with respect to the galactic plane, with a concentration near galactic center, very like that of verified novae.[420] A more serious problem is the possibility of the confusion of tailless comets, and even of some variable stars and meteors with novae; some of the former, too, might aptly be styled "visitor stars." In fact, although the Chinese had separate names for these several kinds of phenomenon, sometimes observers can be shown to have been mistaken—as when they reported that the stranger star had shifted: here they must have been watching small aphelial comets. On some occasions they recognized their own mistake, as when the record states that a visitor star changed itself into an undoubted comet.[421] Probably cases of confusion were few: unlike a comet, which floats along from constellation to constellation like wind-driven miasma, a nova appears *ex nihilo*, and vanishes at the same point in the sky. Novae were also something like planets, as the omen lore attached to them demonstrates: for instance, either a planet or a nova observed in the asterism "Sky Sea" (in Sagittarius) presaged disastrous floods and drownings.[422] Other instances are common in the records. Novae, in short, like meteors, comets, and planets, were clearly distinguished from the patterns of fixed stars, and all of these abnormal stars served as luminous indicators providing information to the wise about conditions in the celestial hieroglyphs. Despite these functional and monitory overlappings, the Chinese tabulation of novae and supernovae forms a unique record which is still of immense interest to astronomers. The outstanding example is the supernova which exploded near γ Tauri in about 3,000 B.C. It became visible during July of A.D. 1054, when it was carefully noted by the astronomers of China.[423] Their modern colleagues have identified this with the Crab Nebula

(Messier 1), an agitated mass of gas and radiant energy which is still expanding at the rate of about 800 miles per second. For the medieval Chinese the Crab Nebula did not exist—or if they had been able to observe its genesis they would have been primarily concerned with its mystic meaning. Not surprisingly, the appearance of a nova was thought to stimulate or signify unpleasant activity in the asterism it visited: loss, war, defeat, sorrow, death, floods, famine, drought, plague, and other disasters are repeated with monotonous regularity in the annals.[424] Occasionally, however, the new star brought good news: its appearance in the Southern Dipper, for instance, signified that retired gentlemen would be taken into government employ and worthy statesmen adequately compensated.[425] Or again, when the stellar visitor turned up among the finely colored stars of Delphinus—which the Chinese styled "Gourd"—it could be expected that gardens and orchards would yield double harvests.[426] One authority, however, takes the view that this apparition means inflated prices of fish and salt.[427]

One thing seems never to have been in doubt: an unaccredited star detected near the celestial palace was always a cause for alarm:

> The stranger star moves the Grand Tenuity:
> The levee departs from the basilica of Lo-yang.[428]

These are the words of Li Po telling how the alien at the wall of the sky palace compels ritual action in its earthly counterpart—austerities and the cessation of state business.

But now to look at the novae reported for T'ang times. In what follows we shall list the stranger stars reported for the T'ang, assuming that they were in fact novae, except for very enigmatic cases, such as the mysteriously complex apparition of 901, which I have arbitrarily placed under "Comets"—as I might also those of 896 and 902—and the shining demilune of 684 which I think I have placed correctly under "Comets" (although at least one authority has surmised that it may be a meteor).

[1-29 November] 829: "A visitor star was seen in 'Water Position.'"[429] This asterism extends from Cancer into Canis Minor. (Arab astronomers are reported to have observed a nova in Scorpius over a period of four months in 827—although there is a little uncertainty about the date.[430] The Chinese apparently did not observe it.)

29 April 837: A visitor star appeared below Eastern Well (μ, ν, γ, ζ^2, ϵ, δ, ζ, λ Geminorum). It faded away on 21 May.[431]

3 May 837: Another nova appeared inside the Straight Gate, which passes between β and η Virginis, close to the [Inner] Screen ([nei] p'ing; ξ, ν, π Virginis). It went out on 17 June.[432]

These last two were both threatening omens for the doomed Li Ang. Strange apparitions in Eastern Well were premonitory of threats to the capital city and of drought. The Straight Gate leads through the gap between the eastern and western parts of the wall around the Palace of Grand Tenuity and represents the forecourt of the palace of the Son of Heaven. A stranger there is a dire portent.

26 June 837: A visitor star, resembling an aphelial comet, was seen in the Southern Dipper (ϕ, λ, μ^1, σ, τ Sagittarii) towards the Lock of Heaven (insignificant stars in western Sagittarius/Scutum).[433] A tailless comet or a nova?

[*9 November–7 December*] *896*: Three stranger stars, one large and two small, appeared approximately on the celestial equator, in a region where Aquarius, Equuleus, and Pegasus come together. They moved together and apart as they drifted eastward, like "women fighting." The two smaller ones disappeared after three days, and then the larger one vanished.[434] Clearly we have no nova here. Split comets—the comet of 1882 is an example—are not unknown, and it may be that was what was seen here.

[*4 February–3 March*] *900*: A visitor star appeared in the asterism Eunuch-Official (Flamsteed 32, 34, 37 Ophiuchi and 60 Herculis). It was the size of a peach, and its flaming light shone through the asterism, so that it could not be seen. It has been estimated that this supernova must have reached at least the second magnitude of brightness.[435]

[*11 February–12 March*] *902*: Another peach-like nova was observed in the Floriate Canopy (in Cassiopeia), but moved on into Leo, where it was engaged in combat by a stream of meteors on 2 March. It remained visible in this region of the sky until 903. The complex apparition was taken as a sign of military combat.[436] (Here we have plainly to do not with a nova but with a slow-moving aphelial comet.)

In the next chapter I shall comment on novae regarded as the transformed souls of men.

And unto the angel of the church in Sardis
write: These things saith he that hath the
seven spirits of God, and the seven stars.

Revelation, 3, 1

7 Embodied Stars

PERSONAL ASTERISMS

A ranking mandarin could—and commonly did—identify himself with one or
more constellations that interacted, *ex officio*, with his life and incum-
bency, the one regarded as fused with the other. The lesser office-holder or
the talented man out of office, whether through removal or through official
blindness to his worth, found it less easy to detect signs of his destiny in the
sky. To potential chancellors and great ministers, kept in collective and
anonymous obscurity through (as they supposed) no fault of their own, was
assigned a rather insignificant asterism called "Lesser and Inconspicuous
[Ones]" (*shao wei*), composed of small stars in Leo and Leo Minor.[1] Not for
them a specific assignment in the hierarchy of heaven, but the same lack of

120

specific identity above that they suffered below. The name "Lesser and Inconspicuous" occurs fairly often in poetry—not surprisingly, since poet-officials were naturally sympathetic with the plight of the unemployed or underemployed members of their profession, their own plight in particular. The following couplet is typical:

> A three-foot sword within a casket;
> The "Lesser Inconspicuous" stars up in the sky.[2]

The poet here uses a pair of obvious symbols—the unused sword and the constellation of the deserving but ignored man, both valuable tools waiting humbly for the day of the proper employment. The same group of four stars was sometimes called "Private Gentleman" (*ch'u shih*) so that a star-minded poet could reasonably compare a worthy but unemployed friend with the "Private Gentleman's Star at the far edge of Heaven."[3]

A star's brilliance or apparent size was naturally related to the importance of the personage to whom it might, however transiently, refer. At the opposite extreme from such conspicuous stellar spirits, with their own names and titles, was the far-flung host of tiny insignificant stars. Undistinguished by any appellation, they represented the mob, the proletariat, the unnamed masses. It could even happen occasionally that the fate of large numbers of common people would be presaged in the sky by the blotting out of masses of these tiny lights by a high thin cloud. Or the anonymous citizenry could be represented by a great throng of fleeting specks in some meteor showers, unmarked by great fire-balls or other noble signs.

It was possible, then, for everyone to find a counterpart, however trivial, in the hosts of heaven, and this sense of temporary or permanent connection was not fanciful or idiosyncratic, but expressed real inner conviction. This feeling of identity was particularly deep and warmly felt in the case of men whose stars reinforced their belief in their own dignity, or which seemed to hold particular significance for them as individuals—at least during their lifetimes. By way of example, consider the case of the asterism which represented every literary man of quality, and, by way of contrast, the asterism of an individual writer who enjoys some fame as a toper.

The literary man's asterism is a cluster in Ursa Major a little in front of the bowl of the Dipper. Its most notable star is θ Ursae Majoris, a brilliantly white double. Its name was "Literary Glory" (*wen ch'ang*). Sometimes it was called simply "Literary Asterism" (*wen hsing*). The former name appears with various other referents: it is a great Taoist deity; it is a river in Kwangsi; it is the name of a royal hall in a poem by Ts'ao Chih; it was the name of a township of the T'ang period on Hainan Island; it was part of the title of the great councillors of state under the Empress Wu; it was the cognomen of the poet Chang Chi. Above all it was an exceedingly common phrase in T'ang poetry.

Ch'ien Ch'i figured the appearance of a literary scholar as "The Literary Asterism let down from the Grand Emptiness"—a kind of cosmogonic apparition, like the epiphany of a primordial deity.[4] Less grandly, a drinking party attended by writers is "this night's gathering of Literary Stars."[5] P'i Jih-hsiu, who wrote that line, elsewhere expressed his regret that he was destined to be a writer rather than a holy adept of Taoism:

> Looking towards this den of the Divine Transcendents
> I despise myself for my dirty vulgar form;
> Once more I hate the Fashioner of Creatures
> Which sent me off to mount the Literary Asterism.[6]

In the following quatrain dark forces cloud the creative spirit in the celebrated critic Szu-k'ung T'u, but he puts no blame on goblin vapors in the Upper Air, obscuring his own constellation:

> Following wind, pursuing white-caps—drive the yellow nuphar;
> Could mere round-head ever analyze the most holy Numen?
> Though brush and palette have lately been freely cast aside
> It has nothing to do with "uncanny pneumas" darkening the Literary Star.[7]

Paraphrase:

1. Unaccountable, chancy winds tear the pond lilies (spatterdocks) loose, impel them (like me) at random;
2. But no ordinary mortal can fathom the workings of the divine spirit through the exercise of reason.
3. I have let the writing tools drop from my fingers recently—why I cannot tell.
4. But I won't attribute it to unseen powers beyond my control: the blame is my own.

It was only natural that the image of Literary Glory should be adapted to flattery in verse. The ninth-century monk Hsi-ch'an praised his fellow believer, the distinguished writer Ch'i-chi, a native of Ch'ang-sha, with the words: "The Literary Asterism lights up the sky of Ch'u."[8] This usage was well suited to posthumous honors, so that when the Literary Stars decline (figuratively) towards the horizon, the death of a great writer is betokened, as Sun Ho wrote in the ninth century, mourning for Fang Kan.[9] Here is an example from the pen of Fang Kan himself, lamenting the loss of his friend and benefactor Yao Ho:

> This night the chilly void lets fall its Literary Star;
> The star may fall, but the literature remains—and his name for
>   a myriad ages![10]

122

Embodied Stars

Frequently (and obviously) this symbol of the civil arts was contrasted with a military constellation, sometimes, as time brought about changes in the human condition, to the advantage of one, sometimes of the other. Here the writer envies the soldier:

> Each will toss out his brush and palette and boast of Yaktail and
>   Battle-axe;
> None sends a Literary Star—they defer to Commander Star.[11]

We have already noted the warlike yaktail banner seen in the Pleiades. There was also a "Battleaxe" (*yüeh*), equivalent to the single star Propus, a binary variable in the left foot of Castor (η Geminorum). But there was another Battleaxe in an undistinguished but powerful constellation in Aquarius. Its name may be associated with the stone-shouldered axes of the Neolithic inhabitants of the far south—the men of Viet (Yüch, *Ywăt).[12] We see Literature and Battleaxe contrasted again in an astrological poem by Tu Fu:

> Axe and Battleaxe [axe of Viet] descend from the blue gulf;
> High-decked ships pass over Grotto Court;
> North wind comes after the bracing airs,
> South Dipper turns away from Literary Stars.
>
> Sun and Moon are birds inside a cage,
> Potent and Latent are water-borne nuphars.
> King's scions and mature men [alike] take action—
> But descending old age is blowing me to bits.[13]

This octet was apparently written in 769 just before the poet's death. The asterism of the two axes in Aquarius represents war, and the execution of war prisoners. The Viet axe in particular is pertinent to the destination of the commander to whom the poem is addressed: it is Canton.[14] The "bracing airs" are General Li Mien's aura—followed by the cleansing wind of the northern army. The ships southbound on Lake Tung-t'ing (Grotto Court) are double-decked warships: this is an allusion to the expedition of Liu Ch'e (Han Wu Ti) against Nam-Viet a millenium before. The celestial powers show no trend; their purpose is ambiguous. They, like the poet, seem to be at the mercy of the winds of chance; while noble youths and stalwart yeomen march forth to war, aging statesmen like Tu Fu and his contemporaries are being swept, like falling leaves, into oblivion, to make way for vigorous new men, as demanded by the nature of the times.

The astrological connections of all literary men were not restricted to such collective asterisms as Literary Glory. Li Po had the distinction of intimate personal connections with at least three celestial entities. His long love affair with the planet Venus is proclaimed in his own name. The legend of his fatal passion for the moon[15] symbolizes a real empathy and lifelong

123

preoccupation. Finally a whole asterism seems to have been created (although we know better) especially in his honor. This was the Wine Asterism, consisting of a small group of stars in Leo, of which the lucida is the double star ψ Leonis. One of this pair is bright orange and the other bluish white,[16] although it is unlikely that the poet was ever clear-eyed enough to distinguish them. The constellation was his real doppelgänger in the sky.[17] When its brightness or color seemed to change, or when such intruders as comets and novae appeared in it, his own life was profoundly involved. Such events were to him as significant as the appearance of an ancestral ghost to the aging laird of a Scottish castle. The asterism was, in a sense, his own property. He felt strongly that its presence among the glittering mass of spirits above meant that wine was a blessed gift of heaven:

> If heaven did not love wine
> Wine Stars would not be in heaven.

These lines appear in the same poem in which he describes wine as a kind of elixir that will lead Him into the mysterious depths of the Ultimate:

> Three cups—I communicate with the Great TAO;
> One dipper—I blend with SPONTANEITY.[18]

In time Li Po came to be thought of as a star spirit himself. There is an anecdote of the Sung period which tells of four mysterious personages who appear at a wineshop in Szechwan, consume considerable quantities of the best beverage, and converse fondly about the Taoist sage Sun Szu-miao. When a magistrate investigates their odd behavior, they disappear suddenly, each leaving a little pile of ashes behind. This strange event was reported to royal Li Lung-chi, who summoned (anachronistically) the learned doctor Sun. That savant informed him: "These were the Transcendents of Great White and the Wine Asterism."[19] In short, it appears that three of the strangers were the major stars of Wine Asterism and their boon companion was Li Po, as the spirit of the planet Venus—but also a familiar of the Wine Star.

Poets of the late T'ang period took this special relationship as gospel. A notable example was the young eccentric Li Ho. In a poem celebrating the cosmic amble of a certain "Prince of Ch'in"—thought to refer to the ruler Li Kua (T'ang Te Tsung)[20]—he envisaged that lord riding through space astride a tiger, his sword flashing through the deep blue abyss:

> At Dragon's head he drains the wine—he toasts the Wine Star![21]

The head of the great Blue Dragon who rules the eastern quarter of the sky is marked by its two gem-tipped horns—Arcturus and Spica: they point towards the Wine Asterism in Leo. Li Ho admired Li Po extravagantly, and this salute with a goblet of celestial vintage is surely directed to his great predecessor's spirit lodged in that constellation.

Embodied Stars

The fact is stated more plainly by another great writer, P'i Jih-hsiu:

I love Li "Grand White"!
His body was the white-soul of Wine Star:
From his mouth spewed verses from highest heaven . . . [22]

And Cheng Ku, writing at the end of the ninth century, was even more explicit. He wrote a quatrain titled "Reading the Li Po Collection":

Why attend either to Literary Star or to Wine Star?
Both lodged at once in Li the Prior Born.
He chanted high, in great drunkenness, three thousand stanzas—
Left his writing among men—a companion to the moon's light. [23]

That a Literary Asterism and a Wine Asterism should enjoy significant status in the sky, resources for apt human analogies, suggested to one writer the possibility of an *ad hominem* construct: why is there no Calligraphy Star? The cursive script of Ma Huai-su, who lived during the reign of Li Lung-chi, was greatly admired by connoisseurs of later generations, among them a certain P'ei Yüeh, who, at the end of the ninth century, made his own apportionment of stars among the heroic figures of the eighth century:

I call out the names of the men of old—
Ghosts and spirits turn their ears and listen:
Tu Fu, Li Po, and with them Huai-su—
Literary Star, Wine Star, and Star of Cursive Writing! [24]

We have alluded briefly to Li Po's kinship and comradeship with the white moon and the white planet we call Venus. Not only did the poets of medieval China take note of this affinity, as when the Buddhist Master Kuan-hsiu pictured him floating serenely, "icily pure," among the magical purple plasmas of the gulfs of space, the very quintessence of Venus,[25] but Li Po himself evidently considered the relationship a simple fact, as when he reported his climb up a mountain peak named for that planet:

Grand White will converse with me;
She will open the Barrier of Heaven for me.
Then I might ride off with the cooling wind,
Straight out among the floating clouds,
By raising my hand I could approach the moon . . . [26]

It is the picture of a homecoming.

SOULS INTO STARS

Quem pater omnipotens inter cava nubila
    raptum
quadriiugo curru radiantibus intulit
    astris.

Ovid, *Metamorphoses*, IX, 271-272

A modern authority has it that the charioteer Phaethon was "catasterized" by Jupiter, to be displayed forever in the dark vault as the constellation Auriga.[27] The Graeco-English verb has a certain grand style to it, but for the phenomenon referred to I prefer the simpler Latinism "stellify" and its nominal forms "stellification" and—even better—"stellation," none of them fortunately unknown to Webster.

That the stars were gods and accordingly had an important role in the history of religions is not a novel idea. The notion that some knowledge of astronomy is even more important to students of the history of mythology (because many ancient legends are unrecognized star-myths) has yet to be widely accepted. Nonetheless, even in Classical Antiquity it was commonly believed that the great myths were astronomical allegories, and that, as Quintillian put it, "acquaintance with astronomy was essential to an understanding of the Poets"[28]—a reasonable assumption in view of the intimate interconnection between poetry and mythology. This relationship has been traced far beyond such simple tales of the apotheosis of Hercules (see the epigraph above) to rather sophisticated mythological exegeses which require a knowledge of the times of the rising and setting of stars at different seasons of the year—the kind of knowledge that seems to be prerequisite to a proper understanding of the euhemerized and bowdlerized texts of both *Shu ching* and *Shih ching*. Here, however, I am limiting my remarks to a kind of star lore that was particularly prominent among the successors of Plato—that souls become stars. This idea survived a very long time. John Donne, for example, was interested in the conception "that souls, after death, inhabited the stars. The blaze of a new star might therefore indicate that the soul of a great man, or woman, had reached its destination."[29]

In China, the stellification of souls never enjoyed the same popularity as the mummification of bodies. Nevertheless, the idea was far from unknown. Sometimes the tales of such splendid events are difficult to distinguish from traditions of the descent of stars—often, no doubt, without any human ancestry—to fertilize antique queens. For instance, the Divine King in Yellow, surnamed Hsüan-yüan, was born after the irradiation of his mother by the Pole Star. Such lusting spirits without doubt often took the visible form of meteors.[30] Moreover, stories of the impregnation of noblewomen by vagrant sunbeams and aggressive rainbows—agencies not very different from stars—are almost too numerous to mention.[31] These presences were most usually seen on earth as dragons of sidereal origin which assumed plainly human shapes. Such were the five old gaffers—the germinal souls of the five planets—who appeared at the birth of Confucius,[32] or the same radiant five who were displayed to a wondering Shun by the aging Yao at a much more remote date.[33] One of these last, the spirit of Saturn, once dwelt a space on earth, having arrived in the form of a yellow stone. His name and powers are associated with the origins of the Han dynasty, with the cosmic activity named "earth," and with alchemy, and he is well remembered in Taoist legend and literature.[34]

But such tales of divine slumming are still a far cry from the belief in the stellation of ghosts. Mythological persons—or persons believed to be mythological—are sometimes to be encountered in the Chinese sky. Fu Yüeh, the shamanistic "minister" of Wu Ting of the Shang dynasty, is an example. He was honored by permanent installation deep in the south below Sagittarius, where he is now called γ Telescopii. This star marks the tail of the Dragon, whose heart is Antares. It is the star of shamankas and inspired sybils, and it seems likely that the "minister" of the ancient king was in fact a *lady* who spoke with the gods.[35] But the Confucian tradition became dominant, and we find the little asterism as a sage counsellor of the male sex in a skeptical or worried poem entitled "Query of Heaven" written by the T'ang dynasty monk Chiao-jan. Here the star is treated as the exalted spirit of a royal counsellor, almost certainly not a woman:

> When ever will there be a lord in Heaven?
> All talk of this is off the true path.
> Who will allege that a true statesman, now dead, ·
> Has become the asterism Fu Yüeh?[36]

I think I detect a bitter note here: a transferral of imperfect conditions at the royal court in Ch'ang-an to its stellar counterparts.

Then there were the two sons of a sky god, Shen and Shang, who have a number of distinct myths: they are sometimes Hesperus and Lucifer, but in a different tradition they are Orion (my Triaster) and Scorpio (Antares)— bitter rivals, each always out of the other's sight at opposite ends of the sky.[37] Still even here we have no clear-cut cases of the stellification of ordinary mortals.

The case is somewhat different with Tung-fang Shuo, the wonderworker and magus of Han times. It is reported that when he died his sovereign asked a crony of his for intimate information about him. This man, it transpired, was an astrologer, and when Liu Ch'e inquired, "Are the stars all present now or not?" he answered in this vein: "All the stars are accounted for. Even the Year Star, which had not been visible, now has appeared again after eighteen years." The theocrat looked up to the sky and said with a sigh, "Tung-fang Shuo was alive by my side for eighteen years but I never knew that he was the Year Star!" and he became joyless and despondent.[38] Another account tells us that the same sage was the quintessence of Venus;[39] if the tale is true he must have had to share the planet with Li Po, or perhaps the latter was the reincarnation of Tung-fang Shuo.

127

A less permanent star was the visible manifestation of another Han wizard. This was Ching Fang, the great authority on omens and divination. When he was languishing in prison because of a false accusation of malfeasance, he told a friend, "Forty days after my death a Stranger Star will enter the Market of Heaven: this will be evidence of my innocence." And indeed he appeared as a supernova in Hercules just as he had foretold.[40]

These two examples of stellation have this in common: both men were deeply conversant with cosmic mysteries—indeed they were virtually boon-companions of the stars. This special affinity evidently implied that the star-essence was already in them when they died, so that their transference to a more exalted plane was no great matter for wonder.

From T'ang times we have many striking examples of the sympathetic reactions of the stars to the deaths of great men. Sometimes this merely took the form of a marked alteration in the normal appearance of celestial fires. So the poetic monk Kuan-hsiu wrote on the death of the statesman Wang Ts'ao that

> The stellar chronograms all have horns;
> The sun and moon are almost without light.[41]

But these are a common and generalized species of evil omen. More specific was the visit of a star to a deathbed, as when a brilliant one descended to hover above the poet Ling-hu Ch'u during his last moments on earth.[42] Such obituary observations occur most commonly in the dirges and poetic laments written by T'ang poets as formal expressions of grief on the death of great magnates—many of them, no doubt, composed on commission. A notable example is a set of three funeral songs composed by Yüan Chen on the death of Li Ch'un (T'ang Hsien Tsung). In the first of these we read: "The Chronogram of the North shifts the God-king's Seat"; in the second, the bow of the deceased monarch, a noble talisman, is invoked to defend the realm against the onslaughts of the barbarians, represented by Sirius, the Wolf Star; in the third, the moon sinks behind the west wall of the Forbidden City in the sky.[43] But even here we have no clearcut stellification, only symbolism or, at most, the assertion that the asterisms vibrate in sympathy with their mundane counterparts. In dirges of this sort there are frequently references to the "Front Star" (ch'ien hsing). This is a small but important star, our τ Scorpii, which the Chinese placed in the lunar lodge Heart (so also for the Arabs, al-Qalb)—that is, the red heart of the Spring Dragon, flaming Antares. This admirable object, which is in fact a binary star with a faint emerald companion,[44] represented the Son of Heaven, and Front Star, which goes ahead of it, represents the heir designate to the Chinese throne.

128

Accordingly, on the death of Li Kua (Te Tsung), and the imminent accession of his successor Li Sung (Shun Tsung), Po Chü-i wrote:

> Front Star accepts the God-king's Seat:
> It cannot leave the North Chronogram vacant.[45]

Here the star displaced far from Scorpio into the polar regions of the north and a completely different celestial environment symbolizes the accession of the prince. But the poem displays only the doctrine of correspondences, not a belief in the actual apotheosis of the late sovereign.

The concept of the "ostracized transcendent" (*che hsien*)—a Taoist spirit condemned to spend a spell on earth for the infraction of some celestial taboo—is a commonplace of T'ang lore and literature. The designation was also used figuratively of writers whose talents seemed to be of a supernatural order. The well-known case of Li Po fits both categories. "With exalted name," wrote Meng Chiao, "we call him 'ostracized transcendent.'"[46] Indeed, in the same poem Meng Chiao stated distinctly that both great writers and great writings augment the brilliance of the stars. The context is a dirge in which the superbeings (*sheng jen*) of Taoism mourn the passing of a writer of distinction:

> The supermen weep for a worthy statesman:
> His bones are mutated; his breath becomes a star.
> His written strophes fly up into the sky—
> The array of Lodgings has added crystal sparkle.

The incarnate spirits styled "ostracized transcendents" were also called—as befitted Taoist beings—"ostracized stars." Li Tung, a poet and scion of the royal family who lived at the very end of that family's rule, described a noted adept as an ostracized star, although we do not know that the person so designated had any literary ability.[47] His sidereal quality was evidently purely religious in nature.

The Great Dipper afforded a perpetual home of outstanding merit, it seems, to a few mortals outside the imperial household. This is the sense of the verses of the monk Ling-i, bewailing a deceased minister:

> His paired guisarmes in Southern Ching: their marks are still there;
> But a lonely cloud-soul in the Northern Dipper now sees far and deep.[48]

(The crossed halberds represent the site of his abandoned military headquarters; Southern Ching is in the central Yangtze region.)

But, unhappily for our traditional conception of the evaluation of human capabilities in early China, a military minister was far more likely to merit stellation than any other mortal below the steps of the throne. Indeed the most prominent among the sky-earth affinities recognized by the Chinese were those that fall into the class of military astrology. Among the most significant constellations were "Great Army Commander in the Sky" (mainly in our Andromeda), and "Army of the Forest of Feathers" (mainly in our Aquarius). Persons anticipating military operations, or actively engaged in them, kept a sharp lookout for aberrations in these asterisms. In doing so, they did not exclude phenomena which are to us purely atmospheric interferences, such as discolorations due to dust storms, or strange haloes caused by ice-crystals in the high atmosphere.

Allusions to the starry counterparts of military men, of armies, of enemies, of encampments, and of victories and defeats are all extremely common in T'ang poetry, where they have (unfortunately) often passed unrecognized. A small sampling is all that can be attempted here. Probably the most usual figure of this class to appear in verse was the image of the victorious—or certain-to-be victorious—Chinese commander striding triumphantly across the sky. Ordinarily this was no mere figure of speech, but an expression of the certainty of cosmic backing for the faithful defenders of the Way of Heaven on earth below. The glorious aspect of these condottieri was delineated, for instance, by Tu Fu, who saw one riding a richly caparisoned horse down the ecliptic and fording the Sky River, while the enemy leaders, metamorphosed into a comet of the "gouging spear" (ch'an ch'iang) variety and the planet Mars in the guise of Dazzling Deluder, crouch petrified by fear.[49] Most commonly the Chinese marshal is styled simply "commander star" (chiang hsing) and his movements on the surface of the earth are confidently transferred by the poets to the depths of the sky. If he is transferred to a new post, his star shifts pari passu, even if invisibly.[50] If he is sent on an exalted mission, his star rises high even if the increased elevation is undetectable to mundane eyes.[51] Looming belligerently over the south, great Antares is a sure guide to his new destination.[52] As his drums and banners approach the sandy Gobi, his star flies boldly through the Northern Dipper.[53] The sky-general is particularly concerned with the protection of the Wall of the Palace of Purple Tenuity, the ultimate bastion of the supreme polar deities. He advances, he retreats, he marches and countermarches—always presumably followed by his anonymous but glittering myrmidons, always with those sacred parapets in view. As a military commander he was also a responsible officer of state, a privy councillor of the Son of Heaven:

> Last night the Star Officer moved in Purple Tenuity:
> This year the Son of Heaven employs his martial might![54]

In the consultation chamber, red-eyed astrologers in rapt attendance, the difference between an actual victory and the ecstatic vision is imperceptible:

Embodied Stars

Yesterday I heard that the battle was over: I was in the Gallery of
    the Pictured Unicorn.
The caitiffs were smashed, the arms gathered in, the tents of war
    rolled up.
Over the watchet sea, I first watched the Han moon brighten—
Down from Purple Subtlety I had already seen the Star of the Nomads fall.[55]

Astrology was, of course, intimately involved with strategy: the magic of battle was above all a sidereal magic, and the "Books of Weaponry" so jealously guarded by the royal court were essentially tabooed treatises of astrological lore. In the following verses we see a civil magistrate called—as was often the case, since all the great arts are interrelated—to military duty:

The Supreme Commander issues from Literary Glory;
The Central Army is immobile in the Boreal Quarter.
He reads the stars—leads out his banners and insignia;
He chooses the day—bows at the altar's precincts.[56]

With all of these crystalline colonels waving their magic batons it becomes difficult to see the underlying humanity of their imperfect counterparts below. Yet sometimes the poets achieved this, as did Hsiang Szu, who wrote of monks with the same facility that he employed to characterize the military of the mid-ninth century. Here he tells of an old soldier, retired to honorable obscurity:

Beyond the Wall the rushing sand impaired the light of his eyes;
Back he came to nurse his infirmities, dwelling in the capital
    in Ch'in.
Now he goes up to the gallery of his high house to watch the
    starry seats;
Or dons white shirt and skirt—walks along, sword in hand.

He often speaks of his aging body—his thoughts of a battle
    command.
Deeply sad at his lack of strength. Ah, to govern a frontier
    camp!
Blue-dyed brow and ruddy face—but at peace with the Uighurs;
Despondent, despairing: the Central Plain does not require his
    weapon.[57]

STAR WOMEN

O'er her fair form the electric lustre
    plays,
And cold she moves amid the lambent
    blaze.
So shines the glow-fly, when the sun
    retires,

Pacing the Void

And gems the night-air with phosphoric
fires.

Erasmus Darwin, "The Loves of the Plants," *The
Botanic Garden*, IV 49–52

In his rhapsody on the Hall of Light, Li Po wrote of "the jade women clinging
to the stars" and "the metal fairy implanted in the moon."[58] The "metal"
—or, if you prefer, "golden"—spirit lady is readily identifiable as the moon
goddess, possibly pluralized, like her companions, the "women of jade."
Whether "metal" or "jade," no one of them requires further epithet to define
her as a shining white creature of the heavens; all are, in short, personifica-
tions of stars. "The Jade Women of the Luminous Star" await the lucky
adepts on Mount Hua, the sacred western prop of the sky, proffering cups of
the "pale liquor" which gives endless life.[59] The most exalted beings enjoy
the privilege of having "jade women of luminous stars" as their housemaids,
as Li Po observed.[60] But by human standards they were among the most rare
and powerful of divinities.

These beings of crystalline whiteness, so closely allied to the stars, are
prominent in Taoist lore, and reveal themselves from time to time to zealous
adepts and alchemists as protectresses and divine guides. In a Taoist characterization they are as insubstantial as snowflakes congealed like ghosts out of
the cosmic breath:

The jade women achieve form by sympathetic reactions to the miraculous pneuma
of Spontaneity. In form and substance they are as luminous and clean, as clear and
shining as jade.[61]

A jade woman has yellow jade the size of a millet grain placed above her nose as a
credential. Lacking this credential, she is a ghost-messenger.[62]

The color of her identification badge is a little surprising, although yellow is
not an uncommon color of nephrite, and also an almost commonplace star
color. (See figure 3.)

Jade women literally immersed in an appropriately mineral environment were seldom encountered in T'ang times. Yet Wang Chien, the noble
founder of the Kingdom of Shu on the ruins of the T'ang provincial administration in Szechwan, discovered such a creature in one of the celebrated
salt wells of that region, and in 907 ordered sacrifices made to her. The
reason for this honor was as follows:

He had seen a woman, whose body was naked, in a salt well. She informed him, "You
shall become the master of the soil and surface of my country, and wealth and
honors shall come to you."[63]

This jade fairy is closer to the alchemists than to the star-treaders, but though
in Taoist ecclesiastical texts her sisters were usually displayed properly
clothed in rich fabrics, she is quite plainly Taoist despite her nudity.

132

FIGURE 3. A Jade Woman (*Wu shang hsüan yüan san t'ien yü t'ang ta fa*.
Anon.; no date. In *Tao tsang*, 103.)

In Taoist tradition, at any rate, these jade ladies were very numerous. A
Taoist classic of the fourth century tells us of the three great goddesses who
were formed spontaneously out of the primordial breath at the beginning of
the world, and were named the "Ladies of the Three Primes" (*San yüan fu
jen*) or, subsumed by their mother, "Mistress of the Three Primes of the
Grand Immaculate" (*T'ai su san yüan chün*), attended in their celestial
palaces by three thousand each of "Jade Youths of the Golden Dawn" and
"Jade Girls of the West Florescence."[64]

In T'ang poetry the jade women are sometimes rather ambiguous fairy
figures, symbolizing the presence of unseen supernatural forces, or evidence

of the approach to divine wisdom—as the sight of gulls signals to sailors the nearness of land. The mysterious ladies appear in large numbers in the untitled Taoist poems of Lü Yen, one of which is translated here:

> Awake to clear apprehension of the scheme for extended life—
> As lotuses in autumn open in place after place;
> Golden lads will climb brocaded hangings;
> Jade women will descend aromatic staircases.
>
> The Tiger Whistle—and volatile souls of Heaven stand still;
> The Dragon Chant—and residual souls of Earth will come.
> Should a man become enlightened in this Way,
> He will become once more an infant—a babe![65]

Such are the powers at the command of the Enlightened Ones: "Tiger Whistle" and "Dragon Chant" are musical modes of invoking revenants and elementals. But what concerns us here is the almost hallucinatory slithering of unearthly beings about the furnishings of the adept's house. They are his familiars and attendants.

A more plainly sensuous version of a jade woman is celebrated in a poem of Liu Yü-hsi. This is one of a number of early ninth-century poems inspired by the alleged visit of a sky woman to a great Taoist monastery of the capital. The site of the story is the T'ang ch'ang kuan, almost in the center of Ch'ang-an, in the An-yeh Quarter, close by a number of other wealthy religious establishment, both Taoist and Buddhist. The friary boasted a garden area named "Close of the Jade Stamens" (*Yü jui yüan*), famous for its beautiful "jade tree" and its fairy-white flowers. One day, when the temple grounds were thronged with flower-loving sight-seers, a lovely maiden appeared, with a suitable following of well-dressed attendants. She was, of course, a Jade Woman, on an excursion to earth to see her own tree. The poem treats her lightly and sweetly, as she awakes to the meaning of time and mortality:

> Jade Woman comes to observe the flowers of Jade Tree;
> Strange scents draw her Seven-scented car ahead.
> She pulls down a branch, touches the snow—
> Marvels—astonished—that here among men the sun so readily
>   declines.[66]

(The "snow" on the branches is, of course, a coating of metaphorical blossoms.)

Beyond the personal attention to the advanced adept, the jade women had their own public cult in connection with the forms of Taoist worship. We see it represented, for instance, in a poem by Wu Jung entitled "The Temple of the Jade Women (or Woman?)."[67]

134

But in what survives of T'ang literature the jade women figure most prominently as mysterious creatures of intensely private visions, lurking in mountain fastnesses—visitors, perhaps, from even more remote mountains, the palaces of the gods. When we read, "The Lotus Platform Palace: the Jade Woman of the Red Cinnamon resides there,"[68] we must not assume that the "palace" is anything but the spiritual emanation of a mountain—and in the end, that the mountain is anything but a secondary microcosm, another version of Mount K'un-lun itself. This mountain-palace identity is implied in the names of many such palaces, such as "The Palace of the Golden Flower: the Jade Woman of the Western Flower resides there."[69] ("Western Flower" is the name of the western Marchmount, the sacred mountain near Ch'ang-an). The theme of the mountain as a disguised or vulgar mis-vision of a sacred palace lends itself naturally to incorporation in a popular sort of tale—the common story in which the hero undergoes marvelous adventures in rich surroundings, only to find himself disconsolate in the wilderness at the end. An almost classic example is preserved in the T'ang book *Yu yang tsa tsu*. It tells of a woodcutter of the Chin period, long ago, who was plying his trade on "Jade Woman Mountain." The reader is not surprised to read that he had a mysterious encounter with divine women in a luxurious palace there, surrounded by trees of jade and similar wonderful appurtenances, and in the end—as usual—he finds that "all was a deserted graveyard."[70] The disillusion goes all the way back to the inconclusive encounters of ancient shamans with the goddesses of air, earth, and water. A famous prototype of all such stories is imbedded in the *Han Wu Ti nei chuan*, a fictional romance, possibly of about the third century, which purports to reveal the magical aspects of the life of the Taoist sovereign of the early Han period. It tells, among much else, of his miraculous birth after the appearance of a red dragon and a divine woman. There is much more in this vein, leading to the later appearance at his great palace of a young woman of extraordinary beauty, clothed in blue. She identified herself as a "jade woman", an emissary from Mount K'un-lun, and one of the many in the service of the goddess Hsi Wang Mu and regularly employed by her as messengers to all parts of the universe.[71] Some jade women attained great distinction: the "Jade Woman of Occult Miracles" (*Hsüan miao yü nü*) became, as has already been observed, the mother of the divine Lao-tzu.[72]

These women of jade—who must, as we have seen, be regarded as transmuted stars—besides their prominent role as divine couriers ("angels" in the original sense of that term), were also the special guardians of the Taoist arcana, at once sentinels and custodians of the mystic archives in the remote palaces of the sky. They are abundantly referred to in this role in the quotations assembled in the "Transcendental Canons" section of the great tenth-century anthology *T'ai p'ing yü lan*,[73] and were especially prominent in the *Shang ch'ing* tradition of Mao Shan. For instance, in the Mao Shan romance "The Esoteric Tradition of Han Wu Ti" they reveal the ultimate arcana of Highest Clarity as a set of secret charms, which, in fact, constituted the canon of that sect.[74] In this exalted capacity of trust they are often connected with the sacred mountains, which are then viewed as great stony vaults and strongrooms for the preservation of the ultimate esoterica. A typical quotation says, "The jade women of Western Hua are placed in the Transcendental Metropolis where they are the guards and wardens of the Canons of the sky-hoards and moon-crypts, and the 'White-silk Strips of the Flying Dragons.'"[75] Or similarly "The jade women placed inside of the Grand Supreme Sextuply Fitted Purple Chamber attend and guard the Cryptic Books of the Grand Cinnabar."[76] The former damsels appear to have specialized in astral lore, the latter were alchemical librarians. They are also the mystic wives of the Taoist initiates to whom they convey, after a ritual banquet, the secrets of immortality which exist in their purest form in the arcana of "Highest Clarity" beyond the sky.[77]

Avatars or subspecies of the jade women were the blue women (*ch'ing nü*), midway between minerals and stars, therefore appropriately the little goddesses of the frost and snow-crystals. The astrological section of the book *Huai nan tzu*, composed in early Han times, contained a list of signs of the onset of winter: "The Hundred Crawlers crouch in hibernation . . . and the Blue Woman comes forth to bring down frost and snow." The standard commentary on this text, written by Hsü Shen in the Later Han period, makes the following identification: "The Blue Woman is a heavenly divinity . . . she is the Jade Woman in charge of frost and snow."[78] Whatever her reality in the cult and beliefs of those early times, to the T'ang poets the Blue Woman (or Women?) offered a convenient personification of the natural forces, traditionally feminine in the metaphysical sense, that converted the waters of heaven into icy crystals—the static, inert ornaments of the winter landscape. The glorious chrysanthemums that best represented plant life in the season of bitter cold survived "an entire lifetime in the frost of the Blue Woman."[79] So wrote Lo Yin late in the ninth century, while his younger contemporary Li Shang-yin compared the glistening whiteness of the apricot blossoms of mid-winter favorably with the best that either moon-fairy or frost-maiden could produce.[80] Li Shang-yin was particularly susceptible to this kind of crystalline whiteness—contrast it with the ghostly pallor that obsessed Li Ho—and more than once expressed it by the juxtaposition of moon goddess and frost goddess, as in the following quatrain, whose title "Frost Moon" invites the interpretation "Frosty Moon," but intends, rather, "Frost and Moon."[81]

136

Embodied Stars

We begin to hear the battle-line of wild geese—and the cicadas
   are gone;
From high in a hundred-foot loft—water joins with sky:
Azure-blue woman, silk-white fairy—both tolerant of cold;
One in the moon, one in the frost—rivals in allure.[82]

From his tall building the observer sees the migrant birds cross the moon, imagines the cicadas dug under the earth; these two realms seem joined on the distant horizon. Two lady rulers: one above, one below (note the chiasmus of the second couplet)—both white, both chilled, both divinely beautiful.

So, by way of the sylphine spirits of the mineral world and the fairies that dip their toes in the ionosphere, we come to the stars.

Our common western mythology makes little of star spirits. Medieval and Renaissance pseudo-science has supplied us with elemental beings aplenty—gnomes, salamanders, ondines, and sylphs—but it was Classical Antiquity that populated the night sky with mythological figures. The female contingent swarms over the heavens: we have Andromeda, Cassiopeia, and Virgo (who is also Leda) among the great constellations, and such groups of individual stars as the Hyades, the moist daughters of Atlas, and the cloudy seven virgins we call the Pleiades. We also have the gauds and glories of antique ladies scattered above the sky: the stephane or coronal band of Ariadne, which we call Corona Borealis,[83] and the shining blonde locks of Berenike, the sister and wife of Ptolemy III Euergetes, which were dedicated to Aphrodite.[84] (Unhappily, the prominent stars of this last group are masculine in Chinese: the Five Barons [*wu chu-hou*].) Although less prominent—the Greeks having preempted the greater part of the firmament—the Virgin Mary has long been allowed a role among the stars. To an anonymous Goliard poet—but possibly it was Abelard himself—she was Venus as the star of the morning, both "gemma pretiosa" and "mundi luminar."[85] In China the gemma would have been jade.

For the conservative Chinese court, the "female" asterisms were typically impersonal and represented offices and incumbencies in the imperial government on earth whose fortunes they controlled. China had few stellified ladies like those of the Mediterranean world. A characteristic example is the single star (ψ Draconis) which bore the name "Female Clerk" (*nü shih*). The title itself can be traced back to the Chou period, when women occupying this office assisted the queen in the performance of her ritual duties.[86] It was much used by later governments as well, but the incumbent was by no means always charged with purely "female" responsibilities. In the office of Overseer of Rosters (*Szu po*) in the imperial palace there were six female clerks; there were four employed in that of the Classifier of

Treasures (*Tien pao*), four in that of the Overseer of Dress (*Szu i*), and many others.[87] Their duties seem to have been mainly those of secretaries, keepers of registers, and storekeepers. Their guardian, the astral "Female Clerk," dwells within the Palace of Purple Tenuity: "Hers is the office of women: she manages the recording of affairs within the Palace. Her auguries are: when she is luminous the clerks who do the recording will use honest language; if she is not luminous it is the reverse of that."[88]

During the reign of Li Lung-chi, stars resided in the imperial seraglio. That monarch called his consorts collectively "Heirgiver and Consorts: the Four Stars,"[89] but we are given no clue to their celestial counterparts, although it is almost certain that they were the four stars of the lunar station "Woman" in Aquarius.

Some constellations, not now plainly designated with female titles, had important female connections and influences. Such a one was Hsüan-yüan, corresponding roughly to our Leo, and usually thought of as representing a fructifying rain dragon. In antiquity it harbored the god of rain and thunder.[90] We should accordingly assume that it had female aspects, as was natural to rain spirits, and indeed some ancient texts make its chief star, which we style Regulus, a queen (*nü chu*), with authority over royal women on earth.[91]

It is even more surprising to discover that women-stars could be discovered in such an unlikely asterism as the Great Dipper. It was characteristic of the Taoists to give women more positions of eminence in the starry sky than did the official star-mappers. They also gave them more individuality and personal character. One pre-Sung Taoist "sutra" identifies female personages in each of the stars of the Dipper. Consider, by way of example, the resident of the Palace Enshrouded in the Nexus (*Niu yu kung*) in the fourth star of that asterism (called Megrez in the West). She is an Inner Consort named "Most Exalted Seven Ardors," and she wears a cloak of purple damask over a yellow-flowered feather skirt. Her hair is done into a "sprawling cloud" chignon.[92] The Taoist arcana regularly provide such pretty details for practical purposes—so that the adept may correctly identify the lady among her many sisters, a matter of some importance. There is a story analogous to that of the voyage of T'ang Hsüan Tsung to the moon, but this one is told about Ming Ti of Han. It is illustrated in Buddhist mandalas, apparently of the thirteenth and fourteenth centuries, extant in Japan. These show the influence of Taoist star cults on Japanese Buddhism, and their Chinese origin must go back to the twelfth century and almost certainly earlier. A version of the story is preserved in the Taoist text "Canon in Twenty-eight Articles of the Most High Northern Dipper," whose date is unknown.[93] It tells how the Han monarch, strolling in the hills south of Ch'ang-an, met a

138

woman, barefoot with loose hair, and clad in white, who introduced herself as one of the seven deities of the northern dipper. (She reminds us instantly of the charming bare-legged maiden sitting under the moon on the now celebrated Han robe recently found at Ch'ang-sha.)[94] Other Japanese mandalas show celestial scenes with various Taoist deities (one of them has the three *hun*-souls and the seven *p'o*-souls all represented as female) and Taoist star palaces, notably the Gallery of the Seven Stars—that is, the Dipper itself.[95]

After this anachronistic, but appropriately suggestive example, we may now turn to an equally suitable story of star-women that cannot be later than the tenth century and accordingly suffers from no such defect:

When Scholar Yao, Notary to the Autocrat in T'ang, was relieved of his office, he dwelt in the left [part] of the city of P'u [old city in Hopei]. He had one son, and two collateral nephews, each with a distinct surname. When they had reached the years of manhood, all were uncouth dullards, and quite unworthy of him. Yao's son was older than the other two youths, and Yao was concerned that he was not studious. Despite daily admonishment and censure, they all wandered about, lazy and unregenerate. Accordingly he plaited floss-grass to make a dwelling for them on the sunlit side of Mount T'iao, hoping that, cut off from extraneous affairs, they would devote themselves to the arts and sciences. There, in the thick depths of the wooded ravines, the din and dust would not penetrate. On the day when he was to send them off, Yao gave them warning: "Each season I will test what you are capable of, and if your studies have not progressed, it will be switching and scourging for you! You must exert yourselves at this!" But when they arrived in the mountains, the two youths did not once open up a scroll, but occupied themselves only with splitting and hewing, painting and plastering. So it went for several months. Then the elder said to the other two, "Now the time for testing has come, but neither of you has attended to his books, and I fear for you!" The two lads paid him no attention at all, but the elder applied himself to his books most assiduously. One evening, as the young man was leaning on a table, unrolling documents by a nighttime candle, he became aware that something was tugging at the back lappet of the cape in which he was clad, so that the overlap on his neck gradually descended. He did not think this strange, but pulled it up leisurely to cover himself. After a while this happened again. On the fourth occasion he turned around for a look, and saw a little pig crouching on his cape. Its color was a very pure white, and it was a bright and glossy as jade. So he pressed a document into square folds and struck at it. The pig ran off hastily with a startled cry. He called the two youths and they took up candles and searched very closely all round in the windows and doorways of the hall, but although they overlooked nothing, they could not learn where the pig had gone. Next day, a blue-gray headed horseman knocked at the door, and entered, note-tablet stuck in belt. "The Lady," said he to the three of them, "has asked me to make inquiries. Last night her little baby, without her knowledge, improperly got into the collar of milord's dress. She is most remorseful about this. When you, milord, struck it, you hurt it over much, but now everything is balanced between you, and milord must not be apprehensive." The three together thanked him with humble phrases—then looked at each other, unable to fathom the situation. In a little while the boy on horseback who had just come approached again, now carrying the baby they had hurt in his bosom, and with him several suckling and swaddling women. The shirt it wore was of white damask and fine cambric, beautifully close-woven and not at all what one sees in the ordinary course of things. Once more he transmitted the speech of the Lady: "The little child bears no resentment." Then he showed it to them. They pressed close to look. From brow to the tip of its nose it was as if streaked with vermilion—the very traces of the blow with the corner of the square fold. The three youths were all the more fearful, but the messenger and the sucklers and swaddlers spoke sweetly to them, and consoled them, and settled them down, adding that the Lady herself would come in a little while. When they had finished speaking they departed. The three youths all wished to get away covertly to avoid her, but were too agitated and bewildered to come to a decision. Now a blue-gray head and several tens of "palace supervisors" in purple clothing came running over the waves. In front they allocated screens and curtains, pallets and mats, and brightly glowing flambeaux with scented vapors—all strange and unusual. Then they saw a carriage with oiled side-walls, blue oxen, and vermilion hubs, as swift as the wind, with several hundred rare horses as escort before and suite behind. When it came up under their gate, it was the Lady that was in the carriage! The three youths rushed out and saluted, while the Lady said with a smile, "I did not intend that my little child should come here, and the injury you did to it yesterday, milords, does not amount to very much. It is because I felt that this might be a matter of anxiety to you that I have now come to console you." The lady's years were something over thirty. In air and attitude she was relaxed and self-controlled, and when she looked up and down like a divinity one could not tell at all what sort of person she might be. "Do you have homes and houses?" she asked the three youths, and when each of the three youths said that he had not, she returned, "I have three daughters, distinguished in manners, and immaculate in virtue, who would be suitable mates for you three young lords." The three youths gave salutations and thanks. But the Lady did not leave—she stayed, and for each of the three youths she founded a cloister. In a twitch and a blink whole halls and extensive galleries were all built and arrayed in order. The following day coaches arrived there. Their guests were resplendent, their suite was gorgeous, surpassing a community of royal relatives. Carts and costumes were dazzling and coruscating: the light flowing from them lit up the land, and their perfumes filled the mountain valleys. Three girls descended from the carts. All were seventeen or eighteen years. The Lady conducted the three girls up into the hall, and then invited the three youths to their seats, where wine and rich viands were set out, with fruits and nuts, of kinds not often found in this world, in plenty and even superfluity—and most of them never encountered by them before. All was beyond anything the three youths might naturally expect. The Lady indicated how each of the three girls was to be paired with her lord. When the three youths retired from their mats, with salutations and thanks, there were then several tens of maids, like divine transcendents, to escort them, and that evening they exchanged the nuptial bowls.

The Lady said to the three youths: "What men treat seriously is life; what they desire is honor. Now before a hundred days have been lost to mankind, I shall bring life to you, lords, enduring beyond this world, and position far beyond that of any mortal magnate." The three youths saluted once more and gave thanks, but were anxious lest their ignorance be a hindrance and their dull wits an obstacle. The Lady said, "Do not be anxious, milords, for this is a simple thing!" Then she enjoined her

140

manager on earth, commanding him to summon K'ung Hsüan-fu.[96] In a moment Master K'ung came, equipped with hat and sword. The Lady approached the staircase, and Hsüan-fu presented himself with a respectful salutation. Standing erect, the Lady asked if she might impose a slight task on him, addressing him thus: "My three sons-in-law desire to study. Will you guide them, milord?" Then Hsüan-fu gave commands to the three youths. He showed the chapter titles of the Six Registers[97] to them with his finger—and they awoke to an understanding of their overall meaning without missing a single detail, thoroughly conversant with all as if they had always been rehearsing them. Then Hsüan-fu gave thanks, and departed. Now the Lady commanded Chou Shang-fu to show them "The Mystic Woman's Talisman and Secret Esoterica of the Yellow Pendants." The three youths acquired these too without missing anything. She sat and spoke with them again, and found that their studious penetration of all the civil and military arts was now as far-reaching as that of a Heavenly Person. Inspecting each other, the three youths were aware that now their air and poise were balanced and expansive, while their spiritual illumination was uninhibited and buoyant—they were in all respects equipped to become Commanders or Ministers. Afterwards Yao despatched a boy of his household with a supply of provisions. Such was his amazement when he arrived that he ran off. Yao asked the reason for this, and he replied with word of the richness of houses, grounds, curtains and hangings, and the abundance of lovely and ravishing beings. Yao spoke to his family in astonishment: "This is surely an enchantment of mountain ghosts!" He hurriedly summoned his three sons, but as the three youths were about to go, the Lady warned them: "Take care against leakage or disclosure, lest you bring about the application of a flogging with switches! Let there be no speaking of this!" When his three sons arrived, Yao too was amazed at the precocious expression of soul and psyche, and the relaxation and courtliness of informed replies. "All of my three sons," said Yao, "must have been suddenly possessed by ghostly beings!" He asked them desperately what was the source of this, but they said nothing. Then he flogged them several tens of times. Unable to bear the pain, they set forth everything from beginning to end. At that, Yao had them immured in a detached place. Now he had an important pedagogue as a long-term boarder, and he summoned him and spoke to him of this. The pedagogue said in surprise, "Most unusual! Most unusual! What were you doing to chastise your three sons, milord? The three youths had been instructed that if they did not divulge this matter they would surely become lords or ministers—in nobility far beyond any mortal magnate. That they have let it out was their destiny." When Yao asked him about the reason for this, he said, "I observe that none of the stars 'Weaving Woman,' 'Minx Woman' and 'Attentive Woman' are lit up. This means that these three woman-stars have come down among men, and were going to bring fortune to your three sons. Since they have now divulged one of the master-keys of Heaven, the three youths will be lucky to escape calamity!" That night the pedagogue took Yao out to inspect the three stars, and the stars were without light. Yao then released his three sons and sent them back to the mountain. When they arrived, the three women were cold and distant, as if they did not recognize them. The Lady scolded them: "You young men did not heed my words, and since you have divulged a master-key of Heaven, we must say goodbye here and now." She then made them drink a hot liquid, and when the three youths had drunk they were as benighted and uncouth as before—they did not know a single thing.

"The three woman-stars," said the pedagogue to Yao, "are still among men, and indeed are not far from this part of the country." He addressed himself closely to his intimates about their location, and someone said that it was the household of Chang Chia-chen in Ho-tung, in which there were afterwards three generations of commanders and ministers.[98]

This rather enigmatic tale, in which the only apparent reason for the little-deserved favor conferred on the three young men is the futile attempt of one of them to cram a little last-minute knowledge into his head, is of interest to us here because of the incarnation of the female star-spirits, and their espousal of three earthlings. The tale distinguishes "hsü woman" (*Hsü* [*Syu* "attentive"] Nü) from "wu woman" (*Wu* [*Myu* "minx"] Nü), although the two terms have been regarded as different names for the same asterism since very early times. The asterism is ϵ, μ, 2 and 3 Aquarii, just below the Equator, south of Cygnus, and close to "Separate Pearls" (*li chu*) on the edge of Aquila. It is also known as "Woman Asterism" (*nü hsing*) and "Woman Lodging (*nü hsiu*). A distinction in the two names accounts for my "Attentive Woman" (*Syu* Woman), who was in charge of textiles and tailoring, and "Minx Woman,"[99] who had responsibility for the destiny of treasures and treasuries.[100] They may here be regarded as twin sisters: "Attentive Woman, the silk-keeper," and "Minx Woman, the jewel-keeper." The Lady herself appears to be the Mystic Woman (*hsüan nü*) of the Nine Heavens, custodian of the occult treatises on the starry chronograms.

The importance of the Woman Asterism in divining the destiny of nations and provinces was often mentioned both in prose and in poetry. In T'ang times she was believed to control political and military events in the Shantung region. But the poets did not forget her purely female character, and could sometimes ignore her sterner role in disastrous geography. As in the prose tale just translated, she was often conjoined in verse with the other divine women, such as those in Vega and the moon. The tenth-century *tz'u* of Mao Wen-hsi has her as a secondary character, shown only through the emotions of the Weaver Maid in Vega—here the unnamed persona—who sees her lover only once each year.

> Seventh evening, year after year—believing it won't go wrong.
> The Silver Ho clear and shallow—the white clouds thin:
> Toad's light and Magpies' shadows—shrikes take to flight.
>
> Ever resentment for the little cicada—sympathy with Minx Maid;
> How often, pretty but jealous, she has put her "duck-tool" down:
> And this very night—the welcome meeting, as rain, gentle and tender.[101]

("Toad" is "moon," of course, and the happy magpies make their annual bridge of wings for the amorous transit. The shrikes—nasty birds—call out, says tradition, in the seventh month [this very time], but they are dismissed.

142

The "little cicada" is a symbol of transience and ephemeral life—it dies at the end of summer, never to see the seventh month. Apparently the Weaver envies it. Is Minx Maid as isolated as herself? The "duck-tool" is an embroidery tool.)

The precious pair, pretty Minx and pearly Moon, are seen together not infrequently, especially when they take sly vacations from the dark empyrean, and turn up at grand soirees in the gilded halls of T'ang.[102]

Sometimes the three great sky ladies appear together. The early T'ang poet Hsü Ching-tsung, in a somewhat erotic poem whose ostensible theme is a divine banquet in the Hanging Orchard of Mount K'un-lun, puts starry Minx and Moon Fairy in somewhat subordinate but suggestive positions, as suitably feminine lights in the stage-set for the all-too-short and frustrating love-making of Weaver Maid (again unnamed) in the second quatrain. Minx applies her celestial rouge in her room, making ready for the big affair; Fairy's carriage is already on its way. But Weaver Maid has other plans.

> Minx's boudoir—a meeting set for this evening;
> Fairy's wheel—adrift on the shallow Huang.
> Welcomed autumn—companioned with evening rain,
> Awaited darkness—a conjunction of divine lights.
>
> In repose on her mat she lowers her cloudy locks,
> In token of her feelings she has loosed her rainbow skirt.
> In the midst of her joy the miserable clepsydra hurries on,
> And after the parting she is resentful that Heaven lasts so long.[103]

The story of the Weaver Maid is widely known and thoroughly documented even in popular literature.[104] It will hardly be necessary here to cover all the familiar ground once more; suffice it instead to stress some aspects of her lore that might prove of special value to readers of T'ang poetry. First of all, consider her many names, some of them, while common enough, almost restricted to poetry. The least charming of these names is one of the oldest—as old perhaps as "Weaving Maid" (chih nü) itself. It is "Grandchild (or Granddaughter) of Heaven" (T'ien [nü] sun), an epithet that occurs in early Han times,[105] but is also known in T'ang poetry, as in the following couplet, in which she takes the form of a royal princess appearing at a party in Ch'ang-an, just as Minx Star takes carnal form in the poem translated above:

> In simurgh car, conducted by phoenixes, the Royal Son is coming;
> From dragon tower and lunar hall, the Grandchild of Heaven appears.[106]

This nomenclature is reminiscent of the use of "God's Child" (ti tzu) as a poetic name for the Goddess of the Hsiang River.[107] But more commonly the loom-plier by the silvery Sky River was gifted with such attractive sobriquets

143

as "Star Fairy" (hsing o),[108] "Star Consort" (hsing fei),[109] "Sylphine Star Charmer" (hsien hsing yüan),[110] and even in double guise, tripping in unwetted socks, like the Goddess of the Lo River, across the sky river's ford among the jade-white leaves of celestial elms, she was simultaneously "Sylphine Charmer" and "Star Consort."[111]

As for the divine lover Ch'ien niu "He who leads a cow by a rope," often called "Herd Boy" in English-speaking lands, although he tends no herds (why not Cowboy?), he is the yellowish star Altair. This is an Arabic name, an-Nasr aṭ-Ṭā'ir "The Flying Vulture," although in Islam the star was sometimes al-'Uqāb "The Eagle." For us the whole constellation is an Eagle, latinized as Aquila. (The Weaver Maid was no lady for the Arabs: to them Vega, and the constellation Lyra, was an-Nasr al-Wāqi', "Descending Vulture."[112]) He had a less common Chinese name, "Great Army Leader" (Ta chiang chün), and sometimes he and the whole asterism were called Ho ku "Ho (River) Drum." It was a martial constellation, reminiscent of Drake's Drum hanging ready at the port of Heaven. There was even a mountain on earth named "Stone Drum Mount" which resonated to the quivering of the heavenly one when the land of Ch'in was in dire circumstances.[113] An early Han source tried to bring this sky cowboy into line with the then relatively new doctrine of lunar stations by transferring Ch'ien niu to the station "Cow," leaving the name Ho ku dominant in Aquila.[114] Some later authorities followed this tradition, but it went against the grain, and many men—probably the majority—rejected the transfer, among them Chang Heng, the great Han cosmographer, who stated plainly: "Ox-leader stands to its left; Weaver Maid resides on its right"—that is, left and right as opposite sides of the Milky Way.[115] Cow Station is in Capricornus, much to the south of Aquila, and is composed of stars of modest brightness—not to be compared with Altair—nor is it on the verge of the Sky River, as Altair is. Let us retain the Cowboy as an alter ego of Ho ku, which itself is not really "river-drum," after all, but probably its homophone ho ku "drum bearer."[116] During most of the year—that is, when he was not making love to the Weaver—we can imagine him leaving his cow (he seems to have had only one, like Beanstalk Jack), and going off to the wars, either as Drummer Boy or as General—and indeed this was an important constellation in military astrology. It appears that he could be unfaithful to his star lady, since some early Sung writers aver that the Li Mu Mountains of Hainan Island, where the Loi "barbarians" live, combine in their name the words for "plough" (li), representing Altair, and wu "minx," the famous star of the lunar station Woman.[117] If such was the case, the inconstant lover was wise in choosing such a remote place for his assignation.

But, for that matter, the Weaver Maid was unfaithful too, and with a mortal at that. We are fortunate in having two pairs of quatrains surviving from the T'ang era that record this love affair. The story goes that the Maid appeared to a certain Kuo Han on a moonlight night and declared her love for him. After that, each night over the period of a year, she descended from the sky and then left him tearfully to return to her sky palace forever. Later a message, mysteriously conveyed, reached Huo Kan. It contained two stanzas of verse. They and the two stanzas he wrote in reply are extant, although it is not clear whether his pair were ever delivered on the bank of the Sky River. Taken together, the four can be arranged alternately to form a kind of antiphon in which the sky girl twice asks her mortal lover to visit her. In each case he replies that the desirable reunion is out of the question. The second pair goes rather like this:

Weaver Maid:
My vermilion gallery stands by a clear stream;
My rose-gem palace holds a purple chamber.
For feelings of delight the reunion must be here—
And this it is that rends my bowels.

Kuo Han:
Your gifted pillow is still fragrant and glossy;
Your wept-on garment remains spotted with tears.
But your face of jade is up in the Empyreal Han,
And vain would be the coming and going of my soul.[118]

(The reader will observe the parallel between "vermilion" and "rose-gem" [the archaic word *ch'iung* has such reddish connotations]. The "clear stream" is the Starry Han, and "purple chamber" is a classical allusion, denoting the private chamber of a great lady.)

Whether longing for her faithful cowboy in the sky, or for a rare mortal lover like Kuo Han, the Maid appears in uncounted poems as the very prototype of the woman who suffers from the endless weary waiting that follows the brief intensity of passion. Another poem by Hsü Ching-tsung will serve to symbolize such desperate ladies. (The reader will note several ways in which this poet's treatment of the theme resembles that in his poem about the fairy feast at the Hanging Orchard, translated above.)

A whole year nursing her grudge—sighing at long separation:
On seventh evening she chokes back her feeling—can speak at last
    of return.
Whirling, swirling, gauzy socks, treading the shining sky.
Vivid and dazzling, makeup renewed, mirroring the moon's brilliance.

Her emotion compels an artful smile, opening a starry dimple,
Unconcerned about showing plain that her cloudy dress is unloosened.
But she sighs when in due course, the clepsydra's toll is done,
And covers her tears, returning to ply the loom as last night.[119]

The obscure and unusual earthly affair with Kuo Han points up a significant quality of the Weaver Maid. She is one of the few star spirits approachable by human beings, and then normally only by the extraordinary means of a sky-raft. Her appearance on earth (other than in certain versions of her original meeting with the Cowboy) is a great rarity—and the Taoist tale translated earlier hardly makes the event less wonderful, since for the Taoists the divinities of the sky are almost as much at home on earth, or under the earth, or even in the recesses of our skulls, as they are in the empyrean. The other sky lady who occasionally receives visitors—though no raft can cross the empty space that separates her from the surface of our globe—is Ch'ang-o, the moon nymph. Perhaps it is not surprising that Weaver Maid and that Silk-White Fairy should be alike in their relative accessibility, since one modern mythographer at least reports them ultimately as the same person—the woman by the river weaves her white cloth out of wisps of cloud, just as a moon girl spins the clouds in an old German tale.[120] Another parallel to the interconnection between Weaving Vega and Silky Luna is provided by the presence in the waters off the coast of south China of the mermaid-like shark women, who weave their shimmering cinnamon-colored byssus on abyssal looms, and, when they are sad, shed not liquid tears but moon-pearls.[121] Indeed the subtle affinity between the two goddesses had an ontological foundation, inasmuch as the Weaver Woman was "a collector of *yin*"—that is, of the substance of the moon—and it was its flimsy beams as much as the feathery cirrus clouds that made the filaments of her divine damasks.[122] In one old tradition she was herself an emanation of the moon: "The quintessence of *yin* from the Toad flowed out to give birth to Weaver Maid."[123]

Weaver Maid resembled Moon Woman also in that festivals and celebrations in her honor were held down below. For one thing, Weaver Maid was patroness of brides. We have a charming but simple example of a T'ang wedding song, preserved in a manuscript found at Tun-huang, in which she figures prominently: it shows the bridesmaids below, peering anxiously at the sky, looking for lucky omens on the shore of the Silver River.[124] She was also a kind of celestial patroness of the domestic arts of spinning, weaving, and sewing, and on her night—the seventh day of the seventh lunar month—it was customary for girls and young women to contend in tests of their skills in the bosom of their family, especially in fancy needlework, applied to colorful and elegant patterns. There were also technical contests in such difficult feats as needle-threading. The goddess herself could normally be expected to be present in the shape of a house spider, and if she wove a web over a gourd it was a lucky portent indeed. On these happy domestic occa-

146

sions, the young men vied in the composition of literary pieces on appropriate themes. As the girls' contest was called "Begging for Skill" (*ch'i ch'iao*), presumably from the goddess, so that of the boys was styled "Composing Texts on Begging for Skill" (*t'i ch'i ch'iao wen*)—a natural interweaving of text, texture, and textile, as all participants were fully aware.[125]

We have a pleasant anecdote about this holiday as it was observed in the beautiful winter palace of Li Lung-chi (the posthumous Hsüan Tsung) at the hot springs of Mount Li:

Each time the night of the seventh day of the seventh month came around, the Divine King and his Noble Consort attended a party at the Floriate Clear Palace. At these times all the women set out melons, flowers, wine, and delicacies. These were arrayed in the courtyard, as they sought the loving favor of the stars Ox-leader and Weaving Woman. Each of them also caught a spider in a little lidded box; in the morning they opened it and looked to see whether the spider's web was loose or tight-woven, as a token of the extent of their skill. If it was tight they said their skills were numerous. If it was loose they said their skills were few. Most common people also do the like.[126]

We are even more fortunate in having a "Text on Begging for Skill" from the brush of no less a stylist than Liu Tsung-yüan, from which the following lines are an excerpt:

> Heaven's grandchild,
> Monopolist of skill in the Sky;
> She interweaves Template with Armillary—
> Warp and woof of starry chronograms,
> Capable of forming patterned figures
> In fretted brocades for God's body.[127]

This is truly cosmic weaving with the archaic astronomical instruments (Template and Armillary) functioning as weaver's reed and shuttle, producing a fabric patterned in the geometric figures that correspond to the numinous asterisms. In effect, the Weaver Maid weaves the sky patterns (*t'ien wen*) that have such a powerful influence on the destiny of men and nations, while the activities of the mortal young ladies on Seventh Evening is a pallid imitation of her gigantic task. So it was said in antiquity:

Weaver Maid is Heaven's daughter. She presides over melons and fruits, and the collection and storage of rare and precious things for the preservation of the Divine Luminosities, and she establishes the set of the strands in the making of garments. Therefore it is the rule in Ch'i that in making patterned embroideries they must correspond to the Ways of Heaven.[128]

It follows from her special competence, therefore, that just as the explosion of a nova in the domain of her lover on the other side of the Sky River is a sign that "in twenty days men-at-arms will rise up," so when her own household is invaded by a stranger star, "silk threads and fabrics will become costly."[129]

STAR SWORDS

And they said, Lord, behold here are two swords. And he said unto them, It is enough!

Luke, 22, 37

In modern times the western world does not recognize any single asterism as an entire sword or as having a unique connection with swords. We have two important sky swords, however, as the armament of two major constellations—gigantic heroes astride the northern firmament. Most of us have no trouble picking out the magnificent sword of Orion, distinguished not only by two triple stars, η and ι, the second of which has components which appear "white, pale blue, and grape red" in the telescope, but also by the Great Nebula. Both of the two stars have been called Saiph, from Arabic *Sayf al-Jabbār* "The Sword of the Giant."[130] The sword of Perseus (i, g, ϕ and ν) presents no such fine spectacle. The Chinese did not recognize either of them. In the seventeenth century there was also Xiphias "Sword", an antarctic constellation which we now style Dorado "Goldfish", near the Greater Magellanic Cloud. In classical antiquity, however, the word *xiphias*, insofar as celestial objects were concerned, was reserved for sword-shaped comets.[131] Few Chinese of the T'ang period could have observed the austral sword, and still fewer would have cared to attach a name or significance to such a barbaric cluster, hovering over the lands of black men and demons. They would, however, have been quick to appreciate the truth that some comets are swords.

In T'ang and earlier times a nicely tapered comet was indeed a glittering blade in the sky, with obvious military implications. The image pervades T'ang poetry. A notable instance can be found in a long poem by Wei Chuang, which, like so many of his, comments dolefully on the disasters of the last decades of the ninth century; in a linguistic atmosphere of secular conflict mingled with supernatural energies—"weird vapors," eagles and ospreys, signal beacons, and magic swords ("the sword in its casket, its tapered blade tempered in spirit-force")—we find, quite typically, "an uncanny star, suddenly flashing its awn-like ray."[132] "Uncanny" stars are normally comets in Chinese literature, and the lofty blade of this lament is simultaneously the source and the spawn of the sword waiting below, trembling in its secret coffer. Occasionally a meteor was a sword—or a sword a meteor: Sun Ch'üan, founder of the third century state of Wu, was said to

148

have had a sword named "Drifting Star," the common Chinese expression for a meteor.[133]

But the sidereal affinities of swords were much more complex than this. Sword and comet were shaped alike, but shape did not count for everything. Indeed swords resembled *all* stars: both gleam brightly, and in poetry are alike compared to snow and frost; both are emblems of winter and death; and both—at least the most important of them—have magic power which brings victory or doom to kings. The following quatrain, written by Chang Wei near the end of the T'ang, weaves an easy counterpoint among the themes of frost, late autumn, white, north (the direction of danger), stars, sword, and the nation's fate:

> Frosted mustaches hug his jaws: he looks out at autumn expended.
> Cloaked in a mantle of white ermine he ascends the watchtower alone.
> Facing the north, gazing off at the stars, sword in hand he stands—
> His life's whole length burdened with care for home and country.[134]

The grizzled warlord depicted so neatly in these few lines looks to the glittering stars for a sign—and it is not by accident that he is shown alone at night, holding a bare blade, far from any battlefield.

In fact, swords were stars.

The great swords of Chinese tradition, like the two swords of the Christian gospel which—to some—represented spiritual and temporal authority, came in pairs.[135] The idea of paired supernatural swords may easily have been suggested by the appearance of comets—those dazzling swords in the sky—with split tails. A notable modern example is Biela's Comet, which split into two segments on its appearance in 1845 and returned seven years later as a distinct pair in the same orbit. Another is Tebbutt's Comet, the Great Comet of 1881, which was distinctly bifurcated. In Chinese thought such double brands were regarded as twins that share the pains and joys of birth and life, or as embodiments of abstract *yin* and *yang* forces, or as brother and sister, often parted. Star swords, of mysterious origin (some were forged by earthly enchanters, some were transformed celestial essences—or dragons), gleamed with purple light or radiated dazzling beams—"awns of light, dazzling the eyes."[136] They also had the power of flight.

The ninth-century poet Tu Mu, describing a precarious military situation, wrote: "An awn of light from a star-sword shoots through Dipper and Ox."[137] Properly, my translation should read "star-swords," the plural, since Tu Mu is referring to the most famous of all pairs—the celebrated swords of Feng-ch'eng. The tale of this pair of metallized dragons—male and female siblings—is one of the best-known of all Chinese stories, and only a brief summary of a basic version (the tale has mutated many times) is needed here. Suffice it to say that the twins were made in the ancient state of Wu, and their numinous essence appeared as a purple vapor in the sky between the lunar stations *niu* and *tou* (Ox and Dipper)—that is, on the ecliptic between our Capricornus and Sagittarius—which together make up the twelfth zodiacal constellation "Star Chronicler" (*hsing chi*). They were detected there during the Chin period by the famous but inscrutable Chang Hua, and on earth passed into the hands of one Lei Huan. They were separated for some time, but finally reunited as dragons when they were dropped into a river. This momentous metamorphosis followed a series of adventures centered in Chekiang, the territory whose destiny fell under the control of Ox and Dipper (the so-called "Southern Dipper," it must be remembered.)[138] This story, and the attributes of the man-sword and woman-sword which constitute its central theme, provide the foundation of almost all references to star-swords in later literature. Indeed the shining pair were so famous that they were subjected to humiliating skepticism by a dragon-woman of the T'ang period, who denied the accepted truth of their tradition on the ground that dragons are by nature antipathetic to metal.[139] I shall not presume to pass judgment on the merits of these allegations, except to observe that their author must be presumed to have had authoritative knowledge of the nature of dragons.

Poets have rung virtually all possible changes on the theme of "Dipper-Ox," the spiritual source and eternal habitation of the famous swords. The phrase occurs in hundreds of T'ang verses, denoting the triangle bounded by the Celestial Equator, the Ecliptic, and the Galactic Equator, south of the Sky Raft, which encloses these two potent asterisms. Together the two lunar lodgings clashed and glittered in the memories of men, they evoked dazzling visions of imperial power and the humiliation of enemies, they were imagined to haunt the land of Wu and simultaneously to tremble on the edge of visibility in the night sky.

In the following double quatrain written in the middle of the ninth century by Li Ch'ün-yü, the Feng-ch'eng pair are lost to this world—but always potentially present, ready for a sufficiently glorious need. The poem is named "Treasure Swords":

> The pool at Feng-ch'eng, where Lei Huan dredged up the swords;
> Years far gone, those events remote, their traces vague and uncertain.
> Mud and sand could hardly cover their pneumas that burst up to the sky,
> But wind and rain have cut off all thought of a time when they might
>   emerge from their casket.

> The night lightning still shakes a shadow at the pond's bottom;
> The autumn lotus spits vainly the flash of a cutting edge;
> But ever since a star cracked behind the Central Eminences—
> They were transformed to make a pair of dragons, and are gone, not
>   to return.[140]

150

Embodied Stars

In short, the spirits of the swords are dragons now, unseen above the clouds; their soul-bereft metal remains hidden in the sediment of the pool, revealed only symbolically in the lightning and the last white lotus petal of the year. (The "Central Eminence" [*chung t'ai*] is the red stars Tania Borealis and Tania Australis, that is λ and μ Ursae Majoris—powers behind the celestial throne.)

The divine sword could sometimes loom in a terrifying way. Wei Chuang, the poetic chronicler of the disasters of war at the end of T'ang, took irritable note of this. After writing of the disappearance of familiar though demonstrably evanescent things, symbolized for him by decayed but conventional animal associations—phoenixes and parrots (Chinese birds *versus* alien southern birds), leaping dragons and flying geese (the dragons appear to represent Chinese power and the northbound ranks of geese the nomadic hordes)—he concludes:

> It should not be—the breath of that pair of swords,
> Forever set beside the Dipper and the Ox.[141]

The celebrated pair have become an endless threat.

In a milder way, the swords represent a challenge to an admirably bold and shining spirit, a young hero, a new Arthur: "a cockerel—like those Treasures, one to burst through to Ox and Dipper."[142]

However, the swords, through their draconic natures, also represented the energy and ascendancy of women. When, in the summer of A.D. 710 the Lady Wei, mother of Li Chung-mao, put her young son on the Chinese throne and declared herself regent, a double rainbow-dragon was seen to traverse the sky. The *Book of T'ang* comments: "a rainbow is the quintessence of the Dipper constellation. The omen reads 'Kingbearers and Consorts coerce the kingly one by means of their *yin*.' "[143]

The Southern Dipper, like the magic swords, had a special relationship with a famous ford in Fukien called "Yen-p'ing Ford" (*Yen p'ing chin*), and sometimes "Ford of the Sword." This was the spot where Lei Huan lost the ancient swords, and where, down under the waves, they resumed their dragon aspect. There, for the imaginative wayfarer, they might be seen to repeat their mutation, as they endlessly do, *sub specie aeternitatis*. A quatrain by Hu Tseng entitled "Yen-p'ing Ford" exemplifies the eternal transformation:

> By the road at Yen-p'ing Ford, the water swells to fullest,
> Craggy walls, a grand sierra—layered a myriad of *li*.
> Yesterday night seven stars were seen at the bottom of the tarn,
> Distinct and clear—divine swords were transformed into dragons.[144]

(The *seven* stars are a mystery here: they belong to the northern dipper—the southern has only six.) Rather different is the archaic vision of divine outrage expressed in Wang Tsun's poem of the same name:

> Three feet of gemlike crystal—shot at Dipper and Ox!
> Never would they yield to common hands to requite a vengeful enemy.
> One morning, at Yen-p'ing—become a dragon's home—
> I watched them carry the wind and clouds to spread over the Nine Lands.[145]

(The Nine Lands are the old Chinese homeland.) The legendary falchion could also come in a dream to haunt the ambitious mind of a frustrated knight. What might not be accomplished if it could only be found again? Such is the vision evoked in Li Chiao's poem "Sword":

> Had I only adamantine swords
> I would look to hurry to a hero's court!
> White rainbow—in season to slice jade;
> Purple vapors—by night to encroach on the stars.
>
> But above their sharp edges the lotus flowers move,
> As frost and snow they light the interior of a casket.
> Trusting to Heaven, I would take them to avenge my country!
> I would mark off my own land and make there a gamecock's name![146]

The two swords of marvel still lie hidden below the waving lotuses, imprisoned in their ancient box, deep in the Ford of Yen-p'ing.

The Feng-ch'eng swords are only the most famous of Chinese magic swords. Other countries knew of such, and other heroes brandished their counterparts. There was a comparable pair, close at hand, not Chinese but known to the early Chinese. The *Book of Chin* gives a legendary version of their story, evidently somewhat recast to fit Chinese conceptions. This tale purports to tell of the foundation of the Kingdom of Champa, in what later became Vietnam, on the southern frontier of the Middle Kingdom. It tells of a slave who caught a pair of magic carp in a mountain stream. The fish, not surprisingly, were disguised dragons, which congealed forthwith into iron: the lucky man forged them into a pair of conquering swords.[147]

We have seen the supernatural blades before in the guise of comets and meteors; here they partake of the quality of lunar haloes, and of frost and snow. These images suggest not only the flashing surface imparted by masters of the art of damascening, but also a real metaphysical connection with the deadly white powers of nature. Accordingly such verses as the following are common:

> The light of his sword like the lightning, his horse like the wind.[148]

This tells of the conquering blade of a general on the northern frontier, but is more than simile: sword and lightning are twins and congeners. The following couplet reverses the imagery, and makes the biting cold of the barbarians' land comparable to that of a sword:

> The Hunnish frost, like the cutting edge of a sword;
> The moon of Han—it might be the ring on my sword.[149]

These last lines were written by Lo Pin-wang, whose poems shimmer with stars and other white celestial lights and surfaces.

Swords trembled in their secret caskets like live things. Light is evidence of energy. What is more, they sang: "Inside its case the treasure-sabre signs both day and night."[150] So wrote Kao Shih in the eighth century in a long poem full of martial frights and prospects. And Lo Pin-wang—as did many others—told of the powerful spirit of the Southern Dipper calling from the sky to the sword in its hiding place: "The breath of the Dipper still calls out towards the contents of the casket."[151] It is like calling to like.

Another characteristic of divine swords is their permanence—a quality that they shared with their sidereal counterparts. Gold might seem to endure forever untarnished—but under the mask of corrosion the steely star-swords retained eternal power, waiting down through the centuries for a great time of need:

> The oyster dies—but leaves a light by night.
> The sword is broken—but leaves a pointed edge.
> The philosopher returns into the great night,
> And for a thousand eons the batons and blazons [of his honor] are
>   handed down.
>
> Batons and Blazons—everywhere among the Four Seas:
> But men themselves are at the caprice of every alteration and change.
> We make odes to flowers—and the flowers have already gone;
> But we make play with the moon—and the moon is still there.
> We shall never know, here between heaven and earth,
> How many times the white sun has been darkened.[152]

This poem, written by Shao Yeh, a little-known southern writer of late T'ang times, contrasts the evanescence of human life, even allowing for the relative endurance of fame, with the perpetual rebirth of other natural things. (The "light by night" left by the oyster is the pearl—the enduring moon in miniature.)

The persistence and incorruptibility of magic swords have been attested by many poets:

> The divine sword does not corrode in the earth;
> The strange linen is not scorched by fire.[153]

Here the natural resistance of the steel of the divine weapon to rust is compared to the innate incombustibility of asbestos. Even the shattered weapon retains its power: "The sword is maimed but still can yield its purple pneuma."[154] Just as heaven-born dragons change only in semblance when they hide themselves in the crevices of the earth, unlike the husbandmen who labor in the fields only to disappear forever, so also the magic sword, like uncorruptible gold, undergoes only phenomenal, never noumenal alteration:

> The divine sword, striking the stars, meets alteration and change;
> The best of gold is formed into utensils, subject to furnace and mold.[155]

Such, then, were the inner qualities of treasure swords (*pao chien* or *pao tao*). Outwardly they were not necessarily jewelled like swords of state, although they might glow like phosphorescent gemstones, and they were revered (even when concealed behind an illusory coating of verdigris) as true national treasures. They were swords of power, often of great antiquity—in the same class as "treasure cauldrons" (*pao ting*) whose immense value lay not in inlays of gold and incrustations of jewels, but in their endowment of supernatural power that connected them time after time with affairs of the greatest moment for the Chinese people. As the virtue of the treasure-sword might be temporarily concealed beneath the rust stains of iron oxide, so the true character of a bronze treasure cauldron was often undetectable—except by a sage—below the blue and green patina of copper compounds.

Treasure swords are often said to be joined or linked to stars, and they are often paired in poetry with powerful curved bows which are compared with the moon, as in the following couplet, in which a sabre is the alter ego of a curved fragment of moon, and is stamped both physically and metaphysically with the pattern of a powerful asterism—probably the Northern Dipper:

> Linking stars enter the treasure sword,
> A half moon rises in the carved bow.[156]

So wrote Lo Pin-wang. Long afterwards Hsü Hun used a similar matching:

> Lowered stars join the treasure sword,
> Shrunken moon defers to carved bow.[157]

These and many other examples make it clear that treasure swords had stellar relationships other than those comparable to or inherited from the notorious Feng-ch'eng pair. There were also swords whose special power was magically induced by runes, as it were—in short, swords engraved with astrological texts and even with figures of the great asterisms: the sword as

154

amulet, the sword as talisman. Of the former sort, a number of literary examples survive from the T'ang period:

> Brandish thunder and lightning,
> Displace occult stars,
> Topple the lethal and evil,
> Offer profit and honor![158]

The longish spell said to have been inscribed on the blade of a sword—somewhat wordy for the blade, but only twelve Chinese characters after all—combines the common heroic reference to honorable victory with the chivalric concept of righting wrongs. Another perhaps imaginary but still likely to have been typical inscription is double: one set of words, one abstract diagram. This is registered in a poem on a "treasure sword" by Li Chiao. The blade is described in anatomical terms, as if it were a saintly body being tattooed:

> Upon its back the inscription was the graphs "Myriad Years";
> On the front of its breast points were made for the pattern of Seven Stars.[159]

The "banzai" characters on the spine certainly had patriotic intent, and indicated the moral purpose of the swordsman; the arrangement of scintillas on the blade represented an asterism of power whose frosty rays might be called on to animate the conquering blade. Naturally, Lo Pin-wang was a connoisseur of star-swords:

> Throughout my life one consideration has weighed with me;
> My purpose—my zeal—are more than enough for three armies!
> The savage sun shares its replica on my battle-axe,
> The stars of heaven are joined with the pattern on my sword.
>
> My bow's string embraces the moon of Han.
> My horse's feet trample the dust of Hun.
> Yet I do not seek a life across the frontier—
> I need only die to repay my liegelord.[160]

This octet should be compared with one written at the end of T'ang by the monk Ch'i-chi, whose verses allot spiritual energy to the sword itself, rather than to the ambitious young paladins:

> Cast by the hands of men of old—a divine creature;
> A hundred times refined, a hundred times tempered, before you were
>   first drawn forth;
> No need for men of today to strive at honing and whetting:
> Your lotus blade—star-patterned—has never failed.
>
> Flick it once, clap it once! Listen: "ching! ching!"—
> And the old dragon wrests its silhouette from the light of the Lamp of
>   Autumn.

When now will you meet a lord, a peerless cockerel,
To take you off for the reduction and ordering of all under the sky?[161]

(Fine swords are often said to have blades "like lotus flowers." The "lamp of autumn" is the moon.)

The stellar diagrams on sword blades are often shown in poetry as endowed with lives of their own. Wang Wei, for instance, has:

He tries to rub his shirt of iron to the color of snow;
He intends to wield a treasure sword—to activate its starry pattern.[162]

This automatic trembling or creeping of the constellation on the polished steel is not merely metaphorical, as demonstrated by the following more explicit description of a treasure sword—one still laid away in its chest:

Within the coffer the starry pattern moves;
Beside its ring the lunar image shrinks.[163]

(The ring, a traditional ornament attached to hilt or guard, and familiar on swords of honor all over Eurasia, is in Chinese poetry commonly figured as a moon symbol.)

The representations of stars on sword blades were regularly circular points of light—as indeed stars are, though we conventionally represent scintillation by adding points around their periphery. If the poets of T'ang can be trusted in this matter, these spots of light were sometimes made by inlaying precious substances in the blade rather than by simple chasing or etching. In the following poem by Wei Ying-wu which tells about two ancient swords whose power of transformation and sky-flight remains, as usual, unimpaired, the asterism is a golden inlay in the shining steel—a technique now sometimes known as "damascening" (in one sense of that word):

A thousand years within the soil—the iron of a pair of blades:
The earth's corrosion has not penetrated, but the stars of gold are
  extinguished.
Deep and remote—a bluish spine, full-set with scaly armor:
A kraken-dragon without legs—a snake with severed tail.

If suddenly it wished to move and take flight, it has within that
  numinous power:
Should a man of mettle win it, it were a match for a nation's treasure.
If enemy household raises its hopes, it will cry out at the mid-point
  of night;
Let no small lad or female child ever approach it!

Dragon and snake, altered, transformed, lie hid within:
When summer clouds run off and away and the thunder peals and crashes—
I fear lest they will become thunderbolts and fly up into the sky.[164]

156

But the sparkling stars in their blue-steel sky appear also to have been devised of other substances; Li Ho provides an example: "Your sword speckled with stars of jade, your yoke of yellow metal."[165] These seem not to have been mere metaphorical gems, although, to my knowledge, no example of such an artifact survives.

A more important question is the identity of the asterisms figured on the blades—assuming, as we must assume, that they were not placed there at random or in decorative but meaningless groups. It turns out that the most important and presumably most common of the stellar images engraved on great classical sword-blades was a set of seven. Perhaps the seven-star figure had its place on even more ancient weapons, but the locus classicus belongs to the late Chou period, and the story of the hero Wu Tzu-hsü (otherwise Wu Yüan), an exile from Ch'u in Wu, who offered the sword to a fisherman to ferry him across the Yangtze, explaining that "in this sword there are seven stars: its value comes to a hundred in gold!"[166] In some degree, all later "seven star swords"—swords of immense spiritual power—were revivifications of this one. The poetry of T'ang—as of other periods—is strewn with allusions to these wonder weapons. A typical example occurs in a verse by Wei Kao, himself a great warrior, and conqueror of the Tibeto-Burman peoples on the southern frontier of Szechwan in the ninth century: "A treasure sword at his waist, with a seven star pattern."[167]

The "seven stars" need not refer to any specific asterism, since many are composed of that number. They are the five visible planets plus the sun and the moon; they are "the seven sweet Pleiades above";[168] they are also the seven stars that make up the lunar station Star in northwestern Hydra—the quintessence of the dominion of the Vermilion Bird in its southern palace.[169] But above all they are the seven bright shiners of Ursa Major. Indeed it is this last group, which to the Chinese was pei tou "Northern Dipper"—not the Southern Dipper of Feng-ch'eng association that we should expect—that was the powerful charm incised on the vibrant blades of treasure swords.[170] In lieu of the representation of the asterism itself, it was sufficient to inscribe the graphs for the words pei tou on the blade, substituting word magic for pictorial magic. The characters for jih "sun" and yüeh "moon" impressed in red on the blade had a similar electrifying effect.[171] Indeed the picture of just such a sword survives as an illustration (reproduced as figure 4) in a short treatise on magic swords and mirrors written in the T'ang period and included in the Taoist canon. This sword, which has the power to slaughter demons, has an elegant form of the character ching "phosphor" on its pommel, while the blade is engraved with symbols of the sun, moon, five planets, and, nearest to the point, the form of the Great Dipper.[172] It may be that although the earlier sources say nothing about stars

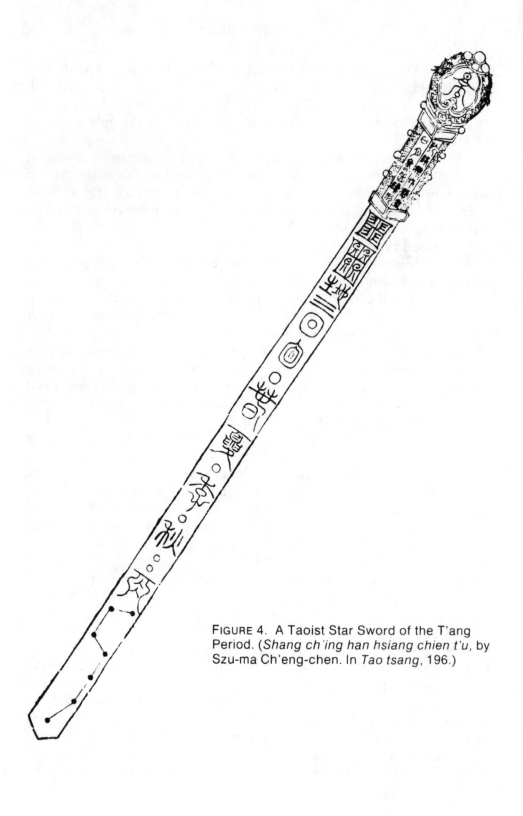

FIGURE 4. A Taoist Star Sword of the T'ang Period. (*Shang ch'ing han hsiang chien t'u*, by Szu-ma Ch'eng-chen. In *Tao tsang*, 196.)

engraved on the blades of the Feng-ch'eng pair, a tradition grew up that they, too, like their cognates, bore such emblems, and when Tu Mu writes " 'Dragon Spring' is waist-hung with its Dipper pattern,"[173] "Dragon Spring" is the name of one of the famous twins, and it is even conceivable that Tu Mu pictured it with the six stars of the Southern Dipper cut into the steel. But normally it was the great Northern Dipper, adjacent to the polar palace of the greatest star-gods, that was alluded to. Here is a typical example from the brush of Wang Wei:

> At your waist: a treasure-sword with a seven star pattern;
> On your arm: an engraved bow—the guerdon of a hundred battles!
> I have heard it said that at Yün-chung [Mid Clouds] you seized the
>   Kirghiz caitiffs!
> Now we know that in the sky there is surely a commander of armies![174]

This poem is addressed to P'ei Min, as illustrious general. It is not exactly what we should expect of Wang Wei—or rather of the aspect of Wang Wei most celebrated by modern anthologists. P'ei Min was skilled at the sword dance, a magical art of shamanistic origin, which was still highly respected in the curricula of military training. The general, with his wonder sword— doubtless an award for valor like Nelson's sword—is revealed simultaneously as a real victor and as a magical triumphator over the dangerous tribe of Kirghiz (described in contemporary sources as large red-haired men with white faces and green eyes).[175] The perfection of his talents persuades us that he incarnates the divine general "Great Army Commander in the Sky" (*T'ien ta chiang chün*), who appears to ordinary eyes as a segment of eastern Andromeda, particularly its splendid lucida Almach (γ Andromedae), a binary and perhaps ternary star, with orange, emerald, and blue components.[176] Perhaps the Taoists saw them separately with their secret eyes.

Happily actual specimens of medieval sword blades engraved with the images of powerful entities in the sky have survived for inspection by our curious modern eyes. Most stunning of these is a sword in the Shōsōin in Nara—very possibly a royal Chinese gift to the pious Buddhist court of Japan. The blade displays several constellations interspersed with the figures of clouds and other magical vapors. The constellations consist of stylized patterns of small circles connected by straight lines. Some are difficult to identify, but one prominent one plainly represents the seven stars of the northern dipper.[177] Similar swords are known from Korea, but are hard to date exactly. One such shows a dipper with a small star on either side of the second star of the handle—as if Alcor, the mystical companion of Mizar, had a mirror image on the opposite side of that star to make a pair of Stabilizers, dear to the secret vision of the Taoists.[178] Such marvels exist in our own day.

159

When H. Boehling visited the monastery on Mao Shan in 1948 he was shown its precious sword "with the gold-inlaid seven stars on the blade."[179]

But there were even greater wonders in the armories of men blessed by the stars. Meteors with brilliant trails were, as we have seen, swords in the sky. These sidereal blades actually came into the hands of men. Among other people who lacked a developed iron industry, the Aztecs, Mayas, and Incas possessed weapons of meteoric iron. This substance was highly regarded—indeed more highly prized than gold—for its celestial origin, and also for the more practical reason that, although malleable, its nickel content gave it the properties of steel.[180] As we have observed elsewhere, iron swords, apparently made in the late Chou period, also gave the ancient Chinese divine weapons created directly—perhaps "restored" is a better word—from the very stars that fell from the sky to provide the material of the magic swords of kingship.[181]

STAR BANNERS AND THE LIKE

... That proud honor claimed
Azazel as his right, a cherub
 tall,
Who forthwith from the glittering
 staff unfurled
Th' imperial ensign, which full
 high advanced
Shone like a meteor streaming
 to the wind . . !

Milton, Paradise Lost, Bk. I

The ensign of Azazel was raised in a righteous cause of cosmic proportions. Early Chinese literature provides us with a comparable banner, but it was the emblem of forces hostile to Heaven. "When a broom star is seen waving and curling in the aspect of a banner, the kingly one will campaign aggressively in the Four Quarters. If it is so long that it crosses the sky, the Lord of Men will lose All under Heaven."[182] This was the serpentine standard of the demon Ch'ih-yu.[183] Its threatening contours flaming through the sky boded evil, and although it was thought to be an emanation from Mars,[184] its physical identity is not surely revealed to us. Some scholiasts have seen it as an aberrant comet, but descriptions of its red and yellow undulations suggest that it usually showed itself as the aurora borealis. Fortunately it was an impermanent apparition. In contrast, fixed among the asterisms, there was a heroic flag, pendent forever from the upraised left arm of the giant we call Orion, streaming down below the red eye of Aldebaran. This royal battle-flag transmits the commands of a great general for the whole world to see, and, as Li Shang-yin observed,

160

Embodied Stars

When high in the sky the flag of Triaster passes,
Among men, all radiant flames flicker out.[185]

Star-patterned banners were known also on earth in antiquity. These normally displayed the seven stars of the Northern Dipper. Such was the "Seven-starred Banderole" (*ch'i hsing chih yü*) which signalized the position of the Chou Son of Heaven,[186] and such was the great banner named "Grand Constant" (*T'ai ch'ang*), a name which suggested the reliable movements of the bright chronograms with which it was emblazoned—the sun, the moon, and the Seven Stars of the Dipper.[187]

But of all the stellar banners the best known, the handsomest, and the most frequently alluded to in the literature of the T'ang period was the yaktail flag of the Pleiades. Metaphorical yaktail banderets are unfurled everywhere, it seems, on the pages of the poets, when the loathsome enemies of Civilisation and the Right Way rush towards inevitable doom. The streamers from the Pleiades fall from the Northern Ho as dispirited captives are led southward towards China;[188] they move threateningly towards the east as the men of T'ang wait confidently;[189] they hang horribly from the sky while the Chinese ponder counter-measures against their menace.[190] And always the young heroes of China dream of personal honor to be gained by cutting them forever out of the sky:

He does not read the books of the "States in Battle";
He does not peruse the "Canon of the Yellow Stone";
He lies drunk in a loft in Khumdan,
Dreaming that he enters the "City of Surrender Accepted."

He dreams also of a joyful moment:

Begs to be handed a Son of Heaven's sword—
To hack down the stars at the Yaktail's Head![191]

This is the ambition of a medieval Billy the Kid, illiterate, unversed in magic and classical tactics—but at least he knows that the attachment at the top of the barbarians' emblem is a solid manifestation of the Pleiades, whose threat is eternal.

The magic pattern of seven stars was seen in many places, swords and banners being only the most obvious ones. The landscape of medieval China boasted a number of "seven star" mountains or hills—normally, it seems, because of some distinctive physical attribute, but doubtless sometimes too because of a forgotten event of supernatural import. Examples are the hills in the north part of Hainan Island, called "Seven Stars" in Sung times: "they were on the shore of the sea, and were shaped like the Seven Stars."[192] Are we to assume that this was an irregular ridge in the form of the Dipper? The same island, in the same period, boasted the Mountain Pass of the Seven Stars, "Close to the sea-coast: its contours were like strung beads."[193] Here it

161

seems we must assume knobs or bumps approximating spheres which represented the seven circles symbolizing the stars of the Dipper characteristic of old Chinese star maps. (There seems never to have been an inclination to represent them as "star"-shaped in our sense.) There were Seven Star Bridges, notably a set of seven bridges in Szechwan, "corresponding to the Seven Stars." Tradition says that they were built by the great Han engineer Li Ping.[194] Some such constructions, however, seem to have been single rather than septuple. An instance occurs in a long poem by Lo Pin-wang: "The image of the bridge is divided in the distance into the contours of the Seven Stars."[195] We also have a poem by Li Chiao called simply "Bridges," with an array of classic examples—a magpie bridge, a rainbow bridge, a moon-shaped bridge—and a bridge "corresponding miraculously to the pattern of the Seven Stars."[196] There was also the archetypical star bridge, a variant of the magpie bridge—not necessarily composed of seven units—which spanned the Sky River on Seventh Evening.[197] All were ultimately the same magic path over divine rivers.

There were also wooden plates, perforated in Seven Star patterns, which, in T'ang times, were buried with the sovereign.[198] There were long seven-holed flutes, popular in those years, whose seven holes were named for the Seven Stars, and had also a side-hole covered by a thin bamboo-skin membrane. They were played in formal court music,[199] and there was also a popular tune named "Seven Stars Pipe"—conceivably played on the flute so named.[200] There were reports of an ancient and miraculous drug, named "Seven Stars Powder," known to that aficionado of alchemy Ko Hung, who tells us that the adept who ingested it would "suddenly be unaware of where he was."[201] The ingredients of this Lethean, if not lethal, medicine are unfortunately lost to us. In a slightly later period, the great T'ao Hung-ching—doctor, scholar, and true founder of the sect of Highest Clarity (shang ch'ing)—was marked as a super-being by a pattern of seven moles on his right knee—the stars of the all-powerful Dipper.[202]

οὐρανοῦ ἐξαπόλωλε, κακὴ δ' ἐπιδέδρομεν ἀχλύς.[1]
Odyssey, XX

8 The Sun

Now, the sun—
Senior among the hosts of *yang*,
Most honored of lords over men!
The cock of heaven chants at daybreak—
The crow of mana hops through the daytime.
At Fu-sang it draws near to the great sea;
At Jo-mu it shines upon K'un-lun.[2]

These lines in Yang Chiung's *fu* on the celestial sphere recapitulate the common lore of the sun, especially its association with the two birds of day and the two world trees, Fu-sang in the east and Jo-mu in the west, through whose branches the sun-bird climbs and descends each day.

The ornithological associations of the glorious sun as it flies across the sky have seemed self-evident to many peoples. For the Greeks, the quail was

163

the forerunner of the returning sun,[3] and its Sanskrit namesake *vartika* was also a solar emblem. But the greatest of sunbirds was Jove's eagle, who was also the symbol of the Nile, the year, and especially of the sun hovering over the equator—that is, of the equinox. He had a number of stellar roles and replicas too, and is for us spiritually enshrined in the constellation Aquila (and, as we have seen, in its chief star, Altair—a name corrupted from the Arabic for "Flying Vulture").[4] The Chinese sky could boast no such noble creature. There the sun contained only the humble figure of a crow, usually said to be awkwardly fitted with three legs. Although modern scientists have deflated him to a lowly sunspot, a different though equally learned version of his nature appears in one of the old apocryphal books on astrology, where he appears as a Trinitarian symbol:

> When the Prime opened up the *yang* to make heaven, it concentrated its essence to make the sun, and dispersed it in proportional spread to make the great chrono-grams. As Heaven is One, the *yang* is realized in Three [approximately heaven, sun, and chronograms—that is, the time-keeping asterisms]; therefore there is a three-legged crow in the midst of the sun.[5]

This sun-crow was the counterpart of the hare and the toad in the moon. Wang Ch'ung assures us that the real existence of this folkloristic sunbird was an article of faith with the *ju*—the orthodox pedants—of Han times. Inevitably he himself regarded the venerable belief as ridiculous, since he believed that the sun was real fire, substantially identical with the common fires of earth.[6] We might be inclined to agree with the famous skeptic's argument that the odd winged inhabitant of the noble sky-fire—for us an atomic furnace—would have difficulty surviving. But surely Wang Ch'ung's point is ill-taken: this was no flesh-and-blood crow, but a creature bred to the torrid life, like the salamander of our own mythology, or—as we might imagine—a crow of unexpectedly crystalline helium in the sun of our astronomers.

Despite his impressive qualifications as a radioactive phoenix, a tripodal crow was more likely, one would suppose, to provoke levity than reverence, and to lend itself most readily to exploitation by comic poets. Nonetheless the actual appearance in the flesh of such a prodigy was occasionally reported. Naturally it was taken to be highly significant evidence of Heaven's approval. For example, the people of Sun-yang were proud to forward the solar harbinger of fortune to the throne in A.D. 559,[7] and another of its ilk was happily discovered in Szechwan in 608.[8] Still another was caught in the capital in 762.[9] Such historical records are probably reliable, at least in the sense that in each case a specimen worthy of serious consideration fell into the hands of the concerned magistrates. Official guarantees of divine tokens of approval were sometimes given in confident innocence, but sometimes they were inspired by guile and imposture. Such a fraud was perpetrated, it is alleged, by a lickspittle courtier (we do not have his name) in A.D. 690. This was the year during which the Lady Wu assumed full imperial titles, slaughtered the scions of the house of T'ang, abolished the dynasty, and restored the ancient name of Chou as that of her own kingdom. Signs of divine approbation of this revolutionary change were welcomed, and the atmosphere of the court seemed right for the otherwise foolhardy presentation of a forged three-legged crow to the usurpatious sovereign. The doubtful authenticity of this embodiment of the divine solar essence was suggested by her "stepson" Li Tan (Jui Tsung), who was later to enjoy the T'ang throne for a brief period. And lo! during a close inspection in the imperial presence, one of the legs dropped off. Another version of this edifying anecdote tells that when the fraud was reported to the throne the "Heavenly Heirgiver" (t'ien hou) merely laughed and ordered the favorable prodigy entered in the official archives. "Why need we examine into whether it is genuine or counterfeit?" she said.[10]

A close cousin of the three-legged crow (no great figure in literature, unhappily) was the red crow, a lucky bird whose history goes back to classical pre-Han times. One such bird—described as "essence of yang," and plainly an authentic sun-bird—delivered a scepter to the future founder of the Chou dynasty.[11] However, one authority, at least, asserts plausibly that the "vermilion crow is a fiery emanation of the planet Mars."[12] Like the rather ludicrous three-legged crow, the handsome fire-red crow has turned up in reliable historical records from time to time. The most famous specimen appeared "like a god or numen," before the gratified eyes of Sun Ch'üan, ruler of Wu, in A.D. 238, in recognition of which the inauguration of a new cosmic era was declared, to which the name "Red Crow" was given.[13] A contemporary poet, named Hsüeh Tsung, celebrated the great event in his "Lauds for the Red Crow":

> Fiery glowing—the red crow;
>   It is the germ of the sun;
> Vermilion feathers on cinnabar body;
>   It is born in an exceptional era.[14]

As for T'ang times, one of these rare birds was discovered in A.D. 776 at Wei-cho, near Ch'ang-an, the location suggesting strongly that it had direct significance for the imperial capital.[15] Other than this (to quote myself), ". . . we observe that a red crow and a white magpie perched in a tree at the home of a recluse named Chu Jen-kuei, signifying his virtue. This seems a rather insipid sign—far from the godly fire-colored bird which appeared to Sun

Ch'üan. But those had been years of a more splendid divinity. No happy affair marks the history of 776. The appearance of the red crow must be taken simply as a guarantee of divine support for the reign of Tai Tsung."[16]

The earth-bound specimens of three-legged crows may have been forgeries, or possibly an occasional mutation, but it seems doubtful that captured red crows could have been anything but genuine fowls—they could hardly have been lacquered. A good candidate for the role seems to be the Maroon Oriole (*Oriolus trailii*) of Central China, a handsome crimson bird with black head and wings, and a close relative of the crows. One may have strayed into north China in 776, where it would undoubtedly have been regarded as a prodigy. But if we ignore birds caught in the wild, there are other possibilities. We have, surviving from the fourth century, at least one recipe for the fabrication of an artificial red crow: "Take a crow that has not yet put forth body plumage and feathers. Mix true cinnabar with ox's blood and have it swallow this. When it grows up both plumage and feathers will be red."[17] But since these artificial firebirds—I am skeptical enough to believe that the outcome of this forced feeding would be a dead nestling—were intended for the alchemist's crucible during his preparation of an elixir, it is not likely that any were carried joyfully to the imperial court.

The gulf between the primordial force represented by the corvine firedrake and the evanescent shadows that drift through the human cortex is great but not absolute. The red crow in a T'ang poem which follows is a symbol of human consciousness, and all the more interesting in that, unlike the lunar hare and even the lunar toad, the solar crow, whatever its symbolism, has been little noticed in Chinese poetry.

Yüan Chen ruminates on sad, dissolving memories. It is an evening in autumn:

> Morning and evening—autumn airs are fresh;
> Wind whirls—leaves slowly lighten.
> Stars multiply—Ho-Han grows white;
> Dew presses in—covers and pillow are clean.

> The cinnabar crow is extinguished within the moon;
> The "papyrus fowl" sings beneath the bed.
> Dim and fast-fading is that which I hug to my heart—
> How much more my feelings for all far-away things.[18]

Here "Ho-Han" is one of the many names for the Star River—our Milky Way. "Papyrus fowl" (*so chi*) is no more a fowl than is the ladybird, but a kind of grasshopper with very long antennae—possibly *Mecopodia* sp. or *Holochlora* sp. Autumn is the season of fading and extinction. The pure whiteness of the spirit world, the world of ghosts, which mantles the first quatrain, becomes active in the second: the tawny sun is quenched by the white moon—no

166

more real than yesterday's candle. So even a recent and cherished memory fades, as do many others that have gone forever into the world of the dead. The sun-crow, then, stands for the lively consciousness, for vivid awareness; the moon—for the cold waters of oblivion.

Separately from the avian associations, the sun rejoiced in other honorable appellations, which emphasized its indispensible role as divine source of light and heat—and hence of life. It shared with the moon the distinction of being one of the Two Phosphors—the blessed lights in the sky.[19] When it sinks in the western mountains, "its light shines back on the east; we call that 'returned phosphor.' "[20] (However this statement may be interpreted by the mythographers or metaphysicians, the reader will probably consider the possibility that it is simply a description of the Gegenschein, the "Counterglow," a particularly conspicuous spot—if any such faint haze can be described as conspicuous—in the zodiacal light, which moves constantly in exact opposition to the sun.) The sun was also called "Radiant Numen" (*yao ling*), "Vermilion Luminosity" (*chu ming*), "Lord of the East" (*tung chün*), "Great Luminosity" (*ta ming*), and "Crow of Yang." "Its charioteer is called Hsi-ho."[21] (This almost forgotten charioteer, a Chinese Phaethon, struck down and split in twain by the arrows of ancient Confucian euhemerizers, survives in respectable texts transmogrified into a pair of ancient astronomers, Hsi and Ho.)

In antiquity the substance of the solar orb was generally taken to be fire, or the germinal essence of fire—a kind of concentrated phlogiston.[22] Not unexpectedly, the doughty Wang Ch'ung adhered to this view in its simplest down-to-earth version: no supernatural or metaphysical fire, the sun was simply a great conflagration in the sky, basically no different from mundane blazes. He admits that, although it is more perfect than its earthly analogues (somehow all are "fire" or "pneuma" or "germinal essence"—not necessarily mutually exclusive), its properties are basically the same. Wang Ch'ung ridicules the true-believers of the *ju* class who would elevate this supernal, but quite comprehensible combustion into an abstract principle—as visible *yang*, for instance. He also disagrees with Confucian conservatives in denying that at night the sun is extinguished by the contrary *yin* principle; rather, he asserts, it disappears in the distance just as any other object going beyond the range of human eyesight. Similarly, the *ju* accept the (to him) naïve belief that sun, moon, and stars are round; not so, says Wang Ch'ung; their roundness is simply an illusion created by great distance which smoothes away the irregularities. This is proved, for example, by the fact that when stars fall to earth (as meteorites) they turn out to be irregular masses—not spherical at all.[23]

The standard opinion—that the sun is a unique species of celestial object, a perfectly shaped manifestation of the warming fructifying essence —led naturally to descriptions of it as a kind of luminescent jewel or gem. This metaphor appears already in the *Shuo wen* dictionary at the beginning of our era: "The sun is a jewel; the embryonic essence of the Grand *Yang*, which does not wane." Similarly, for the Taoist cosmographers it was "fluid pearl"—a term much favored in alchemical writings—and the plaything of the mystic dragon of the east.[24] Here, too, it is a beautiful, divine incandescence—a phosphorescent gem soaring lightly up between the starry horns of the draconic guardian of the dawn.

For the astrologers—and therefore for most people—the sun represented the supreme lifegiving power, with its moral concomitants, and therefore also the person and actions of the Son of Heaven himself. Accordingly, changes in the quantity and color of emitted light reflected parallel changes in the potency or direction of the imperial power.[25] The most spectacular of changes in the aspect of the sun are solar eclipses—total eclipses in particular. The Chinese were as fascinated by these as other peoples of the world have been, and saw in them direct and unambiguous references to the state of the realm. They tried constantly to refine their ability to predict them and so to remove them altogether from the realm of the unpredictable and ominous and to install them among known and innocuous phenomena. Even an approximation of success was always unlikely, not only because, as with lunar eclipses, astronomers were inhibited in some degree by their obsession with simple mathematical resolutions, despite the unpleasant irregularities in the data provided by observation, and also because solar eclipses, and especially total ones, are visible only from a minute portion of the earth's surface. Therefore those visible in China constituted only a small fraction of the total that terrified the denizens of the earth during any given epoch. Even a few days notice—as is possible with weather predictions—would have brought much relief. Indeed this was tried: "When the sun is about to be eclipsed, the color of the Dipper's second star is altered: it becomes a little reddish and loses luminosity. The eclipse will be on the seventh day."[26] The success of this method has not been reported. Indeed, whatever refinements were made in the hazardous procedures aimed at anticipating their reoccurrence, solar eclipses remained terrible events.[27]

Wang Ch'ung reports that the Confucians of his era thought that eclipses were fusions of the abstract or disembodied *yin* and *yang* fluids represented to our eyes by the moon and the sun respectively. He rejected this notion, as well as the to us more plausible one that, in a solar eclipse, the sun is covered by the body of the moon. He preferred rather to believe that these periods of dimming or extinction of the celestial lanterns were temporary lapses of their internal luminosity—much as some Taoists thought of the phases of the moon, superficially a very similar phenomenon.[28]

Whatever their cause, solar eclipses were frightening, and commonly thought, like other celestial aberrations, to be linked with great affairs of state and the fluctuations of the fortunes of mankind. In remote golden antiquity, when all the world was at peace, the sun was never eclipsed. We have this on the authority of Tu Fu, doubtless referring to a very ancient tradition.[29] But in the later, degenerate ages that we assign to "recorded history," the sun was blotted out with alarming frequency. By Han times, if not earlier, there were detailed prognostications available for the curious or concerned, covering variations in the appearance of a particular eclipse, the part of the sky in which it took place, and the day on which the phenomenon occurred. For instance, one of the Han books of omens states that if the erosion of the solar orb begins at the top, "a son will commit violence," but if it begins at the bottom, "wifely violence will arise."[30] Clearly this is a simple-minded deduction from yin-yang theory: "male is top; female, bottom." There was an omen for a potential eclipse in each day of the ever-significant cycle of sixty. For instance, a solar eclipse on a ping tzu day meant a great frost in the fifth lunar month; one on a hsin mao day meant an attack by a vassal on his liegelord; one on a jen ch'en day meant floods and prolonged fogs.[31] An eclipse in one of the lunar lodgings had much the same import as the appearance of a meteor or comet in the same asterism, with the difference that it showed a more direct and intimate relationship with the royal family.[32] A case in point is the solar eclipse of 21 April 646, in the lunar asterism Stomach. This constellation was taken to refer directly to the august stomach of the Son of Heaven, and the eclipse meant "our lord will have a sickness."[33] Just as with the appearance of a comet, a solar eclipse, taken as a sign of divine disapprobation, was the occasion for purges and austerities in the palace. On 11 February 780, the day following a blackout of the sun, a day designated for the great New Year's reception at court, with the usual offerings of token tribute and speeches of congratulation to the sovereign, all reports of happy omens during the past year were proscribed—a dismal restriction which must have shortened the proceedings considerably.[34]

The afflictions that tormented the face of the sun were multifarious, and were minutely catalogued by the diviners of ancient China. We should classify them all in a handful of categories, descriptive of their causes: chiefly turbulences in the upper air, gusts of yellow dust from the Gobi Desert, waves of volcanic ash transported over thousands of miles from Japan or Indonesia. To the Chinese such groupings as these were unknown. They thought rather of the investiture of the sun by malignant auras of various colors, and its defacement by diseased and nauseous blotches which often resembled the shapes of earthly creatures. The T'ang histories are chary of offering interpretations of these disturbing phenomena, but what few comments they do provide indicate that these invasions of the holy disc of the sun were regularly interpreted, through analogy, as responding to manifestations of lèse-majesté or contumacious resistance to the authority of the Son of Heaven.

One species of solar deformity alone is—in our order of things—a truly solar phenomenon. This is the class of mammoth ulcerations we style sunspots, which are comparatively cool cyclones on the sun's surface. If more than 50,000 miles in diameter, these are visible to the naked eye, properly protected.[35] Doubtless the ancient Chinese could observe them through such semi-opaque substances as slices of dark translucent minerals, or perhaps they were visible during great dust storms.[36] It was not hard to see in some of these dark phantoms the fabulous three-legged crow. But there were other possibilities: something resembling a "flying swallow" was seen in the sun in 875. And in 905, when the T'ang empire was in its death throes, an apparition of the northern dipper—that is, a pattern of seven spots—was observed in its disc.[37]

Pareva a me che nube ne coprisse
  lucida, spessa, solida e polita,
  quasi adamante che lo sol ferisse.
Per entro sè l'eterna margarita
  ne recepette, com' acqua recepe
  raggio di luce, permanendo unita . . .
                    Dante, *Paradiso*, II

9 The Moon

The ancient opuscule "Questions of Heaven" posed, among other difficult questions, the following:

> To what are sun and moon attached?
> And how is the array of stars laid forth?

And specifically of the moon:

> What is the virtue of the light of night
> That when it dies it is bred once more?
> And what is gained by a hare in its belly?[1]

There was an easy answer: the moon is anti-sun. As the sun is concentrated *yang*, made visible to the human senses, so the moon is the sensible aspect of the Grand Yin. The noumenal entity underlying the shining disc presented

171

to our eyes was, then, quintessential *yin*—an energetic storage battery (as it were), the necessary complement to that powerful generator the sun.[2] But *yin* was, in all of its aspects, no mere passive entity, as has sometimes been suggested by hasty interpreters of the verbal symbols of the *yin-yang* school. It was not merely derivative, secondary, and passive. Its cold nature was primary, and it was not so much passive as actively receptive. Too much has been inferred by analogy from the supposed social status of women in early China: quiescent, unoriginal, obedient, and placid as they were supposed to be, it was easy to project these qualities upon the glittering, dazzling globe in the sky.

Similarly, it was and is easy to imagine that if *yang* is somehow spiritual, like the exalted human male, *yin* must therefore be "material," like the subordinate human female. By no means. For us of the twentieth century the moon is a solid, non-luminous body—despite its remarkable albedo—a dead mass of cooled lavas, conglomerates, and other end-products of ancient magmas, all overlaid with dust, like an untended mausoleum. The Chinese moon was more truly feminine. Its energy, however, was potential rather than kinetic. Its true brightness was latent, and in this sense only it was dependent on the light of the sun, which, like an ultra-violet lamp, irradiated all or part of its surface to stimulate the characteristic moonglow, as the ultraviolet "black light" evokes the "cold light" in a luminescent object, such as a cluster of fluorite crystals.[3] It is not an accident that the phosphorescence of the oceans was called *yin huo* "yin fire"—for the "feminine" principle has its own kind of cool flame; and that the modern Chinese word for "fluorescence" is "firefly light" (*ying kuang*)—that is, the kind of light shed by naturally luminescent creatures.

"Moon is the germ of water ['germ of water' is also rock crystal] and therefore, although it is luminous within, its pneuma is cool."[4] A Taoist text extant in early T'ang times gives a picture of this inner phosphorescence emanating from mineral substances:

The moon is 1,900 *li* horizontally and vertically. The moon's nimbus is 7,840 *li* in circumference. White silver, *veluriyam*, and water crystal illuminate its interior. The light of flames shines brightly on its exterior. Within it are city walls, fortifications, and human beings. There is a Bathing Pool of Seven Jewels, and the Forest of Waning (*Ch'ien lin*) grows in its interior.[5]

This account demonstrates plainly that the moon is illuminated from within and that its relationship with the sun is purely a matter of cosmological balance, not of dependence. (Perhaps it is also worth remarking that simple calculation from the figures provided by this text shows that the radiant nimbus must extend approximately 100 U.S. miles above the moon's surface.) The same source provides a picturesque explanation of the waxing and

172

waning of the moon, in terms of a lunar population dedicated to the continual refining and polishing of the crystalline moon matter, so that it may radiate to best advantage:

The men in the moon are sixteen feet tall, and all are dressed in blue-colored garments. These men in the moon, from the first to the sixteenth day of the moon, regularly take white silver and veluriyam and refine them in a smelter of flame-light. Therefore when the moon swells to the limit its light is bright and clear and most sheer white; but from the seventeenth day to the twenty-ninth day they gather the flowers of the Three Pneumas below the Forest of Waning to wipe clean the light of the sun and moon; therefore, as the moon fails to its limit, its light is imperceptible, and its luminosity occulted.[6]

One of the Goliard poets expressed the medieval idea of the relation between the moon and the sun very well:

> Dum Dianae vitrea
> sero lampas oritur,
> et a fratris rosea
> luce dum succenditur.[7]

I venture to render this as follows: "When the glassy lamp of Diana rises late, then it is kindled by the rosy light of her brother." These lines reveal the late rising of the full moon in the east, just after the sun has sunk below the western horizon. Phoebus' fire, potent from a vast distance, lights Diana's crystalline lamp—the latter not a bad image from the Chinese point of view. T'ang poetry abounds in lunar metaphors of a similar cast—"silver candle," "crystal dish," and so on.

It may be argued that since *yin* symbolizes dark, shaded things, where the *yang* rays do not reach, the light of the moon is merely a reflection of the lordly sunlight to our eyes, as from a mirror (and in fact the Chinese did sometimes figure the moon as a mirror—but a magic mirror). To this argument we may retort that we prefer to follow the vision of Milton, who, faced with the difficulty of revealing the damned towers of Hell, a place plunged in absolute and supernatural darkness, even to the burning eyes of Lucifer and his host, displays it lit by flames—not by earthly flames, but by a species which shed "no light, but rather darkness visible."[8] Similarly, moonlight was for the Chinese a unique product of the moon—incited perhaps, invoked perhaps, yet of its own, *sui generis*, and self generated.[9]

The moon was also Water—that is, made up of water most refined, most subtle, most perfect, superior to but akin to all earthly waters. This followed from its *yin*-nature—it was the polar opposite to the fire-essence of *yang*. Its unique perfection among *yin* creatures lay in its purity, its homogeneity. It was unmixed, it was not contingent. Unlike camels, marshes, and women, which embodied or exemplified *yin* in larger degree than most beings, the moon was *yin* uncontaminated, a fact which removed it from the realm of natural objects altogether. But it shared more with natural waters than with anything else in the mortal world.

To take the watery nature of the moon as a fact of natural history rather than as a metaphysical truth was fatal to some old but persistent myths, such as the belief that among the denizens of the moon were a rabbit and a toad. In the first century A.D., Wang Ch'ung—otherwise a sensible fellow—used an oversimplified interpretation of the traditional belief to confound the "Confucian" pedagogues of Han, the so-called *ju*, who could not reject even the most fantastic tradition if it appeared sufficiently sanctified by age, or had been blessed by one of the sages of the past. If the moon is water, wrote Wang Ch'ung, the unhappy hare and toad, aliens in such an environment, would long since have drowned. But although the toad, the hare, and other moon creatures were always a source of uneasiness to the literal-minded, they gave no difficulty to the metaphysicians and the poets, whose moon creatures were inorganic. If they had veins, they carried ice water rather than blood.[10] They would have been content in Oberon's "wat'ry moon"—surely no ordinary satellite.

As genetrix of sublunar waters, the moon's behavior—for it was by no means a static object—affected the activities of its progeny below. The Chinese knew very early that "when the moon is full, the tidal surge is great."[11] This congenital affinity showed itself also in living creatures whose nature was watery, and by the same token moon-derived.[12] Accordingly, the moon's miniature replica, the coolly white, water-nurtured pearl, was thought to wax and wane in the oyster's womb,[13] as its divine parent went through her phases, and some pearls of exceptional beauty were styled "luminous moon pearls." Indeed luminescent "pearls"—including the orbs of some sea creatures to which the name "pearl" was also given—shone with the same cold light as the nacreous globe in the night sky, sometimes the plaything of dragons.[14]

Through water, the moon was also connected with inorganic matter, notably rock crystal: both the moon and crystal are *shui ching* "germ of water," and rock crystal was thought to be fossilized ice.[15]

There was also a mystic, perhaps merely symbolic, connection between moon and metal—apparently an outgrowth of *yin-yang* doctrine, which associated Metal with white things, with ice, with cold. The inwardness of this relationship in the actual sky-world is not made entirely clear by the sources, but seems to derive from the standard cycle of the Five Activities (*wu hsing*) in which Metal generates Water, and therefore lies, like a kind of transformer, at the core of the watery moon-substance, where it receives the

174

sun's rays to rediffuse them, like electricity through a light bulb, throughout the body of the moon.[16]

The wateriness of the moon has its western analogues. From early times we have given the name of "seas" to the broad grey areas of that satellite. Among them are Mare Foecunditatis, Mare Nectaris, Mare Vaporum, Lacus Somniorum, and many other wastes of fanciful water.[17] But these imagined seas and lakes, like the supposed canals of Mars, are—as we now know—non-existent, and, in any case, bear little resemblance to the pervasively humid character of the moon as envisaged by the Chinese.

To sum up, leaving such problems as the charting of the moon's journeys and the computation of its perigees and apogees to the historians of science, and focusing our attention on its nature, we have our surest guides in folk belief and literature. Literature does not need to restrict itself slavishly to the findings of science, whether western or eastern, and indeed its metaphors may flesh out the skeletal hints provided by the cautious astronomers. T'ang poetry leads us to a vision of the moon as simultaneously crystalline and aqueous, sometimes a lovely but frigid entity to be personified as a snow queen or an ice maiden—and in fact the Chinese often represented the moon as a woman. Without personification, we are compelled to see it as a crystalline mass, cognate to rock crystal, which, as all the world knew, was petrified ice, eons old, altered to transparent quartz. In his poem "Moon" the tenth-century poet Hsü Yin attributes the crystalline charm of the moon—fashioned, it seems, from glittering icy dendrites—to the special intervention of the abstract and universal entity which (or who) presses concrete objects and specific events out of mere Potentiality:[18]

In a cyan abyss who can detect the authority of the Fashioner of Mutations
Which condensed the frost and congealed the snow to create their unique allure?[19]

We ourselves are persuaded by our television screens that, despite superficial appearances, the moon is stony, and even dusty—something of a disappointment, in short. The Chinese version, however—since we need remote and incredible comparisons—reminds us more readily of the great outer planets of the solar system with their cores of metallic hydrogen and their crusts of frozen methane and ammonia. Or we may wish to compare the Chinese moon to the nucleus of a comet, which has been characterized as "a dirty snowball." The climate of this moon appears to us as wintry and bleak as the climate of Cocytus, the lowest of Dante's layers of Hell, whose cold was truly supernatural. As for the frozen fauna of the Chinese moon mythology, we shall attend to them presently.

A notable feature of the Chinese moon was familiar to Juliet, who conjured her lover thus:

175

Pacing the Void

> O! swear not by the moon, the inconstant moon,
> That monthly changes in her circled orb.[20]

The inconstancy of the moon was so striking that it offered a possible etymology for the Chinese word itself. *Ngywăt "moon" is *k'ywăt "defective": so says a venerable glossary of paranomastic definitions, already a classic in T'ang times,[21] affirming that liability to erosion is essential to lunacy. "This means," says that source, "that once it is full it is once more flawed. [The word 'defective'] is also said to be the name of the moon at conjunction: the moon dies and once more is restored to life."

The Chinese had other names for the moon in its different phases, many of them simple metaphors following the shape of the shining portion—such as "hook," "wheel" and so on—just as the shadowy patches visible on its surface suggested certain familiar figures from nature or mythology, to be named accordingly. But there were also words for various aspects of the moon that attempted to express the true meaning of its apparent permutations rather than their visible appearance. One of the more puzzling of these is the word p'o (*p'ăk), which is most familiar as the appellation of the more earthbound of human souls—the other variety being the soaring hun, a name akin to the word for "cloud." With reference to the moon, the word for the heavier, darker soul appears frequently in such expressions as "moon p'o" and "yin p'o", the latter emphasizing its affinity with the recessive cosmic principle yin, pairing it appropriately with the yang-oriented hun. But if we consult the standard dictionaries in the hope that we will find a firm definition for this usage we are doomed to disappointment. Modern dictionaries yield most frequently something like "the dark or unilluminated part of the moon,"[22] a definition we are inclined to accept without further research—unless perhaps we are a little puzzled at the absence of a corresponding use of hun for the illuminated part of the moon. Our confidence is bolstered by a few direct quotations from early Chinese texts, among which the most complete and convincing appears, at first glance, to be the following from the astronomical treatise in the official history of the Sui Dynasty:

Now the moon is the germ of yin. Its shape is a circle; its substance is clarity. When the sun's light makes it visible we see it as luminous, but where it is not made visible by the sun's light we call it p'o. Therefore on the day of the moon's maximum visibility, when sun and moon look at each other from afar, a man situated between them observes that it is completely luminous, and so its shape is a circle. On the days of the "two chords" [first and last quarters] the sun makes its side visible, and a man will observe this flank—therefore half is luminous and half is p'o. On the day of conjunction [new moon], the sun makes its far side visible, and as the man is on its near side, it will not be seen.[23]

176

The Moon

Clear enough it would seem, and very congenial to our own notions of the dependency of the moon's light upon the sun—but, in fact, a professional view current among court astronomers, and by no means the one accepted by cultivated men generally. A modern dictionary, *Chung wen ta tz'u tien*, citing the *History of Sung* in support, defines *yin p'o* as the moon in its first days after new: before the festival of the seventh day of the seventh lunar month, "the *yin p'o* is already born." This could only refer to the waxing crescent moon, and this definition is supported by that given in the dictionary *Tz'u hai* for the same phrase: "the moon's light before it is fully luminous." The *p'o*, then, is not the dark part of the moon as we were prepared to believe, but the light part of the nascent moon. To compound confusion even further, the Morohashi dictionary gives "alternate name for the moon" for both "moon *p'o*" and for "*yin p'o*," without specifying the quantity of light or dark on its surface. There seems to be no way to reconcile these conflicting definitions: black cannot be white.

But perhaps some help can be found in mythology and poetry—enough help, at least, to give us a fairly reliable guide to ordinary usage of these terms in literature, whatever may have been the usage of the court astronomers as exemplified in "The Book of Sui."

If, in fact, the *p'o* is the luminous edge of the moon, we should expect it to be described as "light" or "white" in early sources, in the same fashion as many other popular moon images, such as "chalcedony hare," "toad light," "toad dazzle," "silk-white moon," "silk-white fairy maid," and "silk-white light," to choose only a few from among many.[24] These are terms from poetry, and others like them are easy to find in T'ang verse. Among them is "jade *p'o*," as in "The jade *p'o* opens out in the eastern quarter," which means that the white moon—since in poetry jade always implies white—rises in the east.[25] We note also, just as commonly, "toad *p'o*," "metal *p'o*," and even "round *p'o*."[26] These phrases, all more or less interchangeable, indicate that the *p'o* is as bright as the moon-toad, that it has the whiteness of "Metal" (traditionally a white symbol), and that sometimes it expands to fill out a complete circle. It should come as no surprise then to find a supporting etymology. An old book of omen lore says outright: "*p'o* (**p'ák*) is 'white' (*bák),"[27] and indeed it is hard to escape the surmise that the presence of the graph for "white" in the character that represents *p'o* (white plus soul) is not fortuitous. Moreover, an old book, well-known to the scholars of T'ang, states: "Before the moon is at its fullness it carries its *p'o* in the west."[28] A fourth century scholiast, Li Kuei, explained this passage in these words: " 'Carries' means 'begins.' '*P'o*' means 'light.' 'Carries its *p'o* in the west' means that the light begins to emerge on the western face, then gradually fills out eastward.' " This is, of course, precisely what happens when the new

177

moon forms its crescent, as it follows close behind the setting sun, receiving light on its western limb: this is the lovely crescent we see just after sunset, hanging low over the horizon, often in close company with the evening star.

But we must constantly bear in mind that the dark part of the moon was not, for the early Chinese, usually what it is for us—a portion of its disc turned away from the solar radiation. What was missing was not sunlight, but moon-soul, and this could pass from moon to sun as readily as sun-soul could take the contrary course. This is made clear in a couplet in a poem by the eighth-century writer and statesman Li Hua:

> The *yin p'o* plunges into the depths of space and time—
> The Grand *yang* borrows its luminosity.[29]

In this poem Li Hua is using the images allegorically—but correctly. As the moon—a spiritual entity with *yin* nature—disappears partly or altogether in the limitless abyss of eternity, the sun, although *yang* by nature, is able to augment its light by drawing from its waning companion.

It is clear that the application of the word *p'o*—for which we lack a good English equivalent—to the early stage of the waxing moon was made by analogy with the growth of the double soul in a human being. For this we have the ancient authority of the *Tso chuan*, which says, "At birth a man begins to form a *p'o*, and after the birth of the *p'o* his *yang* is called *hun*."[30] This is not altogether lucid, but at least it means that the *p'o* constituent of his dual soul comes into being immediately upon his birth, preceding the emergence of the *hun*, which is the *yang* element of his spiritual nature, and that this sequence is identical with the growth of the new moon: "Yang spirit is sun's *hun*; yin spirit is moon's *p'o*."[31] But Chinese texts never refer to the *hun*-soul of the moon, only to a stage of fuller illumination, from about the end of the first week of the lunar month, that is, when a half of the lunar disc is illuminated: this stage they call *kuang* "alight": "On the third day the moon forms a *p'o*; on the eighth day it forms *kuang*," and so on then to the full moon on the fifteenth day.[32] Apparently, although I have not seen this clearly and unambiguously stated, the period of the *p'o* light in the moon, i.e., the solitary embryonic life of its basic soul, lasts about five days, from the third to the eighth of the month, after which the *yang* of the sun takes a hand in stimulating the growth of the moon light to its full glory—a stage corresponding to the growth of the *yang* spirit (*hun*) in the baby after the first torpid week of its life.

We may tentatively align the Chinese terms for this lunar psychogenesis as follows, understanding that no man of T'ang may have devised so neat and rigid a table—still this shows how he used the words:

The Moon

Age of Moon	Chinese	English Equivalent
0 days	*shuo*	conjunction
3 days	*p'o*	primitive soul first visible (as crescent in western limb of moon)
8 days	*kuang*	light (first quarter—beginning of excitation by the solar emanation)
15 days	*wang*	lunar maximum, full moon[33]

But an English equivalent for *p'o* is still needed. Since, as it appears, all ready-made possibilities serve only to deform our conception of the Chinese from the beginning, it is probably unavoidable to use some uncommon word—for instance, "protoplast" or "protonymph" (both from biology, with some added advantage to the latter, which can be taken also to suggest the female character of the lunar substance, and its beautiful goddess Ch'ang-o) —or else to coin a new word with appropriate connotations built in, such as "protopsyche" or "protogene." I shall indulge myself freely in the use of the former, to say nothing of "moon-soul" and "white-soul," rather than the dead syllable *p'o*, hoping that my readers will not be alarmed by translations from T'ang poetry in which "the jade protopsyche follows close behind the sun." The medieval mind peeps through here, however crudely.

Before turning to the lunar soul in literature, a glance at a Taoist view of its nature might not be amiss. The selection translated below is from a metaphysical work by Chang Kuo, whose miraculous powers turned Li Lung-chi (Hsüan Tsung) from a skeptic about Taoism to an enthusiastic believer. The book is based on the conception of a kind of mystic cosmic Ennead, in turn related to the "Nine Revolutions" used by advanced alchemists in refining cinnabar elixirs. The statement purports to reveal the true nature of sun and moon—each with an indispensible germ of the other enclosed within itself, symbolically represented by bird and hare. It exemplifies my terminology in which the buoyant, volatile *hun*-soul is "cloud-soul," and the cool, recessive *p'o*-soul is "white-soul," using "etymologized" equivalents parallel to Chinese paranomastic definitions:

Now the sun's cloud-soul (*hun*) and the moon's white-soul (*p'o*) are *yin* and *yang* —*yin* and *yang* are sun and moon. The sun is of the *yang* and cloud-soul class, and the moon is of the *yin* and white-soul class. There is a fowl in the sun: it is a representation of the Western Quarter, Metal, the lungs. It is of the *yin* class, and so the sun's cloud-soul stores the white-soul of the moon—and the white-soul is fulfilled by the cloud-soul. Therefore the sun is clarified by it. There is a hare in the moon: it is a representation of the Eastern Quarter, Wood and the liver. Liver is of the *yang* and cloud-soul class, and so the moon's white-soul stores the sun's cloud-soul—and the cloud-soul is fulfilled by the white-soul. Therefore the moon is illuminated by it. As to cloud-soul and white-soul—they are the lead and mercury of men . . .[34]

And so on into alchemical analogies. In this view, then, the icy light of the moon's protopsyche is generated by the germ of the solar substance within it—just as the radiance of the sun depends on the germ of the lunar stuff within it.

We have already suggested that the poets, not infrequently, allow the whole shining disc of the moon to be its protopsyche—as if the masculine radiation from the sun had temporarily inflated the aboriginal and puny moon-soul after the seventh day to fill out like a fluorescent balloon. Accordingly we have a pregnant moon in such verses as these: "Its protopsyche fills out—the cinnamon branches round it off."[35] This line, which displays the moon-cinnamon burgeoning out to occupy the whole disc, is preceded by a verse representing the new moon as occupied by a hare, and is followed by one in which it is an impaired wheel and a fractured mirror. In short, the whole period of its waxing and waning is displayed in a sequence of folkloristic images. Or again: two friends play the game of matching couplets as they drift in a boat. One of them makes the following contribution, matching the full moon with the closer lantern of their boat:

> Suspended lamp sends forth flame—
> Distant moon raises round soul (p'o).

But just as often the poet will be strict in his treatment of the newborn psyche—even when (he tells us) he is a little drunk:

> Into the cup—and sudden return to drunkenness;
> Above the lake—the birth of the moon's new soul (p'o).[36]

Are we to understand that Liu Ch'ang-ch'ing, who wrote this second couplet, imagines he sees the first light of the moon in the little lake of wine at the bottom of his cup?

The embryonic nature of the moon's p'o-soul is brought out by analogy in the following double quatrain by Ma Tai:

> Cold-lit Protopsyche emerges above the sea:
> Seeing it afar, I augment my wretched chant.
> A chilly search: but Blackhorse Dragon's chin is heavy;
> A cold plunge: but oyster's embryo is deep.
>
> Candent breath encages all of Hsia;
> Clear light shoots at a myriad peaks.
> Dim and distant, within Heaven and Earth,
> Candid and pure—a Heart fit for us all![37]

A knotty mass of images here—and perhaps we are doomed to be left floundering. But a try at the poet's intent may yield something:

180

The Moon

1. "Cold-lit Protopsyche" (yin p'o) is the new-born moon, despite the mouthful of syllables. I believe the image comes close to what the men of T'ang thought when they saw the new moon.
2. The sight saddens the poet: it appears to be an important personal symbol from which he is divorced. It seems too remote for his poor grasp.
3. It is like the mythic pearl in the heavy jaw of the dark cloud-dragon.
4. It resembles the foetal gem, inaccessible in the oyster's womb.
5. Yet it is visible in the distance, splashing its snowy light over the plains of China.
6. And darting its rays at serried peaks behind.
7. Whether beyond our grasp in the sky (as now), or lying, miniaturized, within the body of a divine animal,
8. It is always the same: Soul of Souls! Pearl of Great Price—eluding us all.[38]

But in these lines the moon-embryo is simultaneously fully realized and darkly latent—we are not concerned with "protopsyche as crescent" but with "protopsyche as potentiality." The former, more "physical," use of the image was more common. We find it in a line by the gifted courtesan Hsüeh T'ao: "Protopsyche is small—in the fashion of a hook."[39] Yüan Chen gives us a pretty picture of the new-born moon-soul, on the third day after new, when the crescent has just appeared (this is the technical meaning of the expression "faint moon" [wei yüeh], where the word I have rendered by "faint" really implies "imperceptible or barely perceptible; hence, rudimentary, infinitesimal"):

> In the blue crystal—the faint moon's hook;
> Shrouded dazzle—the grottoed cold light's psyche.[40]

Or it can change very rapidly, from hook to wheel, as the needs of the poem demand; hence the couplet of Li Ching-fang:

> Hot-lit Crow unfurls its wings at dawning;
> Cold-lit Protopsyche lets fly its disc with nightfall.[41]

More rarely we find the phrase yin p'o used of the vanishing moon at the end of the ninth month—a month traditionally styled "month of murk" or "month of gloom" (hsüan yüeh), because it is the one in which the yin (my "cold light") overtakes the yang in nature:

> The gloomy cold-light invites the falling sun;
> The cool protopsyche terminates in a shrinking hook.[42]

This couplet portrays the eternally alternate flight of round sun and round moon, represented by the hot crow-spirit and the cold moon-soul respectively. But as the moon rises over the eastern sea, it seems to explode instantaneously from the dark recesses below the earth, revealed full-grown. It is as if the non-sentient moon-creature, born as a curved line at the horizon edge, suddenly releases itself as a complete circle: the growth of fifteen days has been compressed into an instant.

The image of the growth of moonlight could be applied to a military situation as an omen or premonition of troop movements, as in the following poem by Szu-k'ung Shu:

> Where cold willows join alien mulberries,
> The army's gate faces a great wasteland:
> Our tent-camp will follow the moon's psyche,
> Our weapons' breath will lengthen the star's awn.
>
> Now a cross-flute hurries the spring wine,
> And doubled furs hold off the night frost.
> The ice has opened—no guard against the caitiffs,
> And green herbs are filling Liao-yang.[43]

These lines, written late in the eighth century, show us a Chinese military encampment on the northeastern frontier, looking towards Manchuria. Its power, its will, and its destiny are reflected in apparitions in the sky. As the light on the moon's face increases from its western rim towards the full, so will the Chinese power extend eastward into Liao-yang. The magic power of the Chinese swords, and the martial "spirit" of the men-at-arms, infuse supernatural light into the comet overhead—a star with an awn—drawing out its tail. There is, obviously, a strong infusion of military astrology in this poem. Most often concerned with such baleful apparitions as falling stars, it is preoccupied with lunar matters here. Most interesting, however, is the dominance of the epigene aspect of the paired Correspondences.

Although its periodic waxing and waning were predictable, the moon, because of another guise under which it was revealed at unexpected times, had also to be classed in the realm of the unpredictable and numinous apparitions. This was a phenomenon to which the name "phosphor star" (*ching hsing*) (with "phosphor" suggesting "spectacular sky-light," comparable to the grandest sights in the sky) was given. Sometimes the same vision was called "Star of Virtue" (*te hsing*),[44] for reasons which will soon become apparent.

Although it may come as something of a surprise to find the name "Star of Virtue" most frequently applied to the moon, not notably a moral symbol, the fact remains that we have many descriptions of this phenomenon,[45] and although they are not entirely uniform, collectively they agree on the following criteria: 1) a Star of Virtue is large; 2) its form resembles that of a half moon; 3) it appears during the period of the new moon; 4) it "assists" the moon's brightness; 5) it is a rare occurrence. In 1901, Franz Kühnert argued—and seems to have established—that these attributes fit only one kind of astronomical event, the appearance on the dark surface of the moon of "earthshine," that is, of sunlight reflected from earth to moon when the earth, from the point of view of the lunarians, is "full."[46] This strange ashen light,[47] does not appear at every new moon, but only under very special conditions, namely, that the sky be absolutely clear,[48] that the moon be as high as possible above the horizon, that the relative positions of sun and moon and earth be just right for maximum reflection of light from earth to moon and back to our eyes, that the sun be a certain precise distance below the horizon, and that the crescent of the moon be not too small.[49] The scene is enhanced by the fact that the additional light added to the crescent, beyond what it normally receives from the sun, makes it seem even brighter and larger than usual. (If it seems strange to us that this ghostly disc should be styled a "star," it should be remembered that *hsing* "star" was a word used to encompass a great variety of celestial apparitions, sometimes quite diffuse ones, such as comets.)

The misty appearance of this untypical lunar spectacle helps to explain some of the Chinese accounts of its genesis. The ancient Chinese seem generally to have felt that this was not a star in the ordinary sense. The common "explanation" for its formation was that it was a fusion of pneumas (*ch'i*), but these constituent pneumas were thought to be less tightly packed than those in a run-of-the-mill star. Wang Ch'ung, following the old glossary *Erh ya*, thought that the Star of Virtue was a fusion of the four pneumas of the seasons, but most later commentators took a rather more complicated view. Typically they saw it as a merging of three yellow stars, two of which had congealed out of a red pneuma and one out of a blue pneuma.[50] In any case, it was a rare spectacle, which seemed to signalize important developments on earth.

As to what these developments were, there was general agreement. To us the apparition is—or once was—a lunar warning of doom at sea, such as that afforded Sir Patrick Spens:

> Late late yestreen I saw the new moone,
>   Wi the auld moone in hier arme,
> And I feir, I feir, my deir master,
>   That we will cum to harme.

In China it was not so. The pallid figure held by the cusps of the shining crescent boded nothing but good. Its appearance was classed, in T'ang times, among the "Great Auspicious Tokens," most important of those signs sent by Heaven to show favor to the nation and its ruler. Also in this auspicious group were apparitions of phoenixes and dragons, the water of the Yellow River running clear, and a perfectly calm sea. Among celestial events, the same category included a quintuple conjunction in which the five planets

183

appeared strung out side by side like so many beads, and the unexpected appearance of Canopus, the Longevity Star, in the haze of the southern horizon, or any "yellow star", probably including some seen through the windborne loess.[51] Although the general purport of all of these fine displays was the same—divine approval of the regime—the Star of Virtue (or Phosphor Star) was said to refer directly to the monarch's "virtue" (te). A typical statement of the sense of the portent is the following: "When the Virtue of the kingly one reaches to Heaven (or, the Sky), then a Phosphor Star will be seen."[52] We have records of manifestations of the Star of Virtue which were supposed to refer to particular monarchs. One, in 110 B.C., was believed to glorify Han Wu Ti, because of his meritorious celebration of the glorious Feng-shan sacrifice on Mount T'ai,[53] while another, early in T'ang times was taken to be a sign that Li Shih-min had been designated future sovereign—surely a rather premature manifestation of divine approbation.[54] Such events, like other desirable portents, were regularly celebrated by the composition of rhapsodies, honoring simultaneously the beneficent monarch whose character and deeds had inspired them and also the divine power which displayed them. In T'ang times these regularly took the form of rhymed fu, with the sequence of rhymes following the words in an appropriate classical sentence, such as "Let down/image/containing/glory/possess/Way/then/seen." which is to say, "[Heaven] lets down an image filled with glorious light; [if a nation or sovereign] possesses the right Way, then it will be seen."[55] One of these rhapsodies begins with the suggestion that the divine and royal powers—if they can really be distinguished—unite to produce the Star:

> Phosphor fixed in the midst of Heaven ·
> Sovereign dwelling over mankind:
>> We scan the high brilliance of the starry figure;
>> We see the vast profusion of the sovereign's virtue.[56]

The power of this magnificent star put that of all other stars, most particularly the baleful stars, in the shade. This power lent itself readily to metaphorical usage: virtue, especially kingly virtue, will triumph. Po Chü-i observed that it will eradicate the horrible influences of comets, just as the pure rain of heaven will cleanse away the foul odor of blood and death:

> The star of virtue melts both Sweeper and Exploder:
> The lasting rains obliterate what is rank and rancid.[57]

The term "Star of Virtue" could even be used in the plural, with reference to gentlemen of much lower rank, as Tu Fu used it to tell of the estimable, even brilliant brothers of the Yen family in Szechwan.[58]

But now let us look at this same ominous moon in its various guises as a divine moon and a magic moon—a moon that is not the simple moon of our

184

daily perception nor the dogmatic moon of the court astrologers. The moon might, for instance, if properly conjured, lend its light and transfer its essence to unlikely places. For example, it could be simulated in thread or paint—an art not much different from that which could create it by the power of magic. The sages knew well that appearances were illusory—that it is hard indeed to say whether a fluttering pattern under a bough was an autumn leaf or a gaudy butterfly. Indeed they might reject the question as meaningless. Absolute being and perfect identity are vain projections of the human psyche. A shining pearl in the folds of a mollusc's body is a true moon. Its empathy with the sky-moon is complete. What of a paper moon?

At the beginning of Lasting Felicity [A.D. 821] Yang Yin-chih looked up Retired Gentleman T'ang, a man of the Way, and was detained there over night. Although the moon was dark, no lamp had been set out. T'ang called his daughter and said, "Why don't you bring us a moon in its lower chord [i.e. its last quarter]?" His daughter took a slip of paper, made the shape of a moon, and pasted it to the wall. She conjured it, "We have a visitor this evening and wish you would bestow light and illumination!" In a moment the entire room was brilliantly lit![59]

Looked at more closely, however, the moon is not simply a shining disc: it has a landscape and inhabitants whose nature may be discovered by curious men and clearly depicted by sensitive ones. Ancient folk beliefs provide some of the elements of these fantastic lunar geographies. It was well known, for instance, that the moon is furnished with a cinnamon tree, and with a man who is perpetually engaged in the futile effort to chop it down. The story of this tree, which has plain affinities with a host of sacred trees, world trees, jewel trees, and fairy trees in all nations, has been told in many books.[60] Its modern organic namesake, one of a number of Far Eastern cinnamons, is the Moon Cinnamon (*yüeh kuei*; *Cinnamomum pedunculatum*), a widespread native of south China. Its valued bark is known appropriately as "*yin* aromatic" in the vicinity of Canton, and it is sometimes called "Indian cinnamon."[61] I have only a few words to say about the lunar prototype here: it shall serve chiefly as a means of introducing the reader to the lunar landscape. We begin with the words of Tuan Ch'eng-shih, that indefatigable collector of curiosa in the ninth century:

From of old it has been said that there are a cinnamon tree and a toad in the moon: thus it is told in books of curiosa that the moon cinnamon is 5,000 feet tall. Below there is a man who chops at it constantly, while the gash in the tree always closes up. The man is surnamed Wu and named Kang. He was a person of Hsi-ho who studied transcendentalism, but went astray. He was ostracized, and commanded to hew at the tree. In the Shakya [i.e. Buddhist] books it is said that there is a jambu tree on the south face of Mount Sumeru. As the moon passes, the shadow of the tree enters into the moon. Some say that the toad and the cinnamon in the moon are the shadows of land, while the empty places are the reflections of water. This assertion is quite close [to the truth].[62]

This précis of lunar lore still needs extension and point-by-point examination.

Most authorities agreed—because old tradition affirmed—that there was a cinnamon tree in or on the moon, in some sense. But opinion varied about the other shadows, reflections, images or entities there. A familiar book surviving from the Chin period had the authority of age, was therefore not to be dismissed lightly:

> As to the vulgar tradition about transcendent persons and a cinnamon tree in the moon: look at it now when it is first born and you will see that the foot of a transcendent has gradually been taking shape. The cinnamon tree appears there as a later growth.[63]

It is not altogether clear whether we have to do with a living foot, or with an impression in some lunar debris. It is at least thinkable, however, that the moon is no globe of subtle *yin*-foam, buoyant in the ether, but is solid enough to support living creatures on its surface—although in some other accounts it seems to be penetrable, like a ball of water held together by its own centripetal forces. This in turn allows the possibility that its surface may be landscaped, somewhat like the surface of the earth—or, more precisely like a chalcedony cameo.

A close study may lead to the suspicion that some of the shadowy shapes on the moon are mountains, as Wu Yung fancied in this poem, called "Staying in a High Building: an Autumn Evening":

> Inside the moon the blue hills are faint as if so painted,
> Within the dewdrops the yellow leaves are filemot-sere—
>   just so for autumn.
> I lean outward from the steep railing, in no way inclined to sleep,
> Wondering only if the Starry Ho may find a way into this
>   high house.[64]

Observe here the pale blue hills set unambiguously *inside* the moon. We can be confident that Wu Yung was not seeing only a surface feature. Similarly —as emphasized by the careful patterning of the quatrain—the faint yellow leaves are reflected in the round dewdrops, which are themselves little moons. The sky moon is penetrated by hills, as moss agate is by arboreal landscapes, and rock crystals by spicules of rutile. So also, with luck, the poet's lofty room may be invaded by the flow of the star river on this brilliant night.

But more prosaic minds could still imagine a dense and solid moon. The tale was told—or at least it somehow reached Tuan Ch'eng-shih—that two men once lost their way on Mount Sung. There they encountered a stranger dressed in pure white: it does not take a quick imagination to recognize him as a moon man. In answer to their inquiries, he informs them: "Are you

186

aware, milords, that the moon is in fact composed of the Seven Jewels?[65] Or that the moon is contoured like a pellet? The images on it are the sunshine on its convexities. Ordinarily there are 82,000 households caring for it. I myself once counted them." Whereupon he gives the travellers a bit of magical jade medicine—undoubtedly moon dust—to ward off illness, and points out the way to the public road, which would obviously be familiar to an all-seeing night-traveller like the moon, or a moon-man.[66] Such a well-tended moon—a mosaic of enamels—would appear to be solid enough, but its gemmy substance is crystalline, and crystals have an ambiguous place in the Chinese imagination. They are allied to water, and so, in a profound sense, are permeable.

But the penetrability of the moon in a different fashion must also be considered. In ancient tradition and in T'ang poetry the moon was pitted and honeycombed, like a volcanic rabbit-warren. From Han times at least we hear of "moon dens" or "moon burrows" (yüeh k'u, *ngywăt k'wĕt), with the second word written in more than one way,[67] and the whole phrase was used ambiguously from very early times. In one sense it refers to the dank and gloomy hiding place of the moon at the western verges of our world—the counterpart of the solar Fu-sang tree in the extreme east.[68] It follows naturally that the expression also symbolizes the cold and remote lands of the western barbarians, notably that of the ancient Kushans, whose name in Chinese even contained the element "moon"[69] and was extended in T'ang times to include other benighted Turks and Iranians beyond the Pamirs. Tu Fu brought out the darkness of these far caverns in a poetic analogy intended to explain a symbolic element in the procedural detail of the great sacrifice to Heaven: "Thus, while the Dens of the Moon are black, Fu-sang is cold."[70] Moreover, whether near the Oxus and Jaxartes Rivers, or penetrating the moist substance of the moon, the caves were not only dark, they were wet. Hence P'i Jih-hsiu, devising a fancy simile for heavy rainfall, wrote:

> Like taking the Moon Dens and draining them;
> Or as seizing the Sky Ho and slapping it.[71]

The dampness does not suggest that these lunar holes made pleasant habitations, but the darkness at least afforded security for some creatures. It will come as no surprise, then, to learn that the "moon dens" below housed not only felt-hatted Tajiks who, doubtless, deserved no better, but also, above, a certain familiar leporine resident of the great sky lantern.[72] His burrow was dark and so was he. There is a hint of Arthur Rackham in Wu Yung's ekphrasis on a painting of a cypress:

> Shaded below—a burrow for the dusky hare;
> Refulgent above—the moon-soul of Ch'ang-o.[73]

(If any lingering doubts remained in the reader's mind about the luminosity of the lunar *p'o*—"moon-soul," elsewhere "protopsyche," and the like—this couplet should finally dispel them.)

It follows, naturally, that lunar bolt-holes provided convenient double images and bivalent allusions, in which it was easy to suggest a connection between earth and sky, mundane and divine. Li Po, a fairly moon-obsessed man, wrote the following about a celebrated hero of ancient times who made his reputation among the nomads beyond the frontier:

> Thirsty, he ate ice from the dens of the moon;
> Hungry, he dined on snow from high in the sky.[74]

While enduring privations, the warrior was still heaven-nourished—and it is hardly possible to overlook his proximity to the Mountains of Heaven, that stretch across Central Asia and, in T'ang times, separated the partly sinicized Iranian oases from the wild horsemen of the steppes. Similarly, in the following century, Liu Yü-hsi, testifying in verse to the nobility of a colleague about to depart on a ritual pilgrimage into the West, wrote:

> The Moon Dens play the host for all of Hsia—
> Cloud Officer descends from the Nine Heavens.[75]

("All of Hsia" [*chu Hsia*] is a collective term for the Chinese people.) Once more earthly and celestial affairs run parallel, and the magnates of our world have their doppelgängers above the clouds: the poet's friend is—as a magistrate of T'ang—an emissary of Heaven going as guest to the caverns of the moon. Is it in the western mountains, or in the lunar hills above? The distinction is fit only for nit-picking literalists.

In all of this there is a mythological strain, even if it is sometimes rather tenuous. The hare's den is a microcosm of the moon's womb—and beyond this universal symbol lie the images of the germinal "sky grottoes" and the underworld to which souls pass and from which they emerge for new beginnings. For instance, to make a meal of the lunar ice from the ice-pits in the sky (which had their earthly analogues) is to be born again, spiritually and figuratively. The interested reader may trace this thread to wherever it may lead.

Where there are secret dwelling places we may expect dwellers. We have already seen a host of them, furbishing the multicolored surface of an enamelled moon. These ant-like and undifferentiated inhabitants of the treasure-orb, however, do not have quite the appeal of the recognizable inhabitants that have been detected there.

The lunar fauna is generally treated by sinologists as if it were negligible, except as nuggets of lore for the earnest researcher in comparative folklore or the indefatigable explorer of the endless trails followed by migratory

188

The Moon

Grand Myths. But for itself the hare in the moon remains only faintly amusing—certainly not to be taken seriously. Or so we imagine. It is enough, we think, to remember the standard tale about his elevation to that chilly planetoid, and to notice with mild interest his repeated appearance as a motif in decorative art, usually grinding away at his mortar like a good tireless beast. That is enough, and by that token he becomes—though we hardly dare admit it publicly—a tiresome beast.[76] I conceive it to be my mission to rescue him from his patient alchemical chores, put him on center stage, and there introduce to the reader facets of his personality that may have escaped the attention of most earthly observers—as they did *not* escape the poets of T'ang. For although many of them saw him as a trivial stock figure, like a moral-monger out of Aesop or a children's cliché from the received Mother Goose, many others saw him as a complex person—by no means as simple-minded and unenterprising as we, in our deluded folly, imagine him to be. Certainly no mere paper cutout, he is a numinous hare—a hare of power. And even more, he is a mysterious and sometimes unpredictable hare.

Metaphysically the hare is an alien element in the cold, impassive moon. The true lunar essence—the *yin* component in the symbolic symbiosis—was the toad, whose damp emanation was neutralized in part by the presence of his rabbity associate, "the *yang* with the *yin*."[77] This, then, was the hare as cosmic force—but even this philosophical eminence is not his most significant attribute. We must turn to the poets to discover the more subtle facets of his nature.

For some of them, the moon-rabbit could be sheer fun. See him now as a rebel—a free-wheeler in the void. The poet is Yao Ho, an eminent official of the mid-ninth century:

Jade rabbit on silver wheel drifting towards the east,
Crystalline clear—it's the third watch, just right for roving!

A slip of black cloud rises—out of somewhere:
Its inky net cages back a ball of water germ![78]

Here we have to do with a delinquent rabbit, not in the guise of grim grinder-out of seasonal decay, a determined time-marker, but a wanderer, a happy hippy, a rebel against clocks. "Silver" and "jade" are commonplace images of whiteness, always appropriate to the moon and its inhabitants. They put the spotlight on the vagrant hare and his silver-penny unicycle. "Why vagrant?" you ask. "It is there, in the first line of the poem," I answer. The beast is pedalling irresponsibly towards the *east*, when he has been programmed to glide placidly towards the west, where he should descend (gracefully but tediously) into the evening sky. He will not. The moon—the

189

vehicular hare, that is—is at its most gemlike and crystalline when high in the midnight sky: the third watch. The roads of heaven are nicely lit, and off he goes on his ramble. But the dark hand of ineluctable fate, in the form of a streak of black cloud, reaches towards him, seemingly out of nowhere. (The physical counterpart of this censorious arrest is the illusion of an east-bound moon caused by the contrary westbound movement of the clouds across the sky—an illusion any connoisseur of moon hares may verify for himself.[79]) Caged in a celestial Black Maria the protean culprit, now plainly revealed in his noumenal form as a crystalline ball ("water germ" is rock crystal, is the planets, is the milky way, is the moon!), is jerked back to his proper path, aimed at the western horizon.

The magic of poetry could also transmute the moon-hare into a figure of fright and inevitable doom, a scythe-bearer, akin to a grinning death's-head. This is how he is presented now by the poet Szu-k'ung T'u, who himself saw the fall of T'ang:

> Fly-y-y crow fly-y-y!
> Hop, hare, hop![80]
> Make mornings come and evenings go; spur on the
> segmented seasons!
> Kua Woman knew only how to patch the blue sky—
> She did not know how to boil a glue that would stick
> sun and moon to it.[81]

These verses have the superficial simplicity of a nursery rhyme, beginning with the smooth steady flight of the sun-crow across the sky, followed by the leaping leporid who represents the variable but irreversible passage of the moon. It ends with the awesome figure of the creatrix Nü Kua, frivolously shown holding a useless paste-pot—is it the same one she used to patch the damaged vault of heaven? She is lost in moody speculation about how to devise an adhesive that might arrest the onward progress of sun and moon—an end to seasonal change—victory over night and death. The verbal images themselves are light in tone—but they tell a message of pessimism and despair: mutability is the only law of nature, and even the gods cannot arrest time in its flight. These are the familiars of Faustus:

> Stand still, you ever moving spheres of heaven,
> That time may cease and midnight never come:
> Fair Nature's eye, rise, rise again, and make
> Perpetual day, or let this hour be but
> A year, a month, a week, a natural day,
> That Faustus may repent and save his soul
> O lente, lente currite noctis equi!
> The stars move still, time runs, the clock will strike . . .[82]

The Moon

Our frost-furred friend is a deadly timekeeper with a staccato thump.

The moon hare even enjoyed power on earth, where he turns up unexpectedly as a symbol of the Son of Heaven, just as the moon occasionally does in an astrological context. A certain Li Hsiao-i, a dignitary at the court of the Empress Wu, was slandered in A.D. 687 by a person who, noting that the character which represented the second syllable of Li's personal name contained the graph for "hare," claimed that he had ambitions for the throne. "It is his assertion," said the calumniator, "that, as there is a hare in his name, and the hare is a creature in the moon, his own lot will be as [Son of] Heaven."[83]

The toad in the moon, although he can hardly be overlooked, has fared less well in literature than the hare. Perhaps it is difficult to make a symbol of dignity out of such a misbegotten creature—and doubtless a venomous one to boot. In one tradition he was a bogle who devoured the moon as he might a robin's egg.[84] Whatever his role, there he was, squatting in the chilly but supernatural moisture, getting what cheer he could out of his official designation of "Germ (or Essence) of the Moon,"[85] A curious and doubtless uncomfortable role for an amphibian-in-name-only, who much prefers the solid comforts of dry land, harassed though he be by harrows, except during his rather specialized rutting season. Indeed Wang Ch'ung saw his lunar apotheosis as strange in the extreme: it simply didn't make sense if the moon—as all the world asserted with confidence—was made of water. The ill-chosen toad would long since have drowned. Despite this unfavorable environment, the creature maintained, in one tradition, sufficient potency to generate, from a decent distance, both the star-studded webster in Vega (who had other good associations with the moon) and the planet Saturn under the name of "Earth Watcher" (*ti hou*), a name which would have suited the big eye in the sky equally well.[86]

Still, in poetry, he tends to be more niveous and marmoreal than numinous and noumenal. He is no crepuscular toad crouching in the shadows, but a toad of jade, a frosty toad, a silk-white toad, a luminescent toad, a metallic toad, a cool toad, even if sometimes a sylphine toad or an uncanny toad.[87] He is revealed now in this dazzling guise as a picnic comes to its end on a spring evening (the quatrain is by Li Chung, who lived in the tenth century):

> The suburb in spring, the drinkers dispersed—
>   evening haze draws in;
> Strings and reeds withdrawn—up they go into
>   the turquoise tower;
> Pink sleeves, songs low sung—golden chalices in disorder.
> But the silver toad flies out atop of the east of the sea.[88]

191

Here the toad is hardly a go-getter like his neighbor the hasty hare, but neither is he static and lumpish. He is a rather serene and soothing figure, his argent form neatly matched with handsome goblets of gold, but contrasted pleasingly with such ephemeral gauds, so irreverently treated by the nocturnal merrymakers. It is a little astonishing that he "flies out," but the emphasis is not on flight as speedy transit, but on flight as a high and tranquil soaring above the fitful affairs of earthlings.

Frequently, however, the toad's place is taken by a frog, presumably because of doubts about the suitability of a lunar environment for a toad.[89] This frog, like his opposite number, is conventionally described as cold, white, shining, and so on. He is also, of course, ultimately composed of the lunar essence.[90] Naturally, when the poet Li Chung was enjoying a quiet evening listening to the sounds of water in a mountain retreat, he heard "frogs in the water, calling to the moon."[91] Like calls to like.

Such terrestrial frogs were seldom frightening, but their celestial counterpart could indeed be frightful. Sometimes he revealed himself, not plainly and quietly as "frog in the moon," but as that creature of terror, "frog who eats the moon." He appears in this guise in the celebrated eclipse poem of Lu T'ung, where he is styled a "weird" or "uncanny" frog.[92] But the poets took a flexible view of the zoology of the matter: in a poem by Li Po the moon is the victim of a voracious *toad*.[93]

If anything, frogs circulate more freely in the worlds of Chinese folklore than do toads. There were ancient frog cults in south China, where the amphibians were represented on aboriginal bronze drums, and boomed, it seems, with the voices of rain spirits.[94] (Indeed, frogs were admired in more than one way in the south: they were favored as articles of diet—a significant stumbling block to sojourners from the north, who could seldom be brought to overcome their revulsion.)

But as moon spirits, it is hard to choose between frog and toad. Both have ancient connections with woman, moon, and pearl.[95] It is even possible that the association with divine women is common to both of them. One form of the name of Nü Kua, the great goddess who mended the sky and created man and is represented as half woman half serpent, was Wa, which is cognate, both graphically and phonetically to *wa* "frog."[96] Other linguistic associations indicate that she may also have been, in another avatar, a spirit of rain pools, puddles, and the slippery animals that frequent them. The toad, on the other hand, may possibly share his name with the great moon goddess. At least one scholar has suggested this, and also that both names are cognate to the Indic word for "moon," *candra*—but the phonological evidence is not entirely convincing.[97]

192

The Moon

We have observed the blue-robed moon-tenders of Taoism, and the hordes of moon-men reported by Tuan Ch'eng-shih. Now it is time for the man in the Moon, more particularly the great rotund face that so many peoples have seen filling the lunar disc. Doubtless it is a cold face, particularly in autumn when it seems carved from a block of snow, as Meng Chiao observed: "The autumn moon—the complexion of a face turned to ice."[98] This is a close-up view of the oldster sometimes seen in the moon, who was styled "a man of ice."[99] His vacuous and anonymous visage holds little charm for us, nor does the traditional figure of Wu Kang (sometimes Wu Chih). This personage turns up in literature from time to time, always a rather undistinguished fellow, a neglected nobody, in striking contradistinction to his female associate Ch'ang-o. Like Sisyphus, he is perpetually punished for a long-forgotten crime by being required to hack away at the immortal cinnamon tree through endless eternities.[100] He is seldom diverted from his fruitless but seemingly spell-binding labor—with the notable exception of the occasion when the supernatural chords contrived by Li P'ing, the harper of T'ang, distracted him, along with many other celestial beings, as recorded in the famous poem by Li Ho.[101] He lacks even the busy effrontery of that guardian of the Great Sky Lantern whose immortal words have been recorded by that specialist in lunatic labors, Ernest Bramah:

Is it not enough that for a wholly illusory crime this hard-striving demon is condemned to live upon an already inadequate sphere and burnish its unappetizing face for the guidance of a purblind race of misbegotten earthlings?[102]

Here, neatly blended, we find both the immense pallid face and the exile condemned to endless labor—the latter, however, a job with some real purpose. Mr. Wu lacks even that consolation. Let us leave this colorless personage.

Fortunately the moon had other inhabitants, chiefly, to judge from medieval literature, moon-maids. They appear under various guises, but are ultimately all sisters, cast in the same exquisite mold. They had been observed at least since Han times—but their lives may not always have been restricted to the moon. A case in point is that of the late-third-century writer Shu Hsi, who fancied the delights of "Lodging at evening in the chambers of the seven fairies."[103] (For convenience and at least temporary consistency I here use English "fairy" to translate o [*nga] as I have elsewhere in "Fairy Radiance," one of the names of the Goddess of the Hsiang River.)[104] Conceivably Shu Hsi had a narcissistic obsession with his own name Hsi (*Sek "bright, shining")—as Li Po did later with his—and thought of these charming young women as a set of lunar heptamerides. But it seems much more probable that they were the feminine incarnations of the seven celestial luminaries—the sun, moon, and the five planets—or possibly even the seven stars of the lunar station "Star" or of the Dipper. But in the course of time the word *nga, my "fairy," certainly did take on plain lunar connotations, contaminated with Taoist ones, so the yüeh o "moon fairy" and yüeh hsien "moon transcendent" came to be used as virtual synonyms. Such at least is the usage of Meng Chiao, a moon-intoxicated poet if there ever was one, who wrote of "The moon transcendents in their high planetoid."[105] Although this name is usually taken to refer to Ch'ang-o, as the sole proprietress of the moon, it could as easily be a collective, especially since the same connoisseur writes of the "Moon fairies, descending pair by pair."[106] Granted that, in this second instance, the poet is using "moon fairies" as a metaphor for falling white petals of the herbaceous peony, the multiplication would be inappropriate if it were not recognizable, and indeed many medieval texts confess that Ch'ang-o was not the only moon fairy.

Perhaps it is superfluous to note here too that the white maidens of the moon had their replicas or imitators in epigene parlors. The literary device of making goddesses out of court ladies and courtesans is probably too commonplace to deserve special attention, and was by no means restricted to the poetic triflers (as they are imagined to be in our old-fashioned criticism) of the late ninth century—misnamed "Waning T'ang." Po Chü-i has generally been allowed the distinction of being one of the few writers of the ninth century who did not disgrace the great clan of true (i.e., moral, philosophic, spiritual, inspiring, edifying) poets—a list sanctified by the authority of almost insurmountable tradition. All the same, he was a fancier of paid-for singing girls, and in one poem describes the swirling dance costumes of a pair of these upper-class cyprians. He compares them to fire and to smoke—they are like snow, and like lotuses. Finally, they are "Divine Women in the rain—Transcendents in the moon."[107] Both together or separately they are the Rain Goddess of Wu Shan and/or Ch'ang-o in her moon. In either case, the flattery is none too subtle.

The abodes of these charming goddesses form an exception to the general rule that the mansions and manors of star-magnates are seldom described in much detail in Chinese literature. We would be hard put, for instance, to specify the supernatural jewels that adorn the high sanctum of "Heaven's Illustrious Great Theocrat" (T'ien huang ta ti), whom we style merely Yilduz or Polaris. The same melancholy observation must be made about the celestial seats of the Grand Heir, the Three Lords, and all the other great dignitaries—to say nothing of the stellar homes of the primordial divinities of Taoism. The great palace of the moon, on the other hand, although it is not the abode of any deity of the first rank, was often rather fully portrayed in both Chinese poetry and prose, especially as a palace of ice crystals—a chilled, angular, brittle abode for a being beyond human conception.[108] Naturally it had all of the attributes of purity and crystallinity that were ascribed to divine dwellings generally, but possibly it was not quite as rarefied and spiritual as it might have been, had it not taken on some of the coarser features of the palaces of devarajas on earth. (Indeed the moon palace had a certain international character, like the jewel tree on Mount K'un-lun which had kindred all over the world. For instance, the Buddha himself had created gem-studded hostels on the moon for the housing of needful devas: "The palace of the moon-devas moves, suspended in the void."[109])

But, although a host of beings were sometimes lodged in it, the palace was above all the divine residence of the goddess Ch'ang-o, and it was furnished with everything needful for her delectation. Other creatures placed in the moon in antiquity were given specific roles there, but they were lesser ones. The palace, of course, had a garden, and, not surprisingly, there was a cinnamon tree in the garden. In a sentimental and perhaps whimsical mood, Po Chü-i, in one of three poems about a solitary cinnamon crowded in under the city wall of Su-chou, suggested that Ch'ang-o might wish to take it as a companion for the famous tree in her palace garden, where there was a superfluity of space.[110] In the higher vision, of course, the two trees could never be proper mates, since the substance of the moon-cinnamon was everlasting. Mao Wen-hsi, a tenth-century writer in the *tz'u* form, composed an ode describing a kind of celestial celebration in honor of the blooming of the moon cinnamon, in which it is plainly pictured as gemlike and crystalline. The form of the strophes is *Yüeh kung ch'un* ("Spring in the Moon Palace"). The poem is divided into two quatrains: the first displays the mineral beauty of the tree, the second describes the admiration of the magnates of heaven, including one of the very greatest:

> Within the water crystal palace cinnamon flowers open;
> Divine sylphs have explored them, oh, how many times!
> Pink fragrance, gold stamens, embroidered in double tiers,
> Down they sink, reversing carnelian cups.
>
> Jade hare and silver toad vie to keep watch and ward;
> Heng-o and Volatile Girl playfully hug one another.
> And, hearing the nine-fold concert "Pivotal Heaven" from afar,
> The Jade Illustrious One comes in person to watch.[111]

("Volatile Girl" [*ch'a nü*] might be translated "Shimmering Woman" or "Mercurial Maid." *Ch'a* is a rather mysterious word, and the goddess, playmate of the Moon Maid, is a rather mysterious person. Her associations are with the Yellow River, youth and beauty; above all, she was the personifi-

195

cation of quicksilver in the arcane language of the alchemists. It is at least a plausible hypothesis that she represents the shimmering, silvery moonglade on the river, and also perhaps the Sky River, "The Silver Han.") In this charming effusion we discover the not uncommon association of music with the moon. The case of the tune recaptured by Hsüan Tsung after his return from his fantastic journey is only the most notorious example—an event well remembered by the poets, as in a recollection of the "Floriate Clear Palace" in its grandest days written late in the ninth century by Cheng Yü. He shows the deposed monarch in his declining years, unable to return to the sparkling moon palace any more than to his lovely winter palace on Mount Li, but hearing the moon's "secret music" faintly and uncertainly, the stony sounds dominant, high in the sky: "Ting! tang! the stones of jade, blended with ocarina and flageolet."[112] Indeed, the moon music inspired much earthly music, including a zither tune named for the moon palace, composed in Kuang-ling (it was paired with a "sun palace" melody—a small tribute to an otherwise neglected sky-mansion).[113] Analogous, but much higher on the social scale, was a song rehearsed in the royal school of music (the *chiao fang* "Instruction Quarter") called "Watching the Moon Palace" (*k'an yüeh kung*).[114] There was other unearthly music, such as the song "Great White Star," whose theme was the planet Venus.[115] We can only imagine how it shimmered and trembled—it is all lost to us.

Earth and anti-earth were interchangeable, and when, in the tenth century, Ch'eng Yen-hsiung wrote of Hsi Wang Mu's make-up mirror left suspended in the sky, he saw this lofty plate of polished lead-bronze converting, by a kind of lunar reflective magic, a wooden house on earth into a house of crystal. Fitly, music was being played at the very moment of transformation.[116]

A somewhat more remote way of transferring the moon palace to the soil of T'ang than this visual one was by the route of metaphor. When Li Hsien-yung wrote about springtime in Ch'ang-an that he saw

Year after year thirty riders
Wheel into the palace of the jade toad,[117]

he was seeing no dragon horses nor ghost riders in the sky, but was adopting an accepted figure for the annual procession of the thirty victorious *chin shih* candidates, who were allowed to ride into the "moon palace" and pluck a branch, emblematic of triumph, from the "cinnamon tree."

These crystal halls were also accessible to earthlings by less conventional means, but with no fewer implications for the political order. A capacious gourd does not, on the face of it, seem a suitable vehicle for a journey to the moon. On the other hand, lacking an adequately powered rocket-ship, it was probably as useful as any other. Such, at any rate, was the mode of transport provided Lu Ch'i when he was invited into the sky to meet the lady of his dreams.

Lu Ch'i was a real person. He was a clever, eloquent man—sufficiently so to become the chief minister of Li Kua (Te Tsung) late in the eighth century—but he was ugly ("blue-faced" like a ghost) and careless of his appearance. He used his high position to amass money and had his opponents ruthlessly exterminated. Accordingly, his biography in the standard history of the T'ang dynasty is grouped with those of other "Faithless Vassals."[118] In the story translated below, written by a T'ang writer named Lu (his given name is unknown but the surname is the same as that of Li Kua's minister), the protagonist is represented as so ambitious and materialistic that he would reject eternal life in a glittering moon palace with the most lovely of goddesses out of sheer lust for supreme power over men.[119] Is it possible that the author is identical with the hero—that this is the minister's satire on himself?

When Lu Ch'i was young he lived in penury in the Eastern Metropolis, having rented a room in an abandoned house. For a neighbor he had an old crone, lone and lorn, named Ma. Ch'i contracted a violent illness, and was abed for more than a month. Mother Ma came and made soups and gruels for him. After his malady was cured, and as he was returning from abroad one evening, he saw a golden calf-cart outside of Mother Ma's gateway. Lord Lu was surprised. He peered at it in wonder, and saw a maiden of fourteen or fifteen years—a veritable goddess. Next day he made a surreptitious call on Mother Ma. "Wouldn't you like to make a marriage alliance? I shall try discussing the possibility with her," said Mother Ma. "Being poor and mean," said Ch'i, "How could I ever dare to have such an aspiration?" "Why, what obstacle is there?" said Mother Ma. And after the night had passed, "The matter has turned out agreeably," said Mother Ma. "You will please purge yourself for three days and meet me in the abandoned Taoist friary east of the city wall." When he reached there he saw ancient trees and wild herbs—it was long since men had dwelt there. As he hesitated, there was a violent onset of thunder, lightning, winds and rain, and all was transformed: high buildings appeared on platforms, golden basilicas with hangings of jade—a pageant of splendid and ravishing objects. A fine coach descended through the void—and it was the maiden of the earlier occasion. When she caught sight of Ch'i she said, "I am a person from Heaven who has received the mandate of the High God. He sent me among men to seek freely for my match and mate. Since milord has the countenance of a transcendent, I sent Mother Ma to transmit my intention. I shall be permitted to see you again after you have purified yourself and fasted for seven more days." The maiden called Mother Ma and consigned a pair of medicinal pills to her. Shortly, with thunder, lightning and black clouds, the maiden vanished, and the ancient trees and wild plants were as before. Mother Ma and Ch'i went home. He fasted and purified himself for seven days, then hoed up the ground and planted the drugs. Scarcely had he planted them when vines were quickened, and in no time at all two calabashes were growing on the vines. Gradually they became as large as a couple of bushel kegs. Mother Ma sliced them down the middle with a knife, and Mother Ma and Ch'i each took a place in one of them. She had also instructed him to prepare three suits of oiled clothing. Suddenly wind and thunder sprang up, and they mounted high into the dark blue empyrean. They heard only the sounds of waves and surges filling their ears. After some time they became aware of the cold, and she ordered him to put on the oiled shirts. It was as if they were in the midst of ice and snow, and she told him to wear more, up to three layers—then he was very warm. Mother Ma said, "We are now eighty thousand *li* from the Lo [River]." After a long while the calabashes came to rest. Then they saw a palace with watch-towers and high-platformed buildings, all with walls and parapets of water crystal. There were several hundred battle-axe men encased in armor. Mother Ma led Ch'i in, and he saw a purple basilica. A suite of a hundred women commanded Ch'i to sit, and they set out wine and good food. Mother Ma stood guard below the praetorians. Now the maiden said to Ch'i, "Milord may have three things from which he is allowed to choose just one thing: he may remain in this palace, living on until Heaven shall come to an end; or, next, he may become an Earthly Transcendent and take up regular residence among men until, in due time, he is permitted to come here; or, least of all, he may become Prime Minister of the Middle Kingdom." "To stay in this place," said Ch'i, "is truly the highest of wishes." The lady was pleased and said, "This is the Palace of Water Crystal. I am the Lady of Grand Yin. My status among the transcendents is already a high one. But, worthy sir, while you have mounted to the sky in the white light of day, you must decide soon, nor can there be any shift or change in you, lest we become embarrassments to each other." She then forwarded a blue paper as a memorandum, and it was presented in the courtyard with a salutation. "Disclose this quickly to the High God!" she said. In a short time they heard a voice from the northeast which said that the envoy of the High God had come. The Lady of Grand Yin and the other transcendents hurried down. Soon there were pennants, standards, and fragrant streamers, escorting a youth in vermilion dress. They took their stand below the staircase, and the one in vermilion dress proclaimed the mandate of the God: "A Declaration on the Winning of the Lady of Grand Yin by Lu Ch'i." It asked what his wishes about remaining in the Palace of Water Crystal might be. Ch'i had nothing to say, but the Lady commanded him to reply quickly. Still he was without words. The Lady and those with her to left and right were greatly alarmed. They hurried back in and took five lengths of "shark pongee" as an inducement for the envoy, wishing to delay him and slow his going. During the interval while he ate something she asked Lu Ch'i again whether he wished to remain in the Water Crystal Palace, to become a Terrestrial Transcendent, or to be Prime Minister among men—this time he must decide! Ch'i gave a great cry, "To be Prime Minister among men!" The person in vermilion dress hastened away. The Lady of Grand Yin said, losing color, "This is the fault of Mother Ma!" And she had him hurried back and thrust into the calabash. Once more he heard the sounds of wind and water. When he came again to his old residence the furniture was visibly dusty. It was now midnight, and neither the calabash nor Mother Ma were to be seen.

This is a plain unsauced tale, free of invention, artifice or any kind of subtlety. Its purpose is a simple moral one. What concerns us here is the presentation of a moon goddess in refined and elegant guise as the Lady of Grand Yin, her fief being the familiar but exalted metaphysical name of the moon. Although the moon is never directly referred to, in a crass material way, it appears in partial disguise in the Lady's title, and many artless signs are posted throughout the tale to ensure that the reader will recognize her true domain, in case its position above the clouds, its richly decorated buildings and the like should not be sufficient evidence. Chief among these tokens are the watery surges, the intense cold, and (above all) the Palace of Water Crystal itself, whose Chinese name combines the regular expressions for the lunar substance ("essence of water") with its affinity to ice and for sparkling transparent rock crystal.

The living creatures native to the moon, ranging from the clammy toad to the radiant goddess, must be sharply distinguished from the occasional visitor from the earth, who could hardly be expected to gain more than the most superficial acquaintance with that chilly orb during his brief stay—shamans and Taoist adepts always excepted. Lu Ch'i was such a tourist. For the T'ang period we have also an even more exalted example—Li Lung-chi, the noble eighth-century monarch known posthumously as Hsüan Tsung.

Many dreams and visions that were to have important public consequences were the portion of that royal Taoist initiate. Among these was the appearance of a phantasm of Lao-tzu, who had been degraded from high celestial rank during the rule of the Buddhist Empress Wu. Now, in consequence of this and other holy apparitions, and also because the primordial god who had once been a sage on earth was regarded as the ancestor of his dynasty, Li Lung-chi exalted him once more.[120]

Elevated to godhead anew—or rather, recognized once more for the great deity he had always been—Lao-tzu took pains to confer special privileges upon his benefactor on the Chinese throne. A most notable one was first reported to the sovereign's personal attendant, the eunuch Kao Li-shih. Here is the story as it was told more than a century after the event by the poet Cheng Ch'i.[121] His account became widely known at all levels of society, and underwent many transformations in later ages:

Once when His Highness sat at the Levee, he pressed on his belly with the fingers of his hand going up and down. On his retirement from the Levee, Kao Li-shih said, "Exalted, on several occasions recently you have pressed on your belly with the fingers of your hand. Surely it is that your Incomparable Frame is a little unsettled?" "It is not," said His Highness. "Last night I dreamed that I journeyed to the Moon Palace. All of the Transcendents entertained me with 'The Music of the Highest Purity,' unimaginably bright and supernally pure—not at all the sort that is heard among men. We drank until we were intoxicated, and after a while they performed a concert of music as they sent me off home. The melody was poignant and moving, but became vague and indistinct in my ears. On my return I recaptured it on my jade flute, but I thought that it might vanish from my memory while I was sitting at the Levee, and so I took the jade flute in my bosom. Then I replayed it, going up and down with the fingers of my hand: it was not that I was unsettled." Li-shih made the double salute, and congratulated him, saying "This matter is not an ordinary one. I should like you, Exalted, to present it once for your vassal!" And the sounds were so remarkable and otherworldly that he could put no name to them. Li-shih gave the double salute again, and then asked its name. His Highness laughed and said, "This tune is named 'The Return of the Purple Cloud.'" Subsequently he had it recorded in the Music Archive, and now it has been engraved on stone by the Grand Ordinaries, where it still exists.

This incident places Hsüan Tsung firmly among the initiates of the "Highest Purity" sect of Mao Shan, even more surely than Han Wu Ti was placed there posthumously in his "esoteric biography" (nei chuan). The tale has also been transmitted to us in the form of a story attributed to Liu Tsung-yüan, although in fact this seems to be a forgery of the early twelfth century.[122] It appears in a collection known as "Register of Dragon City" (Lung ch'eng lu), "Dragon City" being a popular name for the southern frontier town Liu-chou in which Liu Tsung-yüan was a magistrate during a period of exile.[123] The plot is familiar, but is richly and rather conventionally embroidered. It will not be translated here, since it is not a genuine T'ang product. A synopsis will serve. The tale is set in the year 718—that is, early in the long reign of Hsüan Tsung. It is the night of the full moon of the eighth month, the night of the traditional moon-watching celebration.[124] The magical arts of "Heavenly Master T'ien" take a party of three (himself, the Taoist master portrayed in Po Chü-i's poem Ch'ang hen ko, and Li Lung-chi himself) up above the clouds to the cold moon palace, adrift in space. They arrive with their clothes soaked, and observe guards whose weapons glitter like frost, mists like clouds of incense, vast fields of gemmy turf, and deathless inhabitants touring idly about seated on clouds and the backs of cranes. A dozen or so "silk-white fairies" (su o) in white dresses ride about on white birds, dancing and cavorting happily, to the sounds of otherworldly music. After returning to earth, the dazzled monarch composes music for a dance which will imitate the whirling of the rainbow costumes of the divine maidens. This is the alleged origin of the famous dance of the Rainbow Skirt and Feathered Dress, which Yang Kuei-fei performed for the court.[125]

In this tale we are very far from the homely atmosphere that invests Cheng Ch'i's account. We have instead a rather lifeless and conventional picture of a stereotyped Taoist paradise—it could be anywhere. We know that it is the moon because we are told so, and there is a bit of supporting evidence in the dampness, the chill, the white weapons, and white clothes. But little is done with these tokens, and there is no sign at all of the incredible Ch'ang-o, or any of the other ancient moon-beings.

Nonetheless it was natural that the moon should acquire a palace—a mere elaboration of a grotto—and that it should be populated by the super-beings of Taoist mythology. It was also inevitable that this excellently equipped moon palace should also provide a handsome cliché for writers who wished to describe a rich and festive scene at an earthly court, as Wei Chuang did in the following poem. This alludes retrospectively to Li Lung-chi and his parties which, from the point of view of the terrors and desolation of the late ninth century, seemed truly to have been congregations in Fairie:

> A great road, a tall blue building—east of the autocrat's park;
> A jade rail, a fairyland apricot—pink pressed on the branches.
> Gold bells and dogs' bark—the moon is in the sterculia,
> Vermilion-maned horses neigh—the wind is in the poplar-willows.
> Flowing water, flower-girded, pierces the paths and lanes;
> Evening sunlight, tree-blended, enters the blinds and lattice.
> At Gem Pool the party is over—homecomers are drunk,
> Laughing and telling of their Lord the King up in the palace
>   of the moon.[126]

Is there a note of mockery here? The great sovereign, full of mystical raptures and visions of other worlds, is he merely a laughing-stock to his fashionable courtiers who find the court banquets paradise enough?

Now the great moon goddess herself must take center stage. Ch'ang-o (as we now call her) may well have been an ancient deity, but her modern name does not appear to be particularly old, and the legend of how she fled to the moon with the elixir of immortality appears to have been artificially attached to her.[127] But we need not linger over these difficult matters, which still need serious attention by the philologists. Suffice it to say here that by T'ang times she was a standardized figure in popular mythology, and despite the importance of the moon in official religion and astrology, she was not —under that name at any rate—a great deity. But her role in the poetic and popular imagination was immense. The extent of her cult is yet to be determined.

The traditional "moon festival" on the day of the full moon in the middle of the eighth month, whatever its origin, was celebrated in T'ang times, as we know from the direct testimony of a Japanese visitor.[128] This out-of-doors holiday in mid-autumn, with its feminine and erotic associations, could hardly have lacked orisons addressed directly to the divine woman by the people generally. But her status was sometimes even higher: it is known that she attracted the prayers of royalty. A ruler of the independent Fukienese state of Min restored and reconsecrated a great T'ang pagoda in the year 941. Each storey of the pagoda was dedicated by members of the royal family and other great dignitaries to a different form of the Buddha.

The images and inscriptions that furnished it also contained a strong infusion of Taoist elements. To our purpose here is the fifth storey, which honored the Bhaiṣajaya-guru Vaiḍūrya-prabhā Buddha, but also displayed the invocations of three royal princesses to the Moon Fairy (*yüeh o*) and to Starry Wu (*hsing wu*), that is, to the woman-star (ε Aquarii) who presided over the tenth lodge of the moon, and was also protectress of the south.[129] The matching of Ch'ang-o with a female asterism was, if anything, more common in literature that in cult practice, and the glittering pair made convenient symbols for the belles that ornamented the fêtes of the magnates of T'ang. (Elsewhere I have characterized the former in this seductive and fleshly guise as "Sylphine O," or "Constant Luna," or "Rose-gem O" [*ch'iung o*], and the latter as "Precious Wu" or "Precious Stella," or "Minx Star.")[130]

Indeed the moon goddess, understandably, provided the poets with a natural disguise for a bewitching entertainer of men. She had the legendary jade-white complexion which her mundane counterparts could and did approximate with white lead, and she was simultaneously charming and aloof, in the grand tradition of fertility goddesses in the guise of mortal women. Po Chü-i's contemporary Liu Yü-hsi, however, showed a degree of subtlety in lending her attributes to a remembered courtesan: "Heng-o has gone back to dwell," he wrote, "in the depths of the Palace of the Moon."[131] In this treatment he has taken pains not to devalue the goddess in order to make her image more easily detectable in the figure of a Ch'ang-an courtesan. Other poets, perhaps of no greater ability, were able to elevate a hetaira into the realm of traditional fantasy, where she could, like a Dresden china shepherdess, serve transiently as an attractive piece of heavenly clockwork, marking the passage of the months and the fading of beloved images—in short, a symbol of all the depressing residue of separation commonly associated with the moon.[132] Some poets were particularly prone to lend her silver mask to the silk-garbed dwellers in palace halls. Li Shang-yin was one of these secularizers, who—whatever their great gift for words—managed to divest the goddess of much of her charm and mystery.[133] But his example should not make us hurry to discover a white-painted courtesan or white-gowned queen in every ninth-century poem about the moon and its shining spirit.

> That orbed maiden with white fire laden,
>   Whom mortals call the Moon.

So she appeared to P. B. Shelley, and so she appears to us. Indeed she appeared so also to the best T'ang poets, as has already been intimated in our more impersonal discussion of the noumenal and phenomenal qualities of the lovely sky lamp. The goddess could easily assume the physical attributes of

202

the moon—silver train or white fire or crystalline glitter—but it was also possible to make her more human, to discover the ghost of a blush, a trace of maidenly panic, or other fugitive signs of mortal frailty in her. Let us now see her in some of her more equivocal—perhaps really unfathomable—epiphanies, where we cannot say for certain whether she is all mist and mystery, or suffers spiritual uncertainties and even carnal perplexities—whether, in short, her cold flame reddens with latent heat.

The following rather mysterious quatrain is the forty-ninth of a set of one hundred that Lo Ch'iu wrote to honor the beautiful and accomplished courtesan Tu Hung-erh, whom he had stabbed to death in a jealous rage. In them he compares the lovely women of past and present, in this world and the other, to the dead lady—always to her advantage:

> The moon drops, runs off below ground—sun-dark knows
> how to take her.
> Who would say, in her deepest heart, "She wishes to nest alone"?
> Still—affianced to Chiao-fu, she would make no fitting mate,
> And she did not care to pass to the end of life as wife of I.[134]

The commentary which follows may reflect a minimum of light on this little-known example of the moon-courtesan equation:

1. Ch'ang-o sinks below the western horizon and begins her nightly journey eastward, out of human sight. (*An*, translated "sun dark," is the darkness where the sun is not.) It is as if a kind of anti-sun has her in its grip. She gives herself to darkness.
2. Surely she does not wish to pass her days as a secluded widow, hidden from men's eyes?
3. But, it seems, she would hide herself even from such a handsome youth as Cheng Chiao-fu, just like the goddesses of the Hsiang River he encountered in Ts'ao Chih's *Lo shen fu*.
4. And, as everyone knows, she did not care to remain married even to a demigod, the famous I.

It appears that it was the poet's intent to show that Hung-erh must have been so conscious of her matchless beauty that she could not, any more than the moon goddess, bestow herself on any mate, mortal or immortal.

To compare a celebrated courtesan with the moon seems to be a very different thing from comparing a lonely widow with that goddess. Yet, disturbing as the double comparison may be, it is inevitable. In Tu Fu's poem "Moon" the mask of divine radiance conceals a desolate soul. The scene is set in the chilly post-midnight hours, towards the end of the last month of autumn, with the moon waning towards its final crescent. It is the first new moon of winter. All of this is apparent in the imagery of the poem, which is well supplied with fine conceits. Tu Fu's tone, unlike that of Lo Ch'iu, is jovial—even jocular. Evidently he was in no mood to sympathize with a lovely but lonely nymph. "Give her a drink," says he:

Fourth watch—the hills eject the moon.
Shrinking night—the water lights the tall house.
A dusty casket starts to disclose a mirror;
A windy curtain frees a hook to rise.

The hare must suspect its locks are a crane's;
The toad for its part loves its sable cape.
Ladle out a draught for widowed Heng-o:
In the sky's cold she must bear the Autumn Ninth!

Perhaps this is obscure enough to deserve a minimal commentary: the east-facing crescent appears, and is reflected on the dewy pavement in front of the house, where the dewdrops (little moons) wetly reflect the light of their celestial parent. The lunar disc seems to be a shining mirror whose edge is shown above a newly opened casket (the dust refers to the dusty hills where the moon has been hiding); it is a curtain-hook of rock crystal, set free to sail up into the air with the billowing curtain of night. Cranes are bald-headed, and the moon-hare wonders if he too is not too old and hairless for such bitter cold. The shining toad is largely concealed—but happy—in its black, comfortable furs. The poor goddess, with no husband to warm her, could profit from a beaker of good wine. "Autumn Ninth" is the ninth month of the year, the last of autumn.[135]

In marked contrast to this cosy scene is the one revealed in the following quatrain from the pen of an unknown poet of the T'ang dynasty. It is not a very exciting poem, but it exhibits a certain unaffected charm in its personification of a scene from nature. Even so, it does not yield quickly to analysis, let alone to simple inspection. The words are as follows:

Ch'ang-o, within the moon, does not paint her eyebrows;
She just takes cloud and mist to make a netted garment.
It is not known if in dreams she goes off after a blue simurgh;
But she returns holding a flowering branch over her face.[136]

It is possible to treat these lines on several levels. First, as a description of a night scene:

1. The moon shows itself plain and unadorned.
2. Then it is concealed by a filmy screen of cirrus.
3. It has disappeared from mortal sight.
4. It reappears behind the branch of a flowering tree.

But imagine that it is not the physical moon but the goddess in person:

1. The moon woman is not made up for public display.
2. She wears only a simple white costume—a night-dress? neophyte's robe?

204

3. Possibly she dreams of following a divine bird to some fairyland (the "blue bird" is traditionally a messenger of Hsi Wang Mu).
4. She comes back with a spray of flowers—a souvenir of the gem-tree of Mount K'un-lun (and a magic token of initiation).

Can we also take this to be an allegory of a nocturnal assignation—the secret but not supernatural rendezvous of a favored concubine, let us say?

1. Without time to make herself up properly,
2. But throwing on a flimsy gown,
3. She goes off with a mysterious servant, presumably in response to the summons of a high personage:
4. She returns at length, masking her identity from possible observers with a lover's gift—or is it a maidenly blush she conceals?

We can easily admit a synthetic vision that will comprise all of these dream images, and possibly others. Should we press the conventional allusion of *luan*-bird ("simurgh") as "vanity" and as "woman with mirror"—hence, moon woman? I leave it to the reader.

Meng Chiao, the great moon-fancier, took an entirely different tack in a poem that begins with a listing of the ways in which certain beautiful things in nature are shown at their best:

Flowers are most charming when afloat in a springtime font;
Bamboo is most charming when encaged by dawn-glow and haze;
Courtesans are most charming when not overdoing artifice;
The moon is most charming in her natural appeal:
    At the midpoint of night Ch'ang-o attends the levee
        of the Grand Monad;
    Naturally no one among men is her peer in charisma.
    The officers of Han got such royal favor only on
        a few occasions;
    Both Flying Swallow and the Duchess are jealous—
        how they envy her![137]

Here what begins as a connoisseur's pocket-list is suddenly transformed into a paean of praise for the unique loveliness of the moon goddess: what the great beauties of the Han court could not achieve she achieved—and more: a midnight rendezvous with the emperor in the sky! Here, unlike so many other Ch'ang-o rhapsodies, there is no question of flattering a mortal beauty with a lofty comparison—the celestial beauty must always outshine them all.

The theme of jealousy was a natural one, and one which was well-exploited by the poets. Another example, in a rather different tone—that of a trained Taoist with a sense of humor—appears in one of the long sequence of poems on "Saunters to Sylphdom" written by Ts'ao T'ang. I have rendered his charming vignette in rather insipid prose:

Pacing the Void

They quite forgot to tell someone to lock the Rear Palace;
Now the Reconstituted Cinnabar is all lost—the jade cup empty.
If Ch'ang-o had not stolen the numinous medicine,
There would be a struggle for the long life to be achieved
    within the moon.[138]

What has happened is this: the "rear palace" (woman's quarters) of some divine king has been left unattended, and the girls have escaped and appropriated the elixir of immortality ("reconstituted cinnabar") that the god kept in a goblet of jade. These pretty cats, with portions of the supernatural cream at their disposal, wish with all their hearts that Ch'ang-o, their prototype and inspirer, had not gotten to the glorious moon ahead of them with *her* elixir, to reign forever as unique and immortal queen of the night.

In Chinese poetry not only is the misnamed "pathetic fallacy" much more common than is often supposed, but many examples can be found of the comparable "fallacy" in which elements of the natural world are carnalized—or spiritualized, it may be—into parts of divine bodies and supernal costumes. An instance is a poem of Ch'in T'ao-yü in which the blue sky becomes the flowing skirt of the goddess Nü Kua, and there are a number of poems in which the face of the Divine Woman of Wu Shan can be detected, by way of hallucination, in segments of the landscape.[139] We have already seen a black cloud energized as a spiritual being, capturing the errant moon-hare in his erratic flight. In a poem by the tenth-century Szechwanese poet Mao Wen-hsi, whose translation follows, a different cloud becomes a supernatural whitener, who is forever engaged in repainting a fashionable figure in the sky—the eternal restorer of an illusion. In the previous case, the cloud-spirit was the agent of divine forces. In Mao Wen-hsi's poem she is named as the perpetually employed agent of "Primal Harmony," the great balancing and harmonizing power in nature. It is this lady's responsibility to touch up the face of Ch'ang-o each month, like a back-stage makeup artist, with her pot of celestial white lead. The verses are set to the tune "A Single Wisp of Cloud at Shamanka Mountain":

A semblance covers the colors of Shamanka Mountain:
It has just crossed these waves
    that have washed polychrome damask.
Who is it that takes up a brush,
    ascends the Silver Ho,
And limns Ch'ang-o inside the moon?

206

The Moon

Thin-spread she applies the lead powder;
Full-blown she hangs her white-damask net.
The dreaming soul of the servant
    of the sweet-flag flowers is prolific—
Years and eons in the employ of the Primal Harmony![140]

Commentary:

1. A cloudy shape bleaches the colors of Shamanka Mountain.
2. She passes above the Yangtze which, above at Ch'eng-tu,
   is the river in which the Szechwanese rinse their fine damasks,
   and so is there called Chin Chiang.
3. She slides up the Milky Way with her makeup kit.
4. She renews the crystalline moon-stuff as she paints her face.
5. She uses the traditional foundation—snowy ceruse.
6. She drapes herself in fine white damask (contrast the earthly polychrome)
7. So she realizes the illusionist dreams of the nocturnal king of *Lieh Tzu*.[141]
8. She has brought *yin* to its climax in the eternal cycle of *yin-yang* phases.

But all of this sort of thing was froth and folly to the court astrologers and to most of the citizenry of T'ang. What really counted was the moon as plain portent:

And as for the moon—
    the chronicler of the hosts of *yin*,
    the emissary of highest heaven,
    the suzerain of the "other-named,"[142]
    the servant of queens and consorts:
Faced by a mirror basin, it soaks it with shining water;
Ringed with a double nimbus: the wind drives in from the border;
And when it swells in oyster or clam, the caitiff horsemen
    push on their inroads;
And when it makes the unicorns battle the dark tiger takes up
    a covert post.[143]

Here we are far from the intriguing, darkly lit pearl in the sky. We have entered the carefully plotted world of the diviners, where ambiguities are unwelcome, where transient phases, ephemeral glows, and vagrant shadows are either politically significant or they are superfluous. The moon represents the female principle in the cosmos, and its permutations and its journeys among the houses of heaven are reflected in the activities of the *yin* principle in the world of men: we are warned to look out for usurpatious queen-mothers—the "other-named" of our quotation—and outrageous assertions of feminity.[144] But it also sends down the holy dew of heaven into the magic pan prepared for it; it foretells the incursions of the cold wind from the north, whether physical or spiritual; when eclipsed—an event reflected microcosmically in the closing of the oyster shell on the pearl—nomadic horsemen drive in from Mongolia; while the gentle and peaceable unicorns exhibit abnormal belligerence[145] the White Tiger of the western sky crouches in the darkness of an untraceable lair.

Most of all, the passage of the moon through one of the fateful asterisms that spangled the Chinese night sky signified events in the lower world that deserved serious attention. These transits had approximately the same significance as did the entries of planets into the same constellations. In short, power and virtue lay in the concatenation of fixed stars singled out by the moving celestial bodies, not in those bodies themselves. For example, under the date of 21 July 635, the *T'ang shu* reports: "The moon trespassed on Base (*ti*). Base is the place where the Son of Heaven is lodging."[146] The designation of the lunar lodging Base (stars in Libra, especially α^2) as the current dwelling of the ruler of T'ang was the important factor in interpreting the event—the moon was hardly more than a luminous pointer, or perhaps an activating agent, which might, like a fuse, prove to be dangerous. This is somewhat peculiar in view of the fact that not only were the asterisms well mapped, but the motions of the moon and planets along the ecliptic were regular and predictable. Perhaps we may attribute awe in the face of the obvious to an element of conservatism in the realm of the most sacred Chinese traditions having to do with life, destiny and death; science might advance the frontiers of the predictable, but an ominous aura will still cling to the stellar dynamos long after it is understood that their apparent motions are not haphazard.

The appearance of the moon in an asterism was generally taken to be unfavorable for the nation, or for a person important to the nation. Heading the list of these personages was the Son of Heaven himself. Grand Tenuity (*T'ai wei*) was the name of a curved wall manned by high stellar officials, protectors of the royal person, composed of stars mostly in our constellation of Virgo. This wall of powerful stars cut through the autumnal equinox—itself a critical juncture. When, on 6 January 641, the moon breached this rampart, it was taken to mean, "The liegelord is not secure."[147] The lunar station Well was centered on a double star ("crocus yellow and blue")[148] which we style μ Geminorum. When, on 25 April 670, the moon took up its seemingly rightful lodging here, the omen was read as "The lord of mankind is afflicted."[149] Similarly, its entry into the potent asterism Southern Dipper on 22 July 758 meant "The great person (i.e., the sovereign) is afflicted."[150] And so it was down the scale of magnates. The moon covered the chief star of the lodging Heart on 22 June 758. This is the fiery Antares. The omen signified "The Grand Heir is afflicted."[151] Again, "Affliction lies in the Rear Palace (the seraglio)" was the meaning of the moon's entry into the asterism Hsüan-yüan in Leo.[152] But, outside of the palace, the constellations also had their allocations in the disastrous geography, so that a whole region of China might be afflicted when the moon took up residence in one of them. "The moon covered the central star of Ox-hauler (*ch'ien niu*). It is unriddled as 'Bane to Wu and Yüeh.' "[153] Or the disaster might fall on a foreign nation, as when the moon occluded the Pleiades on 25 January 648. This "lodging," in particular, its chief star Alcyone (η Tauri), held the fate of the northern nomads in its power. Accordingly, the omen was read: "The Son of Heaven smashes the Huns."[154] Clearly in this case the royal house of T'ang was favored, not threatened as it was in the other examples cited.[155] Perhaps it is best to understand the relation between the lunar presence and the earthly event not as one of cause and effect, nor as a warning, but as a sheer statement of fact—namely, that the celestial and terrestrial events are not only coincidental but mutually reflective: the moon entered the nomads' sky-sign—the Son of Heaven quelled the nomads. One necessarily involved the other.

However, the moon was most significant in astrology when it was obliterated. Despite the fact that lunar eclipses are predictable with fair accuracy with an astronomy of even modest sophistication,[156] the venerable tradition that attached the profoundest significance to such occurrences was never banished to the dust-heap of mere superstition; the dread of an extinguished moon survived undimmed into T'ang times. In any event, not all eclipses are alike. The moon is often a coppery red when the eclipse is near totality,[157] and indeed there is a whole spectrum of color variations based on atmospheric and stratospheric conditions. There is also a vocabulary for the positions of the vanishing satellite among the asterisms, and so on.

The ancient Chinese were aware that lunar eclipses invariably took place on the fifteenth day of the lunar month, when the moon was full and directly opposite the sun. The early Han book of *Huai nan tzu* speaks of this as the "sun abducting its light,"[158] as if the sun was somehow sucking the luminous energy out of the moon when facing it squarely. This could be regarded as a more rational explanation than the more ancient one, which still lived on, that the moon was being devoured by a celestial monster. Still, belief in the moon-swallowing myth was strong enough in T'ang times to compel the citizens of Ch'ang-an to beat on mirrors, aimed at the moon, in order to compel its disgorgement.[159] This appears to have been an act of imitative magic, the mirror being a replica of the moon, which could be dislodged only by the city-wide pounding. However, many, perhaps most astronomers, under the influence of Buddhist science from India, believed that in eclipses, which take place always at the two nodes where the lunar orbit crosses the celestial equator, the moon was blotted out by the invisible planet Rahu.[160]

The carnivorous language of mythology extended as well to what we would style occultations of planets by the moon, equally significant in astrology. An example is an occultation of Jupiter by the moon in December of 780: "The moon ate Year Star in the allotment of Ch'in. The omen is 'That country will perish.' That month Year Star ate Door of Heaven. Door of Heaven is the central star of Carriage of Ghosts. The omen is 'There are uncanny sayings, and a minor person occupies the See. Our Lord and King loses the pivot; the dead will be more than half.' "[161]

Gleameth of a dusky red
Like the lustre 'mid the stars
Of the potent planet Mars?

Paul Hamilton Hayne,
"Why the Robin's Breast is Red"

10 The Planets

The planets familiar to the Chinese were the "naked-eye" five. They were simply styled the "Five Stars"—or sometimes the "Five Pacers" (*wu pu*) or the "Five Wefts" (*wu wei*). As to us, the planets beyond Saturn were unknown to the peoples of Asia until modern times, and accordingly had no place in their astrology, their religion, or their literature. The little "asteroids"—better named "planetoids"—which orbit the sun between Mars and Jupiter, were just as invisible to them, unless, conceivably, the most brilliant of them, which we call Vesta (magnitude at opposition about 6), "a distinctly orange or pinkish" body,[1] came occasionally into the ken of the ancient astronomers as an unpredictable and therefore ominous apparition.

The standard names of the planets in Chinese indicate either their appearance, their supposed astrological character, or their role in the calen-

211

drical arts. I have tabulated them below, along with the labels of the Five Activities whose concentrated essences they were thought to be:[2]

Name	Identity	Quintessence of
Chronographic Star (ch'en hsing)	Mercury	Water
Grand White (t'ai po)	Venus	Metal
Sparkling Deluder (ying huo)	Mars	Fire
Year Star (sui hsing)	Jupiter	Wood
Quelling Star (chen hsing)	Saturn	Earth

(The Chinese names of the planets have been inadequately studied. Mercury's name appears to be a simple inversion of hsing ch'en, a name for chronographic asterisms that I have regularly translated as "starry chronograms," but it may be that the ch'en of Mercury represents a dragon or similar monster [cf. ch'en "clam-monster," and also the zodiac sign Dragon]; the full significance of the name of Mars has not yet been revealed; the "quelling star" [Saturn] suggests an exorcist.) Despite their mobility, the planets were like other stars in their composition. Some Taoist texts, however, provide interesting glimpses of details of their structure. One of these is a book, apparently of the sixth century, which embodies some of the earliest Mao Shan revelations.[3] In it, Mars, like the other planets, is described as a round mirror, composed of quintessential cinnabar: "Within the star are three gateways, and from the gateways emerge three pointed rays. These pointed rays descend 30,000,000 feet. Inside each gateway there is a red god." Similarly, the mirror of Venus is quintessential metal, but it has seven flashing gateways, behind each of which resides a white god. And so on.

The eccentric movements of the planets as seen from the earth attracted comparatively little interest in ancient China, although the cycles that accounted for them were approximated by about the fourth century. The paths of the two inner planets caused more trouble than those of the others because of the difficulties in observing them when close to the sun. Chinese geometry was never sufficiently sophisticated to map the paths of any of them, either ideally as ellipses, approximated as circles, or even phenomenally as erratic systems of advances, hesitations, and regressions.[4]

Despite a sort of uneasy confidence in these eccentricities, which although undiagrammed were conceived algebraically, the ability to construct planetary ephemerides did not in itself give a sufficient sense of security to permit the neglect of ancient associations of the planets with the sun, the moon, each other, and with the ecliptic asterisms (such as the lunar lodgings) in astrology. As was the case with fixed stars, changes in color, scintil-

212

lation, brightness, and the like were significant—usually of evil.[5] In general, too, it could be affirmed that the appearance of any planet in an asterism—however predictable this advent—portended ill fortune for the place, person, or political agency whose stellar facsimile that asterism was. It was commonly but not always the case that the identity of the planet was a matter of indifference—it was the significance of the asterism through which it passed that determined the omen. (We have observed the same system at work in the case of the moon's passage through asterisms.) It is on record, for example, that when the White Star, Venus, trespassed on the constellation Fixer (*chien*) in Sagittarius on 23 January 649, it was forecast that "the Great Vassals will calumniate each other."[6] Exactly the same prediction was deemed appropriate when Mars, the Fire Star, intruded on the same asterism on 6 October 769.[7] Similarly, it did not matter whether the dangerous lodging of the moon named Well was marked by the presence of Mars or of Saturn—the meaning was "drought" in either case.

Occasionally the planets presented a favorable aspect to mankind. This was especially the case when they all seemed to emit the same color—presumably the result of the interposition of an atmospheric pall between them and the world of men. Such notable apparitions were regularly signalized by the composition of odes of praise by the poets of the royal court.[8]

Before considering each of the planets individually it should be noted that—idiosyncratic mutations aside—each of these stars had a unique anthropoid form it assumed when it revealed itself in human surroundings. So Mercury took on the guise of a woman, Venus that of "a stout forester," Mars "a merry youth," Jupiter "a noble official," and Saturn "an aged gaffer."[9] The reader may wish to compare this list with the representations of the planets on the Tun-huang painting of the Tejaprabha Buddha which forms the frontispiece to this volume: the images of Mercury (a woman in symbolic black at the upper part of the picture) and of Jupiter (the blue-clad magistrate below her) agree well enough. The Sivaite ascetic (token color "Yellow") could be accommodated to fit a wrinkled old man. But the lady in white is hardly a stout forester, and the ruddy demon of Mars (at bottom) is neither blithe nor young. To a Taoist initiate Mars might appear as a more abstract but even more appalling presence; it might, for instance, take the form of a great fire, enveloping the body of the adept.[10]

Mercury, the red runner near the sun (though black in Chinese symbolism) gets little attention in T'ang literature. Traditionally a star of judges, convictions, and dire penalties,[11] its appearance in the judicial asterism signified legal miscarriages, excessive punishments—hanging judges, we may imagine.[12] Occasionally, however, the little planet gave promise of good times: in ancient tradition, when it shone forth clear and fair, it was rea-

213

sonable to expect a proper and seasonable balance of heat and cold, and a fine harvest.[13]

Sharing Mercury's propensity to lurk near the rising or setting sun, Venus, had an advantage over her neighbor in her greater size, greater proximity to the earth, and greater albedo, so that when she is visible at all she is a striking object, well-named "Grand White" and (as sometimes) "Metal Fire" (chin huo).[14] In astrology the white planet tended to be ambiguous, sometimes favoring the fortunes of China, sometimes those of its enemies.[15] In general it was a baleful star, however, and the mild associations that Venus has for us are almost totally lacking in Chinese imagery, for whom white is the color of the ghost world and, in its most lustrous phase, of the flash of deadly weapons. Accordingly "Grand White" is a planet of subversion, plots, cutting edges, and executions. When Venus appeared in Well in the summer of 663 the omen was "There will be executions among the great vassals,"[16] and when the planet intruded on Ghost in 707, the sense was the same.[17] Similarly a visit to Southern Dipper in 744 meant "there will be rebellious vassals," but others took the sign to mean an amnesty.[18] A transit of Well in 719, and of the battle-axe star (Propus; η Geminorum) adjacent to it, obviously portended the employment of battle-axes.[19] In this case the significance was lent by the region of the sky, with the planet—as so often—serving only as a spotlight. So it was also when Venus crossed Barrier of Heaven (t'ien kuan) in 698: the omen was trouble on the frontier.[20] But wherever it was to be found, "Grand White is arms; it is also punishment."[21] The celebrated brightness of the Venerean lamp was, when at its maximum, in itself an omen of importance, but not necessarily a good one. The usual reading of a daytime view of Venus was of the yin principle overriding the yang—that is, of the lower orders showing themselves unwontedly, to the disadvantage of the sovereign, symbolized in this instance by the Grand Yang, the sun.[22] In some situations, however, the imbalance might be a proper one, as when a virtuous vassal rose up against an oppressive monarch. This happened on 24 June 618, when Grand White was seen in the blue sky of day, and the omen, doubtless retrospective, was taken to be "men-at-arms shall arise, and the vassal shall be strengthened."[23] This plainly alludes to the rising fortunes of Li Yüan, founder of the T'ang nation: the hero had formally taken the throne six days before, and the house of Sui no longer existed.

The special significance of Venus for warriors is the focal point of many T'ang poems—perhaps a motif taking second place only to abundant allusions to the planet as the alter ego of Li Po. Wang Wei has shown us a youthful bravo on the Great Wall watching, not enemy horsemen, but the movements of Grand White.[24] Similarly, Liu Ch'ang-ch'ing, in a poem full

214

of somber reflections on war, pinioned captives, gloating nomads, and sweating horses, shows the loyal troops of T'ang:

> Leaning on their swords, watching Grand White,
> Washing their weapons as they approach Sea Gate.[25]

The planet of blood and fire had, besides its more familiar names, that of "Star of Punishment" (*fa hsing*) and "Holder to the Law" (*chih fa*).[26] Curiously both of these are also the names of constellations, the former in Scorpio, the latter a pair of important stars in Virgo, standing like stern Athenian archons on either side of the main gate that gives access to the Palace of Grand Tenuity. The relationship was meaningful, in that in astrology Mars, like its namesakes, had powerful judicial functions, especially in connection with maintenance of ritual protocol and adherence to strict norms of justice and procedural exactitude. Accordingly, when on 23 May 636 Mars made the transit of Hsüan-yüan (in Leo), the omen was read: "Sparkling Deluder is master of the Proprieties: when the Proprieties are misdone, then the punishment issues from it."[27] Similarly, when on 15 March 772 Mars covered the Barrier of Heaven, the doom was pronounced to be: "Disorderly vassals will alter the laws and edicts of the Son of Heaven."[28] Otherwise, however, the advent of Mars portended bane, grief, war, and murder;[29] in addition the planet's hot radiance often gave warning of drought in the region represented by the asterism in which it was a visitor. A typical omen reading for April 797 goes: "Sparkling Deluder entered Southern Dipper. Its color was like blood. Dipper has the 'allotment' of Wu and Yüeh. A blood-like color portends drought."[30] Mars in Well, as it was in 638, 650, and 670, obviously meant the drying up of wells.[31] With a comparable double symbolism the red planet, entering Heaven's Fields (705) and Heaven's Kiang (River) (707), betokened the failure of water supplies in critical places.[32] But otherwise death, and grief for the Son of Heaven were the common omens: typical omens were "The Great Man shall be afflicted" (618), the reference being to the last ruler of Sui;[33] Outlaws will stand by the Great Man's side" (637);[34] "There will be affliction for the suzerain" (675),[35] "The lord will depart from his palace" (754; this forboded the flight of Hsüan Tsung to Szechwan);[36] "Men-at-arms shall arise" (767);[37] "There shall be an abundance of deaths in battle" (888).[38] The movements of Mars could even threaten the life of an individual, as it did that of the great minister Li Chi-fu. A year before his death Mars occluded the star Highest Minister in the constellation Grand Tenuity (Palace). "Heaven is going to kill me," he said.[39]

I have seen no record of the appearance of Mars on earth in human form during the T'ang era, but a notorious case from an earlier age was remembered then. In A.D. 258 (or 259—the sources disagree) a remarkable boy, appearing to be about seven years old, appeared among a group of other lads at play. They were naturally astonished by the brilliant rays darting from his eyes and asked him to identify himself. He said, unabashed, that he was Sparkling Deluder, and that his presence foretold the triumph of the Szu-ma family over the dynasties that ruled the Three Kingdoms. So saying, he shot up into the air, transformed into a trailing ribbon of silk, and disappeared. And indeed, at successive intervals of four, six, and twenty-one years, the three royal families succumbed to the might of the new empire of Chin.[40]

The "year star" Jupiter loomed magnificently over the astrologers of China from the earliest times—sometimes helpfully, sometimes frightfully. As a simple chronometer he had the virtue of measuring out, year by year, with fair precision, the twelve zodiacal positions occupied by the sun in successive months—reinforcing the solar cycle, as it were, at a slower pace. Actually the complete passage of the ecliptic by Jupiter takes a little less than twelve years—a fact to which popular astrology was blind, but which sophisticated astronomy had to take into account to maintain its reputation. Accordingly, studies seeking a persuasive precision on this point were made in very early times. By T'ang times the leap-year of the Jupiter cycle was reckoned at about one zodiacal position each 84 years, which is reasonably close to the requirements of our own astronomy.[41] In the more theoretical parts of astrology Jupiter played a formidable role. The great year star was a concentration of power which irradiated the asterisms through which it passed with productive energy, but detracted from the fulfillment of their potentialities when it failed to make timely appearances among its encampments (tz'u "stages of an army's march").[42] Jupiter, then, was a mighty lord—but his whims could pass as divine law and his movements were those of a Juggernaut.

Examples of Jupiter attuned to the sovereign follow: retrograde in Libra in 629: "the Lord of Men went beyond good rule in the management of his palace apartments";[43] standing in Scorpio in 655: "the Master of Men makes a [mere] 'Guestmaiden' (pin) his Heirgiver (hou)"—this refers to the elevation of Lady Wu by Kao Tsung;[44] in Gemini in 756, while the sun was in eclipse: "the country shall perish" (a clear reference to the An Lu-shan rebellion);[45] and many other similar instances.

Examples of Jupiter attuned to supreme magnates: occluding an important star in Scorpio in 761: "commanders and ministers are afflicted" (and so again in 773);[46] entering Ghost (part of Cancer) in 768: "the Year Star is a noble vassal—and Ghost Carriage governs death and mourning."[47] But Jupiter could be a curse to farmers: visiting "Supervisor of Grotesques" (szu kuai), high in Orion in 696, he promised unceasing floods and droughts;[48] and his presence near Alpheratz in Andromeda in 646 portended that "the Five Crops will be damaged by water."[49] Frequently also Jupiter's message was of war.[50]

Saturn, the yellow oppressor, plays almost as minor a role in T'ang literature as does Mercury. His significance was highly ambiguous—at times baleful, at times beneficent. Like the other large planets he might forbode drought, war, and death,[51] but we also have this report of him: "he is in charge of good fortune, and when he dwells in a lodging in which he should not dwell, the nation will have luck."[52]

Lacking a general conspectus of planetary astrology for this period, and necessarily depending on induction from reported events and their interpretation, we can conclude with no general theory. It is obvious that in many cases the innate quality of the planet itself is in abeyance, and it is the sense of the asterism in which it appears that determines the omen. In others, the reading of the portent depends partly on the meaning of the planet (e.g. Mars is fire) and partly on the meaning of the constellation (e.g. Well is water supply). Then there are various readings which seem arbitrary to us, lacking the necessary astrological manual for consultation. At least one generalization can be made: the planets are conspicuous concentrations of supernatural power; their movements are rapid and irregular, and so tend to inspire a sense of uneasiness and even dread, where more placid motions, neither accelerated nor decelerated, would make for greater peace of mind.

The same uncomfortable feeling was inspired by planetary conjunctions (see Appendix B). One such arrangement was plainly a heavenly design—that is, a conjunction of all five—but others were almost always likely to prove unlucky. All were awe-inspiring.

An official T'ang adaptation of the meaning of conjunctions to "Five Activity" theory identifies Jupiter and Mars (now known to be outer planets) as *yang* planets, governing the central domains of China, while Venus and Mercury (which move inside the earth's orbit) were *yin* planets, governing the marches and outer realms; hence "when *yang* combines with *yin*, Inner and Outer are joined by force of arms."[53] An example cited in the dynastic history was the conjunction of the four planets in Well in 815 which is alleged to have had this general effect—although the specific event is not mentioned.[54] In any case, the doctrine was difficult to apply in practice, and must be seriously stretched to fit the omens as interpreted elsewhere in the same book. Furthermore, with Saturn out of the running, the number of conjunctions explainable by the theory was much reduced.[55] For example, any such lofty doctrine had to combat certain extremely simple analogies embedded in the astrological system, such as the obvious one that "conjunctions" suggest "consultation" and "plotting"; hence, the concurrence of

Jupiter and Mercury on 21 July 626 was interpreted to mean "there shall be revolutionary plots,"[56] and another of Venus and Mercury in late June or early July of 644 was taken to mean "there shall be plots for the use of arms."[57] In some cases at least these unpleasant conjunctions were explained in terms of Five Activity theory—so that a meeting of Venus and Mercury was taken to be a conflict between the agencies Metal and Water. This view is particularly pronounced in the omens of the *Book of Han*[58] Indeed, whatever the metaphysical explanation, conjunctions were generally regarded as conflicts—heavenly powers lined up for tests of strength, usually in single combat, representing analogous contests between mundane powers.[59] There are examples aplenty in the T'ang histories. For instance, when Li Shih-min was campaigning in Korea in the summer of 645, there was a conjunction of Venus and Mercury in Well (part of Gemini), representing as usually the Chinese capital. Here Grand White was said to represent "host," while *Ch'en* star stood for "visitor"—with the conflict taking place in the wilderness.[60] Presumably Venus in this case stood for the true suzerain of these borderlands, and the natives and their allies were regarded as mere interlopers, present on sufferance. The battle between the sexes, so familiar in the popularized astrology—or the mythological symbolism of astrology—of the Mediterranean regions, was quite alien to Chinese belief. If for our own ancestors the pleasant tale of the capture of guilty Aphrodite in bed with Ares by the cunningly woven net of Hephaistos was a disguise for the conjunction of the planets Venus and Mars caught in the filmy net of the Pleiades,[61] the union of the same planets in the same asterism (for the Chinese, in both Net and Mane—that is, in Hyades and Pleiades) in the early spring of 755, meant only armed combat, not a scandalous liaison.[62] The great exception to this principle was a grand conjunction of all five planets, which was virtually always taken as a sign of universal peace, with the implication of the victory of Chinese arms everywhere—a Pax Sinensis.[63] But sometimes—and the refinements of the choice of instances can hardly be guessed at now—a quintuple conjunction could have a quite different meaning. Referring to such a noble apparition in the early autumn of 750, the *Book of T'ang*—noting, however, that Mars, the planet of blood, arrived first at the meeting and was the first to leave—tells us that in such cases "when there is Virtue, it is felicitous; when there is no Virtue, it is disastrous."[64] Although Mars upstaged the other planets in this little drama —which took place in Scorpio, not far from Mars' alter-ego Antares—it is not plain that this priority determined which of the two possible readings was the correct one.

The influence of the moon could also work in conjunction with that of the planets. This cooperation might take place in the same place in the sky, or in different places. An example of the former is reported for 17 November 773, when the moon and Venus entered the house named Base together. This was taken to mean "The suzerain will be involved in weeping and wailing."[65] The other sort of joint action is illustrated by a report for 6 May 650, when, as the moon encroached upon the "Five Barons" (*Wu chu-hou*, composed of stars.in Gemini), Mars simultaneously fell upon the asterism of the Ghost (*yü kuei*), centered on θ Cancri. The double happening was taken to mean "Bane for the Barons."[66] These evil associations were more intense when the planet was actually occulted by the moon.[67] Such events were taken, with great regularity, to mean the eclipse, normally by death, of a magnate of the empire—although, more rarely, a famine might be indicated. We have T'ang examples for the years 647, 670, 790, 807, and 839, the planets so obliterated being Mars, Venus, and Jupiter; their separate identities apparently were not significant in such cases.[68] Typical among these is the occurrence of 26 March 807: "The moon covered Year Star: the interpretation is 'a great vassal shall die.' "[69] Sometimes, however, it was the moon, not the planet, that represented the might man: "Grand White made a raid on the moon; the reading is 'Grand White is a simulacrum of weaponry; the moon is the frame of a great vassal.' "[70] Here, it seems, striking mobility in the sky was interpreted as an act of aggression against a stable personage.

The scholastic philosophers not only
accepted the idea of angelic intelligences
moving the spheres, but also traditionally
named which angel, or archangel,
governed each sphere.

A. J. Meadows, *The High Firmament*

11 Astral Cults

High among the venerated spirits who watched over the doings of mankind
and visited them with appropriate signs or approval or disapprobation were
the beings whose inscrutable forms—if forms they had—were hidden in the
darkness behind the crystal fires they had designated as their visible tokens.
The "star gods" of ancient China were not mere ensouled stars, except,
perhaps, to the vulgar. They were inconceivable beings whose masks and
costumes were always hanging in the Vestry or Green Room of the sky, ready
for occasional use when the formless powers who owned them chose to show
themselves more closely to advanced students of the Highest Clarity than
they ever did to mortals whose vision was more clouded by the obsessive fogs
of ordinary careers and mundane preoccupations.[1] In the icons of all faiths,
it is only glittering costumes that are capable of representation; they are

220

merely useful substitutes and surrogates for representations of the occult powers themselves. This was equally true of religious painting, of sacred hymns, and also of the less respectful if more enchanting forms supplied by the imagination of poets. The latter may be compared, for instance, to the verbal images of archangels—Uriel, Gabriel, and the other souls of the seven planetary lamps—given distinctive personalities by the pens of Milton and Blake. The stars could, in short, be represented by plastic or verbal metaphor in every solemn and reverent situation. The simplest shamankas invoked them as living personages in their homely rites,[2] while the funeral vaults of royal princes and princesses of medieval China were thoughtfully supplied with both the abstract and personified versions of the awful entities who veiled themselves behind the dark vault, which had their distinctive calling cards or signboards permanently affixed to it.[3]

The beginnings of official Chinese worship and propitiation of these remote and sublime intelligences are lost in the roots of Chinese history. In Han times, however, when we begin to have some clear idea of official cult practices and beliefs, star-worship was already firmly established. A prominent place was given to it in the state rituals connected with the worship of heaven carried out in the capital city. An example, under the date of A.D. 26, was the great imperial sacrifice to Heaven, with offerings of oxen to the sky-gods, inaugurated in the southern suburb of Lo-yang. The rite was conducted on a central round "altar" (i.e., ceremonial platform) and external altars to the five paramount gods of the directions. The place of sacrifice was furnished with representations of the purple palace of the pole and with blazons representing the positions of the sun in the east, the moon in the west, and of the Northern Dipper. There were also lesser altars for the planets.[4] These celestial deities were always paramount in the state cult, since they had a special relationship with the imperial house, the earthly nexus of the power that radiated from them. But the star gods also figured prominently in the forms of worship prescribed for the magistrates of the empire, who acted as representatives of the priest-sovereign in provincial rites of lesser importance. An example is provided by a decree of 203 B.C. ordering nationwide sacrifices to the Numinous Stars (*ling hsing*). This was, in part at least, a sacrifice for fertilizing rains made to the stars of the great dragon palace, symbolizing the eastern quarter of the sky.[5] There were also fixed sacrifices at ancient holy places outside the capital, such as those to the sun at Ch'eng-shan, and to the moon at Lai-shan in Shantung.[6]

Allowing for procedural differences and some changes in emphasis since Han times, the same was true of the imperial religious monopoly of T'ang times. Rituals either conducted in or directed from the T'ang palace may conveniently be subdivided into two sorts—private rites for the benefit of members of the ruling family or important courtiers, and public rites for the welfare of the realm.

Elsewhere I have given a brief description of the character of the traditional celebration of the seventh evening of the seventh month, in honor of Weaver Woman in Vega. Here, as an example of a "private" court ceremony, I proffer a translation of an account of this celebration as it was observed by the ladies of the T'ang court:

A two-storied royal hall, a hundred feet high, is constructed out of connected damasks in the palace. Its upper part will accommodate several tens of persons. It is furnished with melons, fruits, wines, and roasts, and arrayed with gear for seating— all for the worship of the two stars Ox and Woman. Each of the royal handmaidens and consorts holds a nine-holed needle up towards the moon and pierces it with five-colored thread. If it passed through, this is a token of artistry to be achieved. There are performance of the *Ch'ing-shang* songs, with feasting and music until morning. Households of both gentry and commons imitate this.[7]

There were many other such pleasant and luck-bringing occasions to interrupt the monotony of harem life, but probably few others, if any, had to do with star spirits.

The high ceremonies of state conducted either by the Son of Heaven himself, or by his surrogates, were an entirely different matter. These were momentous, highly organized, complex affairs, and the number of potent spirits invoked in the course of them was considerable. At the winter solstice, in the most honorable position on the great round altar-platform—that is, the northern one, facing south—the imperial court worshipped the ritual presence of the Supreme Theocrat of the Heaven of Primal Light (*Hao t'ien shang ti*), whose epithet refers to the white radiance of the eternal breath which pervades the cosmos. This supreme deity was also honored by the titles of Radiant Moon-soul Gem (*Yao p'o pao*) and Heavenly Illustrious Great Theocrat (*T'ien huang ta ti*). His outward manifestation is our Polaris. Positions of somewhat lesser dignity on the sacred stage were allotted to such asterisms as Northern Culmen (*pei chi*), the Grand Monad (*T'ai i*), the several spirits of the walled Palace of Purple Tenuity, a number of great asterisms in Leo, notably the seats of the Five Theocrats, the Hall of Light, and Hsüan-yüan—the last two named at a somewhat lower level, along with a host of other glowing spirits, including the Five Planets, the Twelve Chronograms of the zodiac (the Jupiter cycle), the Sky River, and even the Weaving Woman.[8]

This greatest of state altars was two hundred feet broad at the base. It rose in five stages to a height of eight feet and an inch to a summit platform fifty feet broad. It was provided with twelve steps. Nearby were other "altars"—sacred stages, rather—on which the ceremonial dramas in honor of

222

other great deities were performed, among them one to the "morning sun" and another to the "evening moon," both forty feet in diameter.[9] The performances held on them were enriched by splendidly garbed actors of the highest dignity, costly furniture and sacrificial vessels, with carefully selected offerings, incenses, and the holy chants of ceremonial choruses.[10]

The T'ang period, when Taoism was a greatly honored religion, actively encouraged by powerful monarchs, was marked by many innovations in both doctrine and practice. For the star deities alone there were at least two major changes in cult observances. The first of these was increased honor to the great rain deity, identified as ultimately a star spirit. First of all, in 745, the status of both the Sire of the Wind (feng po) and the Master of the Rain (yü shih) was elevated from the miscellaneous group of minor deities worshipped in rites of the lowest grade to join the ruck of star deities in the middle category of ceremonies. They had long been entitled to a special altar in each county of the realm, where the local magistrate offered them officially sponsored honors, and this level of service was continued.[11] Then, in 820, the wind god, to whom the ancient title Master of the Wind was now restored, was shown by classical scholars to be identical with the deity underlying the asterism Winnower in Sagittarius, and was formally recognized at court as indeed that same deity. An altar was erected for his worship northeast of the wall of the captial city.[12] The second case of special honors paid to a stellar divinity took place on 22 August 736. On this occasion it was the Longevity Asterism that was singled out for elevation. As the relevant texts make clear, it was not just the star Canopus, which also bore that name, that was nominated, but also the Jupiter station in Virgo, whose most prominent star is Spica, and which is also represented by the zodiacal dragon. An imperial decree declared that, since the ancient almanac "Commands for the Lunations" (yüeh ling) placed the sun in Longevity Asterism—that is, in Virgo—in the eighth lunar month, appropriate offerings should be made to the divinity on a newly designed altar, and that the correct date for the ritual was the birthday of the sovereign, Li Lung-chi. This glorious day was now styled "Anniversary of a Thousand Autumns," and providentially fell on the fifth day of the eighth lunar month. In this year that date corresponded to 1 September in the Julian calendar.[13] On 5 September similar honors were extended to Canopus—Old Man Star and Longevity Star—obviously in sympathetic rapport with the ecliptical asterism of the same name. The star's rites were carried out on the same altar.[14] Doubtless, despite the traditional respect paid to the star gods in the state religion of T'ang, these new honors were not unrelated to the fact that Li Lung-chi was a devotee of the perfected beings of Highest Clarity, in what we would now call Mao Shan Taoism.

In any discussion of the rather drab accounts provided by the official

223

records of what must have been solemn and dignified but hardly impassioned ceremonies we must always bear in mind that Taoist visions of the deities in the sky agreed much more closely with what we know of popular and poetic—that is, heartfelt—views of them than did the colorless and barren abstractions of them approved and purveyed by court ritualists, even when these were cautiously embellished by court poets. We shall not find the real gods of T'ang in the *Book of T'ang*, that doctrinaire guide to blameless religious thinking. But although true believers in the Taoist traditions of the mystic jewel (*ling pao*) and highest clarity (*shang ch'ing*) could not conceive the ultimate forms—if such there were—of the formless beings behind the stars any more than could the great barons who were required by custom and protocol to show them formal honors, they had very clear ideas about how these veiled entities displayed themselves when they chose to adopt anthropoid masks, and they recorded the points of recognition in their sacred texts—valuable data to which nothing in the official books of ritual corresponds.

It must be reiterated that if we regard Taoist "star worship" as the worship of stars, we misunderstand the nature of the faith as much as if we regarded the adoration of St. Michael and St. Gabriel as bird worship because those creatures of pure spirit are often represented with wings. Although it is still impossible to give an adequate account of medieval Taoist belief and praxis—most of the data we need still lies hidden in the bewildering text-hoard we call the *Tao tsang*—it can at least be affirmed with confidence that in the Taoist system (not completely defined by being styled merely "religious") the stars were not gods but the chosen tokens and guises of cosmic beings, who might assume other guises and reveal themselves in other symbols. They were deities whose location was nowhere, who existed simultaneously in the brain and in outer space, and could exhibit their numinous presence in any manner or place that seemed desirable. It would be demeaning to suggest that the adept could invoke or command their presence—rather he could, by patient study and years of discipline, create wholesome mental and physiological conditions within his person which made it possible for the gods to reach him, or, what amounted to the same thing, made it possible for him to perceive the divine presences.[15] The personal vision, which required as a prerequisite the study of identification manuals,[16] was not the only possibility: the relationship with the divine world could be conceived, not so much as a ceremony of investiture, or the receipt of divine gifts from on high, but more abstractly as the ingestion of astral essences—with the mediation of divine beings—and their conversion to valuable and delightful uses within the human organism.[17] The distinction is merely verbal.

Taoist priests and initiates wore distinctive costumes, which symbol-

ized their spiritual advancement and embodied mana which was revealed outwardly as magical diagrams and talismans. They were in effect
the costumes of gods—and indeed the Taoist divinities are often described in
garb just like that worn by earthly hierophants. Most prominent of these
vestments was the "star hat" (*hsing kuan*), an article alluded to very frequently in T'ang poetry. To a westerner the name may suggest the conical
hat of an Arabian Nights' sorcerer, or of white-bearded Merlin, or of a fairy
godmother. No graphic representations of a Taoist star-hat survive, to my
knowledge, from T'ang times.

Star hats go back at least to Han times—or it was believed in the fourth
century that they did. It was said then that in the year 109 B.C. a divine
personage had descended from the sky in a cloud carriage harnessed with
white deer and alighted at the palace of Liu Ch'e (Han Wu Ti). This
unexpected visitor—who, unfortunately, was thinly treated—was dressed in
a suit of feathers and wore a star hat. He was one of the high Taoist elect, who
had achieved divine status by the ingestion of mica.[18] This being, whose
name was Wei Shu-ch'ing, served as a prototype for a legion of spiritual
brothers and sisters in the poetry of later ages. He (or someone like him) is
even to be found in the *tz'u* of the late ninth and tenth centuries, especially
in song-forms whose titles reveal Taoist associations, such as *Nü kuan tzu*
("female hatted one," used in T'ang of Taoist priestesses) and *Wu shan i
tuan yün* ("One strip of cloud at Shamanka Mountain"). One poem in the
former style by Wei Chuang features a lovely Taoist lady in her handsomely
appointed temple—hardly distinguishable from a temple in the sky. She
wears "a star hat and a cape of pink clouds."[19] One in the latter mode, by Li
Hsün, pictures an immortal being riding down from the sky on a magic
bird:

> Moon cloaked—thin as auroral pink-clouds;
> Star hatted—clustered with jade stamens.[20]

("Auroral pink clouds," or "sunrise clouds of morning" [*chao hsia*] however suggestive of filmy mists suitable for the dress of a dawn goddess,
was a name actually applied to a pink cotton fabric in T'ang times. The
material was associated with the people of Southeast Asia, and sometimes with the cassocks of Buddhist monks)[21]

A few details about the figures on at least some of these astral head-
pieces are preserved in the Taoist canon. An apparently early text, for
instance, refers to a divine hat, suitable for an adept of the "Five Pecks of
Rice" sect, in the mode of "The Twenty-eight Lodgings," that is, showing
(presumably) five dipper patterns and rimmed with the shapes of the whole
cycle of lunar lodgings.[22]

In other Taoist liturgical and meditative works there are references to "golden flowered" hats showing the "Seven Planetoids"—i.e., the sun, moon and Five Planets. Such a splendid tiara was particularly suited to crown the heads of great cosmic deities who preside over all of the Nine Skies, such as "The Supreme Illustrious Jade Numinous Lord," who was also girded with "a purple sash with cinnabar figures."[23]

Not surprisingly, considering their importance in T'ang astral lore, especially that of the Mao Shan sectarians, polar figures were especially prominent in these hat designs. We have an example from lyricized folklore in a stanza by the eighth-century poet Li Ch'i. This is a rhapsodic song about Hsi Wang Mu, the divine woman who presides over the Taoist earthly paradise in the mountains of the far west. She is shown riding in a "unicorn"-drawn carriage, which flaunts a shining rainbow banner and is shaded by a canopy of feathers. The goddess holds the pears (sic!) of immortality, and wears "a nine-star hat" on her head.[24]

The association of the star hat with some kind of a feather cape or canopy or skirt is constant; it is reminiscent both of the bird people or feathered men whose figures are the earliest representations of Taoist divine beings we have, and also reminds us of actual feathered costumes—symbolic of magic powers of flight beyond the world—worn by both male and female adepts from Han to T'ang times.[25] An example from T'ang poetry shows an aging imperial concubine regretfully leaving the palace, but "costumed in feathers, hatted with stars, her mind fixed on the Tao."[26] The eighth-century poet has placed this scene of retreat from secular life far back in Han times, but the description is more suitable to the T'ang. Similarly, Han Wo, writing early in the tenth century, reveals a man, obliged to retire from the court, already suitably dressed to take his place among the seekers of eternal life: "Crane cloak and star hat—the guise of a 'feathered visitor.' "[27]

Indeed feathers and stars went together so obviously and naturally that a suitably crested bird was the inevitable image for a Taoist sylph:

> Petalled hat—star-dotted—the dress of the gentry of the Tao:
> A woman from the Palace of Purple Sun-power, her body transformed
>   for flight.
> She has power to transmit to the Uppermost Zone tidings
>   of springtime's surge and ebb;
> If she should come to Mount P'eng—they would never let her go home.[28]

This is a little ode to a hoopoe, that finely striped bird with the spotted panache who—in western tradition—played the go-between for the Queen of Sheba and Solomon. In Chia Tao's pleasant conceit, the bird is the form assumed by a Female Perfected (nü chen) in the Mao Shan tradition of Highest Clarity for a visit to the world of men to observe the play of passions and affairs in the springtime. She should beware of alighting on the lower paradise of P'eng-lai, however, since the gossip she could transmit would so delight the bored transcendents there that she would be caged up forever—far below her proper home beyond the stars.

Star emblems and star patterns evidently figured prominently on other parts of the adept's paraphernalia, but none of these is frequently alluded to, as is the star hat, which must have been a striking and memorable part of the initiate's vestments. We note, however, that both the gods and the high-level initiates carried seals of power, many of which, at least, would yield astral impressions. On the side of the deities, for instance, the Jupiter deity possessed "a seal of penetrating luminosity" and the spirit of Mars was equipped with a "seal of Grand *Yang*"—that is, of the sun conceived as a great cosmic force.[29] On the side of humans, proper discipline might make one worthy of "a seal of the Penetrating Numens of the Five Stars, which seals with the talisman of the Numens of the Five Stars."[30] Evidently these potent instruments were used to create amulets and apotropaic tablets by imprinting charms instilled with the vigor of the stars and planets, and were carried ready on the belt.

These men who could draw on the energies of the stars spent much or all of their lives in the private establishments—which were often large and splendid—misnamed "Taoist monasteries" or "friaries" in English. The ancient Chinese name for them was *kuan*, properly "watch-places" or "observatories," a name which, whatever its remote origins may have been, was well suited to the predominantly astral character of Taoist religious exercises in the T'ang period. These well-appointed retreats were most frequently placed in remote mountain areas, not only because of the peace and quiet afforded to facilitate the meditations and disciplines of the residents, but because of the obvious benefits of the proximity of supernatural beings, both those who held court deep within the caverns of the mountains themselves, and those above, for whose spirit-places the high hills provided convenient courtyards and vestibules. Most prominent among these holy mountains, on which Taoist sanctuaries tended to cluster in special abundance were the five I have styled "marchmounts" (*yüeh*, M.C. *ngauk*), which marked the center and the limits of the sacred kingdom in early antiquity. The hallowed fanes that dotted their slopes were familiar to the writers of T'ang who, like other intelligent men, were often disposed to visit them, to participate in philosophical discussions with the resident priests, to undergo periodic purification of their bodies and cultivation of their spirits, and to enjoy such esthetic charms as murmuring brooks, aged pines, and the close and invigorating approach of the high asterisms on clear and radiant nights. Their memories of these visits, or the desire to give thanks and praise to their hosts, frequently inspired elegant or wry verses, as the occasion or writer's temperament demanded.

The widely admired poet Meng Chiao, who had artistic temperament to spare, has left us ten verses reminiscent of his stay at the "Cloud Platform Observatory" (i.e., Taoist friary) on Hua Shan, the western marchmount, near Ch'ang-an:

> Hua Marchmount—unique for strange and numinous beings;
> Its herbs and trees always new and fresh.
> Throughout the mountain—five-colored stones;
> Throughout its waters—single-colored springs.
> Transcendents' wine—it will not make men drunk;
> Transcendents' polypores—they extend the years of all.
>
> By night I hear the hostels of the numinous stars;
> The hours are attuned to the chords of "female bindweed";
> Respect for this place will not let me sleep:
> I burn some cypress—chant from books of the Tao.[31]

A scene of magic stones, holy water, life-giving food and drink, music of the spheres and of nature closer at hand—all inspiring to profound thoughts.

Hsiang Szu, like Meng Chiao a poet of the ninth century, gives a different view: he reveals the Taoist retreat as subject to the same mortality that its former residents had hoped to overcome. The poem is entitled "Old Observatory":

> The observatory's founding stele is broken now.
> The years spent watching the pines are indistinguishable.
> Who in this grotto can recognize the herbs?
> Outside, daily, more cairns are added.
>
> The turtle, allowed its freedom, heads for the water;
> A deer, called hither, is apprehensive of my melilot.
> Beside the altar I see ashes and fire—
> How many burnings here—offerings to the patterns of stars?[32]

Here the memory and traces of the past are gradually being obliterated, and the barrows of dead priests proliferate. The wild animals no longer feel at home on the grounds of the friary. But someone still makes his little oblation to the star gods—a vestigial relic of centuries of the great astral ceremonies of the past.

Religious life in the "observatories" revolved around sacred precincts or stages, where the divine dramas were performed. We call them loosely "altars." Along with the star hats, the star altars (hsing t'an) were the feature of the Taoist sanctuaries most frequently alluded to in T'ang poetry:

A pair of banners with twinned rainbows—an apron sewed
 with feathers;
At the Altar of Seven Stars he salutes the Primal Ladies on high.[33]

This couplet forms part of a poetic account of a Taoist ritual witnessed at the
very end of the T'ang period. The "Primal Ladies" (*yüan chün*—the word
chün actually lacks gender) are the female deities of the highest rank, especially in the favored Mao Shan Taoism of this age. (It may also be, of course,
that the celebrant alluded to by Wu Yung was also female.) The "seven stars"
were almost certainly the seven prominent stars of the Dipper. Their place in
the sanctum was as crucial as it was on the head-coverings of the initiate.
There can be little doubt that in the course of many significant rites the
priest stepped through the pattern of the constellation on the altar, thus
reproducing or inspiring the experience of actually treading the stars. In one
of a series of untitled pentasyllabic poems, as full of alchemical longings as of
astral aspirations, written by Lü Yen, the famous Taoist of the late ninth
century, the poet suggests the possibility of progressing beyond this ceremonial imitation:

Let me but have a staff of power on the true and perfect road—
How then would I trouble to worship the moon or tread the altar
 of stars?[34]

The true Tao is available to the man with right instruments—no need for
wishful exercises on the ground.

Indeed the poetry of star altars sometimes is invested with an air of
frustration—even of disillusion. Salvation and eternal life do not come easily
or early. An example may be found in one of a handful of poems from the
pen of Li Chao-hsiang, who wrote towards the end of the ninth century.
Most of them are in a Taoist vein. This one is roughly in the tradition of the
Sorcerer's Apprentice. It presents the persona of an alchemist, regretting his
vanished youth and irritated at the trials and frustrations he has experienced
in his attempts to gain access to the secrets of his life:

I have in mind the original tradition of the Nine Conversions'
 formula;
Under peaks in the clouded dark blue I invoked the Splendid One
 in Space;
My granular cinnabar did not mature; my heart was rent in vain;
Hard put to stay in the bright white sun as my temples turned to gray.

I have no road into Grotto Heavens to follow up Mu Man;
I have at times, with mortal men, thought well of the young Liu;
Although the grace of the Transcendents has been redoubled—
 I have no means to repay;
All burned up on the Altar of Stars is the incense of the
 midnight hour.[35]

229

(The Nine Conversions is the arcane procedure for the ultimate refinement of the cinnabar elixir; "The Splendid One in Space" is a cosmic deity in the Mao Shan tradition; Mu Man is Mu Son of Heaven, a mythical prototype of the successful Taoist ruler; young Liu is Liu Ch'en, one of a pair of youths who had mystic encounters with divine maidens on Mount T'ien-t'ai, and simultaneously Liu Ch'e, that is, Han Wu Ti.)

Occasionally a poet might use the star altar as a means of conveying a sense of the transitory nature of human lives and institutions—but even then the hint of final transcendence can be detected. Li Chung was a talented writer who lived and worked in the tenth-century state of Southern T'ang. The quatrain which follows describes his visit to an eminent adept in a Taoist friary. He finds him gone:

> I came, at leisure, to the Transcendent's Friary—
> asked for Hsi-i;
> Clouds filled the altar of stars—water filled the pond.
> The "feathered visitors" did not know whither he had gone:
> It was the season when, all at once, flowers fall in abundance
> before the grotto.[36]

(Hsi-i is a sobriquet derived from a pair of words signifying "mysteriously invisible and inaudible" from the book of *Lao tzu*; the clouds cover the mystic altar where the star rites are performed, symbolizing that the spirits have withdrawn into the unknown; "feathered visitors" are the Taoist adepts themselves—transients on the stage of earth; the flowers fall in autumn, season of decay and death; the grotto is the entrance into a Taoist paradise. The object of his inquiry has died—or rather, gone up to a better world.)

The rites performed on the star altars were as complex as the manifold documents which comprise the Taoist canon. However, the story of the outward manifestations of Taoist piety or ambition in T'ang times is, if anything, less well known than those which are visible on the island of Taiwan in our own day. Here no attempt will be made to remedy this deplorable state of ignorance—only to hint, or suggest, or typify, by less than a handful of instances. First in importance: the gods stand, invisible to all save a few, *behind* the stars. This basic fact—as important as the knowledge that the images of Christian saints are not the saints themselves—is exemplified in a quatrain by the late eighth century poet Ch'en Yü. In it, we are projected into the past and are shown Liu Ch'e, seen through the eyes of a medieval writer. The cloud-painted carriage is an authentic attribute of his religious innovations, but there is no contemporary evidence of his personal adoration of the Northern Dipper:

230

> The Martial one of Han, purified, purged, has read the Tripod Texts;
> The Grand Functionary helps him up into the cloud-painted carriage;
> High on the altar the moon shines clear; the palace halls are closed;
> He looks aloft to starry Dipper—does reverence to the empty void.[37]

The final verse here seems to show something more than the simple recognition that the stars themselves are not gods—that the gods are enthroned invisibly beyond them in the deep void. It reads as if it is space itself that is being honored. Is it possible that Ch'en Yü was being ironic? That the gods are—nothing?

In T'ang times there were many formulae for appealing to the "star functionaries." Their variety can only be suggested here. In common with other religious ceremonies, at any rate, these involved the careful choice of auspicious days and hours, and were accompanied by sacrifices of edibles and valuable goods. For instance, in a text of the early seventh century, a ritual of prayer asking wealth and fortune was directed towards an auspicious star in response to a dream. Seven cups of wine were required on the altar, and seven strips of deer charqui, seven sheets of "memorandum paper" (the fine yellow paper used in the chancellery), two feet of red cloth, seven pieces of paper money, two lamps, and a stick of incense.[38] The plurality of the number seven in this prescription suggests the prevalent Dipper symbolism. From approximately A.D. 900 we have an invocation to the planet Jupiter, beseeching his aid in ridding the celebrant's body of the "corpse-worm" which gnaws at the vitality of us all. This will serve as an example of the spoken appeal to the deity—perhaps not necessarily made on the star altar itself. Choosing the correct cyclic day (*chia i*), in the spring, the celebrant "rectifies his heart," fixes his gaze on the planet, and conjures it thus: "I desire that the luminous star in the eastern quarter support my cloud-soul, unite with my white-soul, and make my longevity like that of pines and cypresses—a life extended through a thousand autumns, a myriad years. . ."[39] As my first instance suggests some of the trappings of a group ceremony, even if not a public one, the second seems more likely to have been an individual one, the prayerful act of an adept seeking to purify the substance of his soul with the help of a star-masked divinity.

The iconography of Taoism, which is in considerable part the iconography of star spirits, has not been adequately studied. This is another deficiency in our knowledge of medieval China that I do not propose to rectify here. Rather, I propose only to suggest faintly the character of the great celestial deities as they were conceived in T'ang times. We know that they appeared to adepts wearing the specific countenances and costumes that made their identification possible. Indeed they were represented in the religious art of that age. We know, for instance, that the late ninth-century painter Chang Su-ch'ing, who had marked Taoist leanings, did images of "The Seven Planetoids" and of "Longevity Star."[40] But unfortunately no examples of significant Taoist art survive from T'ang times. At most we can surmise the astral identity of some ceramic tomb-figurines, or speculate about the true age of the wood-cut delineations of star deities accompanying the printed versions of early texts in the modern version of the Canon. Our best visions of those supernal magnates must in the end be based on descriptions of the phantom visitations written down by ecstatic initiates, or accounts confidently handed down by ancient and therefore doubtlessly reliable traditions. For instance, a medieval text, otherwise undated, provides an exact description of the divine being who lurks behind the second star—named "Light" (*kuang*)—of the Southern Dipper. This personage rejoices in the title "Lord of the Tao of Supreme Perfection in the Jewel House of Cinnabar Quintessence and Occult Luminosity." He looks like a small child dressed in a robe of feathers embroidered in purple, and is girded with bells of "fluid gold." He sits on a golden couch, holding the six stars of the Southern Dipper in his hands, "like fluid pearls threaded on orange cords."[41]

However, it was the arcane spirits of the Northern Dipper—a stellar oligarchy of supreme importance in medieval Taoism—who attracted most of the fervent attention of poets, liturgists, and presumably of iconographers as well. We have given literary examples of the worship of the stars in Taoist friaries, but they alluded to no specific stars. When the effusions of lyric poets specify the recipients of the priests' entreaties, they almost invariably refer to the Dipper, and more specifically the Northern Dipper. Writing in the middle of the ninth century on Taoist themes (as often), Ma Tai described a hopeful neophyte partaking of a life-giving dose of "cinnabar frost"—that is, a mercury sublimate—contemplating the sacred texts while seated on a simple stone bench, and (above all) giving his devotions to the starry dipper at midnight.[42] In another stanza, the poet's characterization is quite simple:

> A remote dwelling—a cave in the white clouds;
> Focusing, purified, on the Canon of the Red Pine;
> Now and again, up above the Dragon Tarn,
> He burns incense, and honors the Dipper's stars.[43]

A pre-T'ang text, well known to the Taoist enthusiasts of T'ang, has described the correct procedures for invoking the separate star-spirits of the Northern Deities to induce their descent into the adept's body, where they will aid him to form the embryo of his future divine and immortal body. It provides descriptions of some of these divinities, especially of their costumes and distinctive jewelry. Many of them were female: such was the "Supreme

232

Consort of the *Yin*-numen Jade Purity of the Nine Quintessences," who sits within the Dipper, with purple mists swirling about her divine figure.[44] The account of her appearance, however, is rather conventional. A more intimate and revealing view of the person of one of these great ladies is provided by a hagiography of Wei Hua-ts'un, better known as Lady of Southern Marchmount—the chief inspirer of Yang Hsi, whose dazzling visions ultimately provided the celestial foundations of the Mao Shan sect. She has not been identified with a particular asterism, but the description shows plainly that her body is composed of stellar stuff, and she must have struck the fortunate observer as clearly one of the company of radiant beings who populate the reaches of outer space:

> Empyreal phosphor, glistening high;
> Round eye-lenses doubly lit;
> Phoenix frame and dragon bone;
> Brain colored as jewel-planetoids;
> Five viscera of purple webbings;
> Heart holding feathered scripts.[45]

This characterization, full of celestial imagery, yields a transparent effigy, a goddess of crystal, lit from within, her vital organs plainly displayed as phosphorescent jewels—not sources of vital juices, but of electric rays. We see her in eighteenth-century guise in Dr. Darwin's philosophic verses which appear above as the epigraph of the essay on "Star Women."

My soul, there is a country
Far beyond the stars.
Henry Vaughan, "Peace"

12 Flight Beyond the World

Tours of space were a commonplace of ancient China. For persons who lacked the resources to make such journeys in person, fully documented reports, dictated by the cosmonauts themselves, were readily available. The most popular of these flight-logs survive in the familiar and much commented on classical collection of the *Ch'u tzu*, a title which may be rendered as "The Word-hoard of Ch'u."[1] These richly and elegantly phrased rhapsodies were based on the god-seeking soul voyages of the Chinese shamans. At one time, these ecstatic medicine men had been dominant at all levels of Chinese religion, but were gradually degraded to the lowly status most of them occupy today after centuries of persecution by a hierarchy that was totally intolerant of mysticism—even one that sprang from purely Chinese roots. This historical aspect of the matter has now been investigated with

234

some thoroughness, and further reportage would be redundant here. Suffice it that priority among the literary versions of the primitive shamanistic narratives is usually given to the *Li sao*, attributed to the always official and now popular hero Ch'ü Yüan, possibly because of its traditional susceptibility to "Confucian" paraphrase. Despite differences in plot and style, settings and stage business with which these long poems are furnished, they all have much in common. An important difference among them, however, is the mode or vehicle of flight they describe. Sometimes the sky-traveller projects his soul over the rainbow and beyond the clouds. Sometimes he seems to go in the flesh, like John Carter on his celebrated trip to Mars, or as we all shall do at the sound of the last trumpet. (Often it is not easy to be sure which of these methods the poet intends, or if he wishes to be precise on this crude point at all.) Sometimes the protagonist flies off on the back of a beast powered by some undescribed metaphysical fuel—the dragon being a prime favorite. Or he may even glide aloft in a chariot, either drawn by a divine animal or else invisibly propelled by a primitive anti-gravitational device. In either case, he skims the housetops of the star-gods, as easily as Santa Claus, and hobnobs with the porter at Heaven's gate—but his ultimate goal is always the city of God, the pure and holy center of the universe where all antinomies are resolved and all anxieties evaporate.[2] These evolved shamans and refined medicine-men were very familiar with the stars, and their familiarity was no mere nodding acquaintance. They treat them in an offhand, neighborly way, and even exploit them, much diminished it would seem to us, for trivial purposes. The reader is sometimes reminded of Hollywood cartoons—Bugs Bunny shuffling about purposefully in the moon dust. Our ghostly heroes of Ch'u wield comets as falchions—or as pennants; they drink perfumed wine from the Dipper—or twist off its handle to use as a magic wand.[3]

Such visions as these were refined, probably by some time in the Han period, along with other elements of folk belief and religious practice, into the sophisticated system we call "Taoism." Today we can still see superbeings astride winged dragons, as if they were horses, in the tomb reliefs of that period,[4] and in the same era other media of transport, for instance the magical Manchurian crane, began their long careers, as indispensible attributes of the immortals.[5]

MECHANISMS OF SKY TRAVEL

Up in the skies
His body flies
In open, visible, yet magic, sort.
  Thomas Traherne, "On Leaping over
  the Moon"

Pacing the Void

By the beginning of the Han period the ideas of mystical flight, of magical flight, and even of mechanical flight had begun to crystallize into commonly accepted patterns. Men remembered that the divine king Shun had been taught by his two wives to fly like a bird.[6] Strange tales were told of the bird people on the eastern littoral—oviparous and feathered sky-climbers.[7] Sometimes they were called "feathered people" (*yü jen*) and sometimes "ascending, or transcending ones" (*hsien che*).[8] In both visions were the germs of the mature Taoist conception of a fully realized superbeing capable of traversing the mysterious wastes of the world above the clouds. He bears a startling resemblance to the ostracized angels of Persia, the peris, whose name means "winged ones." (". . . Tearless rainbows, such as span//Th' unclouded skies of PERISTAN!")[9] Flight with wings—actual or symbolic! Since antiquity, royal princes wore, on suitable ceremonial occasions, high hats styled "far-roving hats" (*yüan yu kuan*). Although a princeling so capped might have little inkling of its antecedents and once magical purpose, there is no doubt that these aspects of the headgear loomed importantly in the intent of other wearers. I give two examples from the tenth century. When the last ruler of Southern Han, soon to be absorbed into the growing Sung empire, took the throne in Canton in 958, he let the actual reins of power fall into the willing hands of a god-possessed woman, who presided over her enchanted court clad in such a "hat for rambling in the distance" and a "stole of Purple Aurora"—the latter being plainly a man-made facsimile of the magical mists which envelop the heavenly cities of the gods.[10] Here the hat was no bit of antiquated regalia, holy only because it was mentioned in approved classical books. It was meant to work —or at least to give visible show to the true powers of the queenly sorceress. That the wonderful hat had actual lines of force tying it to stars—such would seem essential for correct performance—is evidenced by the following couplet by Hsü Hsüan, who wrote in tenth-century Chekiang when it was ruled by the kings of Southern T'ang. The occasion is happy and festive, a princely wedding illumination which attracts a multitude with free wine and other great inducements:

> Flowering candles—babble and hubbub: the Minister-chancellor's
>   Office;
> Stellar chronograms—shifting and shaking: a
>   Distance-wandering Hat.[11]

Each burst of artificial stars (it seems likely that the "flowering candles" and the "stellar chronograms" here were pyrotechnic displays)[12] outside the buildings that house the magnates of Southern T'ang elicits a responsive roar—just as the power of a far-wandering hat induces violent activity among the asterisms fixed along the ecliptic.

236

But we can descend even lower on the scale of devices providing space transport. Much in the tradition of the seven-league boots of Hop-o'-my-Thumb are the faintly comical "Flying-cloud shoes." The tale that Po Chü-i himself owned a pair while he was engaged in alchemical pursuits in Kiangsi is doubtless aprocryphal.[13] But there is no doubt that people were sometimes actually shod with such dynamic footwear. Li Po, mocking at the classical pedagogues of ancient Lu, Confucius' home state, described them in these words:

> Ask them about the strategy of "designing for relief"—
> They are as vague as if shedding haze and fog.
> On their feet they wear far-wandering shoes,
> On their heads they bear square-mountain kerchiefs.[14]

("Designing for relief" is something like "planning for economic assistance." "Square mountain kerchiefs" were worn by celebrants in ancestral temples.) In short, these respected upholders of the traditional morality had no useful counsel to give, only blather—and they affected archaic costumes that only made them look ridiculous. But in T'ang times such magic shoes could be taken as seriously as the "pace of Yü" in lifting their wearer beyond the grime of this world. The following poem, translated in its entirety, is also the work of Li Po. He wrote it in honor of his friend the Taoist priestess Ch'u the "Thrice Pure," on the occasion of her departure for the southern marchmount (Her head kerchief is no square mountain!):

> Lady of the Tao from the Kiang in Wu,
> With lotus flower kerchief borne on her head,
> And a rainbow dress unwetted by the rain—
> Unique and rare—a cloud from Yang-t'ai.
>
> Under her feet far-wandering shoes
> To skim the waves, raising pure-white dust,
> To find the transcendents, heading for South Marchmount—
> Where surely she will see the Dame of Wei.[15]

There are a number of rather conventional allusions here—almost all to divine women of the past—flattering comparisons are inevitable. Lady Ch'u would surely be diminished if we were to style her a "Taoist nun," a practice commonly but poorly employed when writing of these proud and often liberated women, who remind us rather of the daughters of medieval French and English kings—who might also do turns as secular abbesses. This "nun"—no timid drab recluse—wears splendid silks and a real or artificial lotus on her massed black hair. The cloud and rainbow dress symbolizes the rainbow woman of Wu Shan, the ancient fertility goddess and consort of kings, celebrated in the eighth century in the notorious fairy dance of Yang Kuei-fei. Yang-t'ai, as everyone knows, was her sacred home. A goddess who skims the waves is herself so pure as to stir up little dust, and that a dust of holy whiteness, in her wet path—a kind of candent moonglade. The prototype of this snowy vision is the goddess of the Lo River, whom Li Shang-yin invested with the garb of the moon-goddess Ch'ang-o to tread, not the water of a river of earth, but the icy surface of the moon.[16] Lady Ch'u is undistinguishable from her as she crosses the Yangtze River ("the Kiang in Wu"), departing for the high home of the gods. Li Po's friend, then, combines the attributes of both ancient rain goddess and ancient river goddess in his picture of her, and he concludes that she will inevitably meet another sacred woman-figure, Wei Hua-ts'un, the protectress of the mountain. The anti-gravity shoes lift her off the ground—ready levitation is an attribute of divine personages—and symbolize the weightless and tenuous nature of her person, fit for floating through the rarefied atmospheres of divine worlds.

Tales were also told of wooden flying machines, said to have been invented already in Chou times.[17] Soon, too, hopeful sky-treaders could experiment with the powerful elixirs of the alchemists—such as the mineral compound "the color of hoarfrost and snow" which levitated the body itself by a remarkable chemical action.[18] Both of these techniques, mechanical and chemical, had something crude about them.

It had become possible for Taoist initiates to gain mastery over the stars and realms beyond them by wonderful feats of self-projection. Allusions to these arts can be found even in pre-Taoist literature; an example from the old songs of Ch'u tells of "treading the array of stars to the limits of light."[19] This perilous adventure is now our theme.

STAR TREADING

Go, wing thy flight from star to star,
From world to luminous world, as far
As the universe spreads its flaming wall.

Thomas Moore, "Paradise and the Peri,"
Lalla Rookh

While still earth-bound the Taoist initiate—following the precedent of his ancient shamanistic predecessors—could walk among symbolic stars (or, in some sense, actual stars) tracing out their formations in order to achieve mastery over them, or to draw power from them, as a sorcerer can enchant a person by reciting the syllables of his name or manipulating his image in a manner approved by wizardry. Since as far back as the Chinese records show, an instant injection of supernatural energy was the reward of the adept who danced the step of Yü, the divine king, sometimes called the "shaman's step."[20] This halting but potent pace is thought by some to have originated in the shambling of a bear, impersonated in an archaic animal cult, and by others to represent the hemiplegic gait of Yü himself: the demigod had sacrificed part of his physical powers to gain heavenly ones.[21] It is still performed by female shamans in south China.[22] But its most sophisticated and dramatic development was in the esoteric practices of the Taoists. A good account of the technique survives in a Taoist classic of the early post-Han period, although the modern authorities are not in complete agreement in interpreting details of the recommended procedure.[23] The same book confides many circumstances when performances of the hopping ritual can be profitable—as when calling for the attendance of a fairy, entering a haunted forest, or approaching a ghost-infested rock.[24]

The Great Yü's mother, it is said, conceived him after seeing a shooting star in the constellation Orion.[25] It is possible that this extraordinary pregnancy is not unrelated to the power of Yü's pace, which derives its energy from the stars, and in turn provides authority over them. Yü's pace was important in Taoist enchantments and exorcisms. For instance, a Taoist text of about A.D. 900 tells us that an important part of the procedure for expelling the "Three Corpses and Nine Crawlers" that drain man's vital energies consisted of performing the Pace of Yü while shouting a spell for aid addressed to the sun and the moon, which are themselves similarly infested —but with hare, frog, and three-legged crow.[26] The pattern on the steps always followed, it appears, that of a powerful asterism, most prominently the sequence of stars in the northern Dipper. Doubtless it is the same "art of treading the Dipper" which, in primordial times, was taught to the Yellow God by the Mystic Woman of the Nine Heavens, enabling him to overcome the monster Ch'ih-yu.[27] (See figure 5.)

The Grand Supreme Perfected Men (t'ai shang chen jen), the most exalted class of Taoist super-beings, can summon the high polar deity Grand Monad (T'ai i)—he is also known by more elaborate titles—by "pacing the road of the Nine Stars."[28] These are the nine stars of the true dipper, the dipper of the Taoist cabala. The identity of the two stars other than the familiar seven is somewhat ambiguous. A Sung dynasty commentator on the "Nine Songs" states that they are the Sustainer (Fu = Alcor, attached to Mizar) and Far Flight (chao-yao = γ Bootis—the tip of an extended Dipper handle).[29] But medieval Taoist texts regularly present another view. A map in the canonical version of one such work shows two ancillary stars in its chart for dipper-treaders: one is Sustainer and the other is Straightener. The former is our little Alcor; the latter is said to be attached to topaz-yellow Phecda—but in fact Straightener is an invisible star, and one of its names is Void. It is an anti-star, a "black hole."[30] Another authority, however, makes

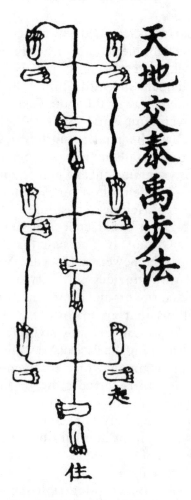

FIGURE 5. The Pace of Yü around and through the Dipper.
(*T'ai shang chu kuo chiu min tsung chen pi
yao*, by Yüan Miao-tsung [A.D. 1116].
In *Tao tsang*, 987.)

both of these accessory stars invisible: "Of the nine stars of the Northern Dipper, seven are seen and two are hidden. The eighth and the ninth are the embryonic essence (*ching*) and actualizing spirit (*shen*) of the Grand Reverend Godly Illustrious One (*Ti huang t'ai tsun*)."[31]

Polar star-magic took many forms and employed many techniques, such as breath control, whose nature is preserved for us in the books of the Taoist canon.[32] One T'ang work on Taoist practices advocates an internal approach to the dipper: the walk involves a simultaneous vision of the star Alcor, seen as a spark behind the closed eyelids.[33] Another recognized technique, orig-

240

inating it seems before T'ang times in the Mao Shan tradition, requires that the adept repose himself at night on a diagram of the dipper laid out on his bed, with its bowl like a canopy over its head, and his hands and feet pointed at major stars. He is to recite the names of its stars, picture them in his imagination, recite prayers, and in the end bring their sublime embryonic essences into his body where they build up, in the course of time, an immortal body which will ascend to heaven in broad daylight.[34] There were still other techniques.

So, in the end, it was within the bony planetarium of the skull that the divine asterisms are spread—identical, through a kind of supernatural topology, with the constellations we think we see above our heads. Travel by trance is easily the best method, far superior to either dragon or crane.

The outline of the Dipper was only one of many sidereal paths available to the adept. He might, for instance, choose to make the circuit of the Five Planets, hopping along the ecliptic.[35] There is also a pattern for treading the Three Eminences, a powerful constellation below the Dipper. To achieve this the adept must hold his breath while he makes the step of Yü in the transit from the Jade Mainstay to the Gold Mainstay.[36] These Mainstays (kang) are the cosmic meridians which bind the stars to each other and to the poles, above all to the little star named Knot (niu) which marked the celestial pole in the T'ang period. As an early (pre-Sung) Taoist text says: "The common stars are all bound to the Northern Chronogram."[37] Some say that there were twelve Mainstays, corresponding to the twelve annual stations of the Jupiter cycle.[38] Their southern nexus is an asterism which is itself named "Heaven's Mainstay," and is represented by our δ Piscis Australis, near the great reddish star Fomalhaut. This is exactly in the antipodes of the Northern Dipper. Perpendicular to the Mainstays are the Strands or Filaments (chi) of the sky, lesser bindings which have a symbolic counterpart in the constellation "Heaven's Strands" (T'ien chi) whose lucida is ξ Coronae Borealis and which extends horizontally across the northern part of Hercules, connecting ε, ζ, and other stars of that group. This strand is in the latitude of Vega and is perpendicular to the Sky River. This may explain the statement in the Book of T'ang that the "strands of earth" (sic; as contrasted with the stays of Heaven) extend from the "Cloudy Han."[39] The Mainstays, then, are the sturdier elements in the cosmic net, and depend from the horizontal Dipper; the filaments are the slenderer elements, attached to the stays, and feed out from the vertical edge of the galaxy. They might be compared with the standards and threads of basketry, whose meshes are closely analogous to those of a net. Sometimes the initiate actually climbs or skims along the Mainstays themselves to proceed from star to star. This appears from the following passage:

241

Pacing the Void

The Black Stars are the cloud-souls (*hun*) and white-souls (*p'o*) of the Northern Dipper. When you wish to tread the Mainstays, first hold in your breath, and turning left outside the seven stars, travel three circuits above the Dipper's cloud- and white-souls. Only then may you ascend to Solar luminosity.[40]

That is, here begins the walk along the Mainstays of the Dipper. ("Solar luminosity" [*yang ming*] is the true but esoteric cloud-soul name of α Ursae Majoris—that is, Dubhe, the cloud-soul whose phenomenal aspect usually was called "Pivot of Heaven" [*t'ien shu*]. There are five "cloud-soul" stars and four "white-soul" stars underlying the nine phenomenal stars of the Dipper. Collectively they are the "black stars.")[41] There were some dangers inherent in this undertaking. The sacred text says: "The method for treading the Mainstays: do not proceed horizontally, infringing on the Mainstays. To cut across—transgressing on the Mainstays of Heaven, is to hack and cut at the *Way* of Heaven, and is absolutely taboo."

In the T'ang dynasty there were poets to testify to the power of the shaman's shuffle over the dominions of the night sky. Yüan Chen, relaxing in a Taoist friary, where many such great traditions were still observed, wrote:

For binding oaths—their golden sabres are weighty;
For beheading trolls—their precious swords are keen.
When they pace like Yü the starry stays are moved;
When they burn their spells the furnace spirit does homage.[42]

(The "furnace spirit" is the deity that presides over the alchemist's fire and hermetic egg.) But many ways that might lead to the star-spirits were explored. The possibility of flight was—so to speak—in the air, and the best space flights were to take place through the cosmos of the spirit, and to find their Baedekers not among the historians and the compilers of alchemical manuals, but in the poets.

THE POETRY OF SPACE

The faery beam upon you,
The stars to glister on you;
A Moon of light,
In the Noon of night,
Till the Fire-drake hath o'er-gone you.

Ben Jonson, *The Gypsies Metamorphos'd*

The form in which this theme was most fully exploited by the medieval poets—a form or manner that has been very fully explored by other scholars—is the one commonly styled "*yu hsien* poetry." This expression means "relaxed journey to or in the transcendental realm." It might be paraphrased in some of the titles of poems in which it appears as "A Saunter to Sylph-

242

dom," "A Picnic in Paradise," or "A Trip to Transcendence"—possibly even (with a nod to Victorian iconographers) "A Stroll through Fairyland." Other variations might be rung on this theme in accordance with the treatment afforded it by the poet, the tone of his language, or possibly the whim of the translator. Some of these versions might be expected to express by one nuance or another the pigeon-holing of the particular poem in one of the easily defined slots to which the majority of this genre (to use a much-abused word) have traditionally been consigned. These would include such plausible categories as "rejection of dominant social values" (i.e., "alternative life-style"), "social protest," "escapism," "romance and erotic fancy," "retrospection and historical nostalgia," "religious practice and belief," "mysticism, visions, and ineffable personal experiences," "yearned-for worlds of the spirit," "idle or transitory dreams," "imitations and pastiches," "Parnassian trifling and word-play," "Taoist transmutations, alchemical and otherwise," and so on. Doubtless the reader will be able to add to the list and supply the names of some classical examples from the writings of such unavoidable "masters" as Ts'ao Chih, T'ao Ch'ien, and Li Po. What all of these conceptual modes share, other than the jargon of their typology, is the theme of journeys "out of this world."[43]

The linguistic imagery of the medieval sky-tourists was, of course, derived from a rich variety of sources. It will serve our purposes sufficiently to mention only two of them here. These are especially worthy of attention because of the handy phrases supplied by their titles alone. The first is the first chapter of the received version of the *Chuang tzu*, entitled *hsiao-yao yu*—the familiar one that begins with the parables of the giant bird and the enormous fish. The word *yu* is not consistently used—as is quite proper. When the eponymous author of the *Chuang tzu* uses it of "travel in the interminable" (*yu wu ch'iung che*), he seems to be using it in an abstract, almost intellectual sense, quite alien to the concrete figures of speech that dominate the rest of the chapter. But then again he uses it in a highly poetic and visionary way in the following:

> Inhale the wind, drink the dew,
> Do not eat the Five Cereals:
> Mount cloudy vapors,
> Rein flying dragons—
> Then wander (*yu*) beyond the Four Seas!

There is a flavor to these images that is immediately recognizable as authentically "Taoist" in the special sense of that word when it is used to describe the techniques for finding liberation from the bonds of ordinary perception and sensation. *Hsiao-yao* is employed in the parable of the great but immobile tree whose very uselessness—its inexploitability—guarantees its survival. The phrase, then, has strong overtones of movement without effort, of travel envisaged as sleep-walking, even of lying idle and letting the movement take place in the hills and furrows of the brain. Related, but somewhat different, is the phrase *yüan yu*—perhaps "far roving"—whose *locus classicus* is the title of the poem of that name in the songs of Ch'u, which tells of a journey through the stars, beyond even the Pure Metropolis and its Godly Palaces within the starry circumpolar wall, finally even

> To leap over the Unworked to attain the Pure, oh!
> To join the Grand Antecedence and be its neighbor.

To summarize: *yu hsien* implies the attainment of some paradisiacal vision or ecstatic state; *hsiao-yao yu* implies effortless transition; *yüan yu* implies passage beyond the bounds of space, time, or reason. All of these phrases—all too often rather stale—and many variations of them, were available as the stock-in-trade of T'ang poets, some of whom found new subtleties with which to sauce them.

Not surprisingly this body of interstellar imagery had a special appeal for poets whose temperaments inclined then towards Taoist belief. The most sublime instances of poetic space flight are provided by the verses of Wu Yün, a friend of Li Po, whose name is associated with the Taoist court of Hsüan Tsung. A mountain recluse early in life, his reputation for sanctity earned him an invitation to the palace, where he was installed in the Forest of Quills among the most successful writers of the realm. Somehow he could not live permanently under these worldly circumstances (he is known not to have been interested in fashion and current affairs)—or perhaps the opposition of the powerful eunuch Kao Li-shih, a partisan of the Buddhists, made life uncomfortable for him, or, conceivably, he did not provide the desired kind of advice to his starry-eyed imperial patron. It is reported that when Li Lung-chi asked his views about such great matters as alchemy and the gods, he replied:

These are matters of concern for rustics, who must dedicate the efforts of months or years in pursuit of them—they are not so deserving that the Lord of Men should give his attention to them.[44]

Not very encouraging counsel for an all-powerful monarch. Whatever the causes, he returned to the life of a hermit in the middle of the eighth century. (Some say that Wu Yün was primarily motivated by prudence, since he had a premonition of the success of An Lu-shan's armies.)

The poems of Wu Yün abound in astral flights, and although he plainly owes much to the linguistic resources of the *Ch'u tz'u* and to Taoist writings in the broadest sense, which he plunders freely—like so many poets of the late T'ang—no one who reads his work carefully would confuse his poems with those of any of his predecessors. Above all, he is concerned with paraphrasing clichés that suggest passage into the Infinite. The emphasis is not on providing such convenient metaphors as "farflight" in place of "alienation," "departure," and "escape" as it was for many of his materialistic contemporaries, but on suggesting, however imperfectly, the ecstatic bliss of empyreal being. He yearns for the ultimate but featureless sublimity, which becomes a kind of white glare in some of his poems, of the state where even the stars are grimy particles far below his feet. In the end, he aims to transcend what is already transcendental. All of these longings are expressed in radiant language, spangled with the very distinctions which he hopes to negate. To suggest this ultimate transition he rings the changes on all conceivable phrases for otherworldly travel. These are displayed at their most typical in a series of ten poems collectively styled "Treading the Void" (*pu hsü*), an old *yüeh fu* title also favored by other T'ang poets. In these he points the way through the "myriad heavens;"[45] he spirals upward, "footloose and fancy-free" (*hsiao-yao*) above the Grand Aurora (*T'ai hsia*);[46] he is irradiated by the silvery brilliance of Venus; finally he achieves the passage of the "Grand Emptiness" (*T'ai k'ung*).[47] (Wu Yün is particularly fond of the word *hsia* "pink-tinted clouds of dawn," which I have just rendered "aurora." The phrase "luminous aurora" [*ming hsia*] occurs frequently in his writing. For him the phrase seems to connote a realm glowing with more than human splendor, not violent and overpowering, but subtle, gentle, and calm, like the divine touch of *rhododáktylos ēós*.) Here is a prosaic version of one of Wu Yün's more sparkling journeys:

> Gently, easily, away from man's domain,
> Leaving, quitting, drawing near the town of God,
> Up far beyond the strands of starry chronograms,
> Looking down on the lights of sun and moon.
> Swiftly now past the Great Tenuity:
> The blaze of a sky dwelling—gleaming, glittering![48]

This stanza is one of twenty-four entitled "*Yu hsien*," but the language is of the class of that of the void-pacing poems (*Pu hsü tz'u*). The eager collector of golden streets and pearly gates will have a happy time with Wu Yün, but will have to temper his enthusiasm for jewelled imagery with a due appreciation of abyssal plunges and cruises through black or blinding gulfs. Along with the vocabulary of "strolls in purple courts" (*yu tzu t'ing*), "leaping climbs to the seats of the Nine Sylphs" (*ch'ao teng chiu hsien wei*), and "effortless motion in the precincts of divine luminants" (*hsiao-yao shen ming yü*), he must come to grips with more abstract and equivocal phrases such as "stroll in the great emptiness" (*yu ta k'ung*), "breathless dash into mysterious gloom" (many words for *ch'ung hsüan*), and "calmly set in the obscure void" (*t'ien-jan tsai hsüan hsü*).

A century later the youthful Po Chü-i penned his now familiar words about the search for the soul of Yang T'ai-chen. His account is drab and stereotyped when compared with that of his less known predecessor:

> He pushes clouds aside, harnesses the pneumas, courses like
>   the lightning,
> Mounts the sky, enters the earth, looks for her all around,
> Exhausting the cyan gulf above, the yellow springs below.[49]

Even the master recognized later that this was paltry stuff, and that his interests and talents lay in other directions—as the world has to come to recognize. The world does wrong, however, in supposing that "Taoist conceits" and "courtly eroticism"—it often pleases the world to find the one inevitably a disguise for the other, and in any case mere trivial word-play —are apt characterizations of some of the finest writing of the ninth century, a century heavily populated with good poets, unrecognized because their themes are no longer popular. This was the era when the seeds planted by the shamans of antiquity bore their fullest fruit and flowers. The merits of these skilled dreamers, although in common estimation heavily outweighed by the mighty reputations of Han Yü, P'o chü-i, and Liu Tsung-yüan, are gradually coming to be recognized, or at least searched for.

Fortunately the heritors of the sky-journeys of Wu Yün multiplied. One of the most eminent was Lü Yen, who failed the great examinations during the second half of the ninth century, and thereafter haunted the bistros of Ch'ang-an until he discovered his own Tao. Then he vanished, later to achieve apotheosis in the popular company of "Eight Immortals." He left a long series of untitled poems in seven-syllable lines on alchemical and transcendental themes. Sometimes he does not disdain the plain, familiar imagery that Po Chü-i made so widely popular, but intermingled with it there are usually phrases of cosmic and philosophic import conveying an entirely different flavor: in these poems we leave romantic fiction and endless love and find metaphysical devotions and the lure of invisible, almost inconceivable worlds:

> Girding the lightning, coursing the clouds, I fly to sun and moon;
> Goading the Dragon, urging the Tiger, I pass beyond both Potent and Latent.[50]

This couplet begins, as it were, on Po Chü-i's youthful ink-slab, but we soon see the stars shining in the great houses of the eastern and western skies, where Dragon and Tiger reign, and leave them quickly for the mystery beyond. Or again:

246

Flight Beyond the World

Without image, without contour, lying beyond the Formed Mutations:
There is a gate, there is a door, placed in both Potent and Latent.[51]

The created, phenomenal world counts for nothing here. Certainly this journey bears no resemblance to the hunt of Po Chü-i's Taoist detective for a woman's spirit—the search is for the road that leads beyond such illusory distinctions as the world we know provides. Outer space has become inner space.

At the very end of T'ang and throughout the tenth century, the flights of poets' minds into immensity underwent continuous development and writers ransacked both minds and thesauri for words to express the inexpressible in new and untarnished ways. Happily the name of Kuan-hsiu, the magnificent monk of the court of Shu in Szechwan, who made new and exciting discoveries in these fantastic word-worlds, is no longer as unfamiliar as it once was. We shall encounter him again elsewhere, building palaces of otherworldly gems. Suffice it here to take note that he, like Hsü Hsüan and Han Wo—but not exactly like them—made a specialty of dream travel. Conveyance by dream is a fairly obvious way of solving the problems of the transit of interstellar space, particularly since it was the common old Chinese view that dream experiences were true experiences. The soul, completely endowed with personality, went abroad, met friends in distant towns, bedded mistresses, and carried out all kinds of other "normal" activities—as well as many that were rather difficult when encumbered with carnality. For Kuan-hsiu, however, what the soul saw and encountered in dreams were by no means everyday things, or even extensions of them, but imperfect shadows of the truly real and perfect entities that are concealed from us during our waking hours. For him, the inner eye sees, in the phantasmagoria of dreams, only wavering segments of the ultimate vision. (Dream travel among the stars could also serve a purely prophetic purpose, as when Kao Huan, the future founder of the dynasty of [northern] Ch'i in the sixth century, "dreamed that he went about treading on a multitude of stars. He rejoiced within himself when awoken."[52] This was the most important of several signs of his coming eminence.)

SKY PALACES

What if within the moon's fair shining
    sphere,
What if in every other star unseen,
Of other worlds he happily should hear,
He wonder would much more; yet such
    to some appear.

Edmund Spenser, *Fairie Queen*, II

247

Pacing the Void

In T'ang literature, the final goal of the dreamers and treaders of the void was, as in earlier writing, the buoyant but solidly-set paradises where white-limbed, angelic beings lived their endless lives. (It is easy to think of them as "Taoist immortals," but they were part of the general heritage of Chinese lore.) They and their castles in the air appear as if seen through wavering mists or semi-translucent glass, but occasionally they are sharply defined, like the cunningly constructed copies of them built by royal architects in the great cities of Asia. In the hazier versions, they were only partly revealed to the profane eyes of the poet through amethystine fogs, imponderable clouds high above the clouds that drift over the fields of earth, and magic mists charged with energy analogous to the pink swirls of the aurora borealis, billowing far above the familiar world of rain and snow. The poet Wei Ch'ü-mou, an adept of the end of the eighth century (later converted to Buddhism) described, in one of a long set of "Pacing the Void" poems, the divine city as a "town of white clouds" (or "in white clouds")—*po yün hsiang*—near the Two Phosphors (*erh ching*), the two great sky lights of the sun and the moon.[53] The eminent Taoist metaphysician and mythographer Tu Kuang-t'ing has the palatial structures actually carved and twisted out of cloud-matter: "Formed of knotted pneumas, framed of clotted clouds in the void."[54]

Usually the divine mansions are represented as uncertainly placed in the blue depths of the sky, or floating serenely on banks of clouds, encircled by colored mists. Even in these lofty locations they had the aspect of hillsides covered with splendid buildings, and indeed whatever their location—a relative, illusory matter in the first place—they shared the attributes of divine mountains. Important among them were distant sea-girt isles, hovering on the horizon where sky and water are hardly to be distinguished, and therefore simultaneously earthly and heavenly dwellings. Occasional apparitions of these enchanting lands must have been appearances of those rare mirages, seen over expanses of water, which are called *fata Morgana*: "landscapes with towns and towers and parapets, rising above the horizon, transforming, crumbling, fairy-like scenes, producing a deep sense of happiness and an endless longing."[55] Then, too, some houses of the gods were placed on the summits of the five sacred mountains, once thought to have supported the canopy of the sky—although by historic times they could readily be identified within the confines of the Chinese empire, retaining their holy power for all that. Most remote of all was Mount K'un-lun and its jewel-like residents beyond the snowy summits of Serindia, one of many analogues of Indian Mount Meru, which the Chinese identified so readily with the numinous massifs of the west and south, whether known of them directly or reported by travellers in the flesh or in the spirit.

248

Flight Beyond the World

Sturdy and stable above the clouds, paradoxically weighted with abyssal stores of rich metal and gem-pregnant magmas, the magic mountains were also stepping-stones to the sky. They appear to earth-bound men rising above low banks of fog, as if afloat, however solidly anchored—and indeed we now know that their granite roots float on the plastic basaltic crust which lies like a black ocean below them.[56]

The paradox of weighty and weightless, so characteristic of mountains, could be expressed poetically by such antinomies as "flower palaces in pebbled piles" (*hua kung lei-lo*), or, in the same stanza, "jade peaks on cerulean clouds," set among the high asterisms.[57] Not merely flowers on stones, or stones on clouds—there are no prepositions in the Chinese original—but flowers as stones, and stones as clouds, level with the constellations: such are the habitats of divinity.

As foci of sky-derived energy, sometimes made visible as gathering clouds and flashes of lightning, the great mountains of China rightly had their places in the official pantheon, although their numens remained anonymous, except in popular lore. But however familiar their craggy exteriors, their hidden depths contained mysteries. Among them were the mighty chambers which housed not only divine beings, but whole universes. These adyta could be found only by way of haunted tunnels and secret adits, often recognizable near the surface as glittering caverns, hung with limestone stalactites—themselves sources of potent drugs. The Taoist adepts claimed to possess the cabalistic lore that provided access to these divine labyrinths and antechambers. Here were the "grotto heavens" (*tung t'ien*), whose capacity could not be guessed from the apparent size of the mountains that formed their outer husks. Sometimes it was said that there were ten large grotto heavens and thirty-six smaller ones, but other sources report only thirty-six all told.[58] Each constituted a distinct cosmos—a duplicate universe completely furnished with sun, moon, and stars, and with "the palace towers of the numinous sylphs."[59] In any event, their size was completely illusory—a matter of absolute relativity, like the "true" size of dragons. Accordingly they were often compared to worlds concealed in pots and calabashes—themselves symbols of the cosmic egg from which all created things had emerged. Each had its own secret name and its presiding deity. Taoist literature provides the necessary data for the adventurous sight-seer. For example we have Lo-fou Mountain in Kwangtung, one of the most celebrated of the sacred eminences for believers in the Tao. Its particular grotto heaven is named "Vermilion Luminosity Shining Perfection"; it is five hundred *li* in circumference, and its genius is "The Prior Born of Blue Potentiality." Or again, consider Kou-ch'ü Mountain, the holy place of the sectarians of Mao Shan (its other name). Its interior heaven is the "Grotto of

Golden Altar Flowered Solarity," but only one hundred and fifty *li* in circumference. Its god is "The Perfected Person of Purple Solarity," who in life bore the name of Chou I-shan.[60]

These worlds with mineral shells are frequently alluded to in T'ang poetry, most often perhaps by the so-called "Taoist" poets with a taste for the great mysteries. An excellent example is Lü Yen, of whom we have already taken note as a pathfinder of outer and inner space. He is particularly fond of the macrocosm-microcosm illusion. The following couplets are typical:

> Wind and thunder inside the grotto—returned to my palm's grasp;
> Sun and moon within the pot—placed in my breast lappet.[61]

Or,

> Starry chronograms flash out from the bud of a blue lotus;
> Sun and moon have power to hide in "white horse's tooth."[62]

Or,

> Starry chronograms now and again are seen inside a pot;
> Sun and moon time after time are hidden within a cassock.[63]

The implications of these images were carried to the obvious conclusions. The visible world evolved within a macrocosmic hermetic egg—the *hun-t'un*; *ergo*, the first causes and ultimate forces that made the visible world must also have been generated—if that is not too forceful a word—within the mystic egg. Lü Yen has something to say on this point too: "Inside a pot —only Potent and Latent are self-generated."[64]

It was possible to regard our world beneath the sky-dome as only a kind of pallid duplicate of the sphere of a grotto heaven. Indeed all of the grotto heavens can be themselves envisaged as an infinite regression of mirror images, one the reflection of the other, and all equal in the end. The translation which follows is of a poem written by Tu Kuang-t'ing, our chief T'ang authority on grotto heavens. He worked and wrote during the ninth and tenth century, climaxing his career at the court of Former Shu in Szechwan with the title "Heavenly Master who Transmits the Truth" (*Ch'uan chen t'ien shih*). He died, full of honor, at the age of 85, leaving us an eloquent account of the multiplicity of unseen worlds, their origin, and their present condition:

> When Potent and Latent were split apart,
> When Clear and Turbid were first divided,
> They made by fusion the Kiang and the Ho,
> They made by knotting the hills and holy mountains;
> Now alloted above as solar and lunar asterisms,
> Now hidden below as the skies in grottoes.[65]

250

The underground skies and their population of lights is, then, a replica and twin of our own sky and its array of asterisms. Moreover, each of these worlds has its own great deity to watch over it. Sometimes it is even suggested that the grotto heavens, conceived as the residences of mighty gods, existed before the general creation of distinct entities—or at least were formed simultaneously with the universal sphere, before the "big bang." This conception contains an inherent contradiction, which itself may be illusory:

> In the years of the Grand Culmen when the cosmic sphere (hun-t'un)
>   split open,
> This mountain too was the home of the divine transcendents.[66]

In short, a grotto heaven, concealed within a particular mount, existed prior to the world we know. Were there, then, many cosmic eggs, or was each simply a reflection or echo of the other—and so in the end only One?

Tu Kuang-t'ing's near contemporary (mid ninth century) Tuan Ch'eng-shih gives a rather different account of the hollow worlds:

The Six Palaces of the Grotto Heavens are a myriad li in circuit, and 2,600 li in height. These Six Palaces of the Grotto Heavens are the very palaces of the spirits and divinities of the Six Heavens . . . All men when they die go among them . . . The Six Heavens range across the Northern Road, and here are the dwellings of the spirits and divinities. There is a total of two myriads of worlds which make up only the names of the Sky Palaces. If you read them inaudibly during the night, you will keep ghosts and goblins away.

Here Tuan Ch'eng-shih is reporting Taoist lore that came to his eyes or ears from an unknown source.[67] The grotto heavens are here only six in number; they are placed in the vicinity of the celestial pole instead of in the roots of mountains; and they are the future homes of the souls of the dead. But this view is at least compatible with the accepted belief that the transcendent beings were basically classifiable into several sorts—for instance, those of heaven (t'ien hsien) and those of earth (ti hsien). Mystic worlds under mountains were suitable homes for the latter.

In the same passage of Tuan Ch'eng-shih, that bibliophile refers to the "little heaven of k'ung-tung " (k'ung-tung chih hsiao t'ien). In many graphic variants k'ung-t'ung or k'ung-tung was the name given both to the vast subterranean caves and also to the mythical mountain in the far west—a place of rebirth, initiation, and enlightenment where Huang Ti is said to have been given instruction about the nature of ultimate reality.[68] It is hard to overlook the tantalizing but unprovable suggestion of a linguistic affinity with K'un-lun, the name of the world mountain of the west, and with hun-t'un, the primordial sphere, to say nothing of a large word family, apparently derived from an archaic root like *klung, whose members have such mean-

251

ings as "vault," "cavern," "canopy," "dome," "roof," "cage," "chamber," "rainbow," and the like.[69] That some such relationship existed in the minds of men of T'ang who liked to speculate on this theme is amply demonstrated by the "Song of the Hollow Grotto" (*K'ung-tung ko*) of Chang Chih-ho, a cosmologist who lived during the second half of the eighth century. By ingenious division and recombination of familiar words and phrases, he suggested the identity of the grotto of his title with the cosmic sphere from which all things came, and whose nature is not susceptible to the ordinary processes of rational analysis:

> It is as it is, though not of itself—
>   the Prime of itselfness.
> It is transformed by no shaper—
>   the Alpha of the shaper of transformations.
>
> Patent-like, guileless-like
>   its form sphere-clumped.
> Turn away your sight!
> Turn away your thoughts!
>   So is it viewed![70]

Here "itselfness" is "self-generated"—but "it" has no self; "it" creates structures, but lacks structure.

Conceived in terms of myth and ritual—that is, of myth expressed in symbolic action—the thirty-six grotto heavens offered the possibility of an orderly journey through the spirit worlds where, following the precedent of Hsüan-yüan Huang Ti, the adept might make a circuit of the spirit realms in which, "within the grotto, sun and moon and starry chronograms are linked in order."[71] Ultimately, like Yü the great, he would be handed the magic talismans that unlocked the universal mysteries,[72] or simply, in the vacuity of a symbolic closed space, like a chick in an egg, meditate in absolute quiet and experience some sort of mystic revelation.[73]

The high significance of caves has been noted in many times and places. For some peoples they are symbols of underworlds and abiding places of the dead, simultaneously symbols of rebirth into the eternal chain of being. They are both tombs and wombs. In them, the shamans and their successors could communicate with the ghosts of the dead and other spirits.[74] There was an obvious connection between the starry palaces of the great gods in the blue grottoes and the richly furnished tombs of ancient kings. Folklore, fairytale, and fantasy everywhere have been enriched by grave-robbers. The successful plundering of the tombs of the Pharaohs has fertilized the magical legends of Aladdin and Ali Baba and their cognates in many lands.[75] Such also was the case in China, where medieval tales tell of richly furnished palaces, like those of ancient kings, accessible usually with supernatural help through curious tunnels and grottoes.

Sometimes these subterranean worlds opened in places visible only to the inner eye of a true initiate. They had no crudely perceptible limestone cavern for entrance. An example could once be found under the Altar to the God-king of Heaven (*T'ien ti t'an*) on the Taoist sacred mountain and cult center Mao Shan: it was the centermost window of such a grotto heaven.[76]

To sum up: underground caverns, whether natural or artificial, were never simple. Their roles were related to the purpose, the hope, the training, or the destiny of their visitors. They could appear as sumptuously jewelled burial chambers like that built by Ch'in Shih Huang Ti, or the refined grave-palace of Chao T'o at Canton;[77] they might be recognized as spiritual kingdoms presided over by Taoist gods and equipped with complete starry domes; or they might serve as ideal retreats where shamans or hermits held converse with spirit-beings. An example of the latter vision is provided among the long series of quatrains, written by Ts'ao T'ang at the end of the ninth century, each of which opens a little window on some part of the transcendental world:

> Inside the grotto the mists and auroral clouds never cease to appear;
> Within the grotto—a Heaven and earth well suited for golden polypore.
> The moon shines on, clear and bright on a tree at the head of a gorge,
> Where white-haired old men face each other over a game of go.[78]

(The "golden polypore" is a divine variety of the sacred pore fungus, one of the rare plants traditionally held almost equal to mineral drugs in its magical powers. The last line of the poem is reminiscent of scenes of winged figures at gaming boards familiar in Han art.)

All of these versions of globe-universes were identical at bottom. Their true nature was withheld from the profane, but they held the meaning of life and death within them. They might be revealed in part to adepts but the full revelation of their nature must be awaited with patience:

> In what place shall I search for the Mystery's solution?
> Among men there are the Grotto Heavens:
> All my zealous activity is for the Tao,
> And, ostracized here below, I am still a Transcendent!

> Occluding the Phosphors, I shall ride a vermilion phoenix.
> Pushing back the Void, I shall harness the violet mist,
> And not offended by the arrogance of the lackey at the Garden,
> I am willing to wait in front of the Jade Sanctum.[79]

(My "I" here is purely gratuitous. The persona of the poem could be any-

253

one.) This double stanza seems to represent a humble seeker for ultimate wisdom. He has been banished from the inner sanctum and denied access to the great mysteries for some subtle fault; but he is still zealous, still devoted: sent down from the sky dwellings where he once resided, he can find his way once more to heaven by way of an underground tunnel, and will sit patiently at the entrance to the garden of the gods until judged worthy of readmission. Like him, many of us may have had the worldly terminus of our mystic journey in sight, only to see it dissolve in the Nothingness of the Absolute. The stony mountains of earth are all about us—but we know well that they are no more than opaque encrustations which conceal the interior of the cosmic geode from vulgar eyes. We still yearn for the crystal spires within.

MINERALS OF PARADISE

Thus, cavern'd round in CRACOW'S
    mighty mines,
With crystal walls a gorgeous city shines;
Scoop'd in the briny rock long streets
    extend
Their hoary course, and glittering domes
    ascend;
Down the bright steeps, emerging
    into day,
Impetuous fountains burst their
    headlong way,
O'er milk-white vales in ivory channels
    spread,
And wondering seek their subterraneous
    bed.
Form'd in pellucid salt with chisel nice,
The pale lamp glimmering through the
    sculptured ice . . .
Far gleaming o'er the town transparent
    fanes
Rear their white towers, and wave their
    golden vanes;
Long lines of lustres pour their trembling
    rays,
And the bright vault returns the mingled
    glaze.

Erasmus Darwin, "The Economy of Vegetation,"
The Botanic Garden, II, 125-142

Northrop Frye has shown the mineral world as a mine of fundamental symbols for the world of myth: above all it yields the vision of the ideal city, the perfect palace or temple, glowing like a precious stone.[80] The universality

254

of the metaphor is inherent in the nature of the mineral world itself, in which the elemental form is the crystal—a disposition of atoms in patterned lattices around planes and centers of symmetry. The world of minerals is a world of pinacoids and prisms ranging in complexity from the simple cubes and octahedrons that manifest themselves in native copper and fluorite, up through the complex bipyramids of sapphire and the trigonal alps of quartz to the radiating clusters of green dioptase. These are the building blocks of Ratnapura, the City of Gems, or the microcosms in which are reflected the houses of the gods. All is perfection in them—and therefore all is changeless and endless. They represent the eternal inorganic universe, not subject to the tremors and tumors that mean change and growth in the world of organic life.

In Chinese literary imagery the sky as a whole, the remote paradises of the gods, and even holy tracts on earth appear in mineral form. For instance, in a very widely held Taoist tradition, the uppermost of the three spiritual spheres—even above the zones of Highest Clarity and Grand Clarity was the realm of Jade Clarity (*yü ch'ing*), attainable only by the most highly evolved of transcendent beings.[81] Its name reveals its nature: "Jade Purity is like jade in that it is so tough as to be indestructible, and so pure as to be impollutable."[82] In poetry the wide roof of heaven is jade, swept by snow crystals;[83] but even the land consecrated for a Buddhist temple is "a precinct of crystal,"[84] and the pure lawns of a Buddhist paradise may be reflected in an earthly language on a "berylline" or "glassy" frontier.[85] A divine world, parallel to our own, might be mineralized from top to bottom: " 'fire germ' was suspended to make a sun, with carved black jade to make the crow; 'water germ' was used to make a moon, with blue turquoise to make the toad and hare."[86] ("Water germ" is our rock crystal, but we can only surmise the identity of the carbuncle represented by the expression "fire germ"—a ruby or pyrope perhaps?)

If the very skies and their inhabitants could be crystalline, it is not to be wondered at that the gardens and trees and flowers of holy places should be crystalline too. It is as if the realm which breeds the sky-blue prisms of celestine[87] or the blue quartz crystals once called siderite, or star sapphire —also called asterias—provides also for the existence of pseudo-organic forms like moss agate, "desert roses" of gypsum, arborescent gold, dendritic silver, and mammillary malachite. Indeed the earthly presence of these sparkling illusions of a kind of divine protoplasm must have furnished the minds of the early Chinese with ample raw material from which to create the magic palaces of Taoism and Buddhism, with much to spare. Some writers, certainly, exploited these resources more fully than others. The tenth-century Buddhist prelate Kuan-hsiu, celebrated for his poems and paintings alike, was an outstanding creator of word-paradises based on the metaphor or

255

actuality of mineral worlds of the spirit. He set his paradises—or some of them—on silver land whose trees bore gemmy fruits and flowers that chimed attractively in the wind. He furnished this dazzling realm with objects of crystal, coral, red salt, ice, sulphur, amber, diamond, orpiment, cinnabar, pearl, jade, and glass in many colors.[88]

What was true in fact or as symbol of the secret worlds of the gods was also true in the fancy of the parks and pleasaunces of holy abbots and almost deified kings in the world of T'ang. Kuan-hsiu himself—as did other writers—composed a birthday poem about the garden of the King of Shu, in which we are not surprised to find a golden tree with jade foliage.[89] But there was little to distinguish the ideal of a royal palace from the conception of a paradise hovering on a cloud over the sea. A self-contained mineral cosmos seems to our short vision as unlikely in either place. Wei Ying-wu put a crimson coral tree on the fairy sea-mount of P'eng-lai; Kuan-hsiu furrowed one of his nacreous Edens with sandy-bottomed golden canals; Ts'ao T'ang reveals dainty lotus-lands strangely built of such volatile and soluble minerals as quicksilver and white alum—the reagents of Taoist wonderworkers and the very stuff of their strangely mutated elixirs, glittering like tiny crystalline landscapes in newly opened crucibles.[90]

The kiosks, pavilions, and reception halls that represented humanoid activity in these splendid lands were, if possible, even more wonderful than the groves and gardens in which they were placed. Palaces, like temples, are not only microcosms, but also miniature, idealized mountains, their prototypes seamed with precious metals and lodes of gem-bearing pegmatites. The pseudo-mountain, however, could be more homogeneous: Kuan-hsiu wrote unambiguously, "The mountain is a palace of water germ (rock crystal)."[91] The palaces in the hidden bubbles deep within the sacred hills of the Taoists were, in a sense, made from the raw materials close at hand—as accessible to the subterranean builders of China as to the deep-delving gnomes of medieval Europe.[92] Divine buildings tended to be constructed of metals. (It must be remembered that metals are crystalline too, even though comparatively few men have troubled to inspect the cubes of native gold, the octahedrons of native silver, or the dodecahedrons of native copper.) And although massive minerals like marble, jade, and chalcedony, whose crystalline structure is rarely obvious, were sometimes made into walls, floors, and roofbeams in these happy halls, patently crystalline minerals, such as quartz and garnet, went into their ornamentation, gauds, and objects of vertu.

So Kuan-hsiu invested the buildings of paradise—we are hard put to distinguish Taoist from Buddhist in his poems—with the charm of old bronze and the mystery of incorruptible gold.[93] Even Lo Pin-wang, three centuries before him, was familiar with inorganic architecture, and pictures a vast Taoist palace with buildings of silver and watch towers of gold.[94] These two writers represent the whole three centuries of the T'ang dynasty—bracketing innumerable other poets who conjured up such metallic visions, if not always so skilfully as they.

In such imperishable surroundings only immortal beings could dwell. Where there is no change there is no death. Or if there was change, it was imperceptible and death a remote shadow at the end of inconceivable eons. As I once wrote elsewhere, such a mineral world displays at worst "the quiet and slow fluidity of petrological process. [It is] a world which shares the qualities of such fluid materials as quicksilver and the soft gold of the alchemists . . . the immensely slow life of rocks."[95]

The stars themselves were palaces and had the gemlike quality appropriate to such sources of sparkling light: "The stars are the palaces and homes of many devas." So says a standard source of Buddhist wisdom transformed into Chinese concepts,[96] and a poem of the early eighth century tells of the "radiant beams of the serried star palaces."[97] All were flashing diamonds in the sky. The Five Planets, certainly the abodes of powerful spirits, were properly conceived as jewels of well-defined species. A Taoist work of unknown date, and possibly therefore post-T'ang, advises the adept to attune himself to the cosmos by evoking a picture of Jupiter as a blue gem in his left hand, Venus as a white gem in his right hand, Mars on the top of his head as "a fire orb," Saturn in his heart in the form of yellow gold, and Mercury as "black jade" (normally a term for jet) under his feet.[98]

Especially distinguished among the changeless mansions was the crystal palace of the moon. We have already observed the gemmy character of its inhabitants: a turquoise (as it may be) hare looked up from its floating disc,[99] and a toad of nephrite squatted sedately on its surface.[100] Its special character was its accessibility to persons of no great spiritual distinction. To some extent the same condescension was allowed by the great sky river, whose ageless foam sparkles on the beaches of Heaven.

THE SKY RIVER

Wynken, Blynken, and Nod one night
  Sailed off in a wooden shoe—
Sailed on a river of crystal light
  Into a sea of dew.
  Eugene Field, "Wynken, Blynken, and Nod"

The glimmering tattered ribbon visible high in the night sky from places free of neon lights and layers of smog is better known by name than in actuality

257

in the western world. We call it The Milky Way. An early version of the standard tale that purports to explain its origin may be found in a catalogue of constellations attributed to the Alexandrian astronomer Eratosthenes. It tells how baby Herakles, kindly suckled by Hera (all unaware that her husband Zeus had fathered the sturdy brat on Alkmene), nipped the divine nipple so lustily that a jet of milk spurted across the sky. There it remains hypostatized as visible plasma, rich in hydrogen atoms.[101] The edifying tale—a warning to queenly wet-nurses—has been reornamented endlessly in literature, from Eratosthenes to Disraeli, by way of Rabelais and others, and splendidly depicted by Tintoretto. The pleasant details of this remarkable history may be read elsewhere.[102]

The Chinese took a rather different and, it must be admitted, a more refined view of the origin of the misty light. For them it was a kind of crystallized water—a celestial complement of the great rivers of the Middle Kingdom—retaining the fluidity of its earthly counterparts the River Ho ("the Yellow River") and the River Han. This divine, subtle, nebulous, quintessential liquid winds its way mysteriously along the surface of the sky dome. Originally, like all of the stars, it was the finest product congealed out of the Primal Pneuma at the beginning of time.[103] Some ancient sources identify the Sky River, equally or alternately, with metaphysical Water, with the quintessential germ of Metal.[104] This association seems partly based on its whiteness which, in "Five Activity" theory, puts it under the metaphysical rubric of "Metal," and partly because of its silvery aspect, congenial to other beliefs in the ultimate mineralogical character of all entities above the sublunary world.

Both water and silver figure in the many traditional names of the galaxy. It also had its nebulous aspect, and so was called "Cloudy Ho." For its affinity with the stars, it was also styled "Starry Han." Prominent among its attributes were, appropriately, "luminous," and "heavenly," and "of the sky." The name "Silver Han," or "Silver Ho," was seasonal. It was deemed suitable after the retreat of the damp summer monsoon and before the advent of dusty winter monsoon, that is, during the early months of autumn when the clean, clear sky permits high luminescence to the Sky River. It was especially fitting that the seventh month of the year should be the very time when the shining sky waters, "unruffled by windy waves," to use the words of Tu Fu,[105] made easy the annual transit and reunion of the divine star lovers, parted from their homes in Vega and Altair, to the endless delight of earthly observers.

Although the Sky River was passable in season, it also had its permanent ford, marked by Deneb and other prominent stars in Cygnus. Accordingly, the focus of spiritual power at this crucial point controlled the des-

258

tinies of the fords and bridges of the great rivers of China, and in its own sphere, "it was the means whereby the deities communicate with the Four Quarters."[106] This high crossing had its counterpart below in a celebrated bridge over the Lo River at the Eastern Capital—Bridge of Heaven's Ford (*T'ien chin chiao*). It is at its most natural when, as portrayed in an evening vision by Meng Chiao, it crosses a river of ice, the earthly equivalent of the crystalline sky river. Human beings disappear, but a holy mountain—a residence of the gods—appears in the distance:

> Under the Bridge of Heaven's Ford the ice begins to form;
> From over the lanes of Lo-yang human activity breaks off.
> Dreary desolation sounds through elms and willows, between high-decked
>   halls and galleries;
> But by the moon's glow—a direct view of Mount Sung's snow![107]

Indeed, the Sky River, not unexpectedly, had spiritual connections with water of every sort. It might even signal the prospects of precipitation to watchful astrologers: "When its stars are abundant, water is abundant, and when they are few, we have drought."[108] The importance of this interaction was recognized, as was necessary, by the divine sovereigns of China. Accordingly, in 100 B.C., not long after the Han court had adopted the custom of declaring the inauguration of new eras and cycles whose names magically induced desirable results in the realm, the new era "Sky Han" was proclaimed, to bring an end to a long period of drought.[109] The most potent source of the divine water was a kind of hydraulic nucleus near Aldebaran and "the rainy Hyades"—that is, the lunar lodging Net. When the moon—a great energizer of celestial activity—approached this center of concentration, it stimulated a buildup of the divine fluid, and consequent leakage on the fields of earth, as was repeatedly noted by the poets of T'ang.[110] Other forms of precipitation had their origin in the Sky River, too. It generated both the dew of morning and also "the mystic frost" of heaven.[111] So the divine river blessed all of the plants of earth with its life-giving water:

> Lustrous and clear the Starry Ho comes down,
> Wets the lichens and also sprinkles the pines.
> It fades the floating herbs—so bright its moist light;
> It pales the quaking peaks—so halcyon blue its stone.
>
> At this country friary—moon blended in clouds,
> In these autumn walls—bells from the clepsydra;
> I know milord is intimate with this remote place:
> He is seldom to be met on the Nine Streets of Town.[112]

This poem by the ninth-century poet Ma Tai tells of a meeting with a friend in a Taoist friary, appropriately named "Friary of the Quelling Star," that is, of Saturn. The stars clearly made their influence felt in such a holy spot, much more than in most places. (A note: the "halcyon blue stone" must, of course, be the dark blue dome of the lapis lazuli sky against which the starry river shines.)

On a larger scale, the Sky River was, as has been noted, intimately connected with the rivers, seas, and lakes of earth. The interaction was not confined to the two great rivers, the Han and the Ho, for which the heavenly stream was named. The great circumambient ocean itself received its waters. Men had long surmised that where the sky river passed beneath the two horizons its numinous waters fed the hidden springs of the seas, and, more particularly, where it divided below Vega and Altair to begin its downward flow, the surplus added to the already brimming ocean produced the phenomena we style the tides.[113]

In short, the Sky River was a cosmological fixture of the utmost importance, and occupied a central position in the archaic view of the world. The story was told that when, long ago, Ch'in Shih Huang Ti laid out his capital as a microcosmic reflection of the supernal world, he based it on the structure of the purple palace around the pole star, and diverted the Wei River through his divine city as an image of the Sky Ho. He built a bridge over that water as a shadowy replica of the causeway that allowed the annual meeting of Cowboy Star and Weaver Star.[114]

The supreme significance of the Sky River was celebrated in literature as well as in fanciful architecture. An example which may serve for the whole genre has survived from the pen of Hsieh Yen, a learned gentleman of the early seventh century. This is a rhapsodic treatment of the Sky River, which he titled *Ming Ho fu* "Rhapsody on the Luminous Ho."[115] It flows somewhat more freely than others of its ilk, but the seams between its segments do not always fit together as neatly as we might expect, although we could, if we chose, attribute these unheralded transitions to poetic genius rather than slipshod technique. For instance, about midway through the poem (it is baffling that so many scholars still refuse to admit that a *fu* is a poem) the spirit of Vega, the protector goddess of the River, appears as the faintly discernible image of a beautiful woman, composed, as in a hallucination, out of fragments of the sky and air. She is not plainly identified as the familiar weaving maiden, however, and, after her brief and dreamlike appearance, she vanishes, leaving the scene entirely to a brilliant but impersonal display of the glittering colure for which the poem is named. Disjointed or not, the piece seems to me to be a competent and sometimes a beautiful composition, and therefore worth rendering in full, although I cannot come close to a faithful reproduction of its fine figures and elegant patterning. Indeed, in a

260

few places, I doubt my own judgment about the best interpretation of the words' intent:

> The moon begins its return in the *yang*-light of evening,
> The sun makes its nightly plunge down the Mainstays of Heaven.
> Afoot on the courtyard flags, on a leisurely walk,
> Conscious of the empty expanses of the cloudy empyrean,
> Where airy shapes in a myriad varieties stand all aligned,
>> farflung along the Starry Ho—
> The whole road brilliantly alight:
> The Numinous Elders netted in the Silver Han—
> Alone I watch.
>> Splendent, oh! as if burnished,
>> Immaculate, oh! as if polished.
> When luminous moon shines on it, it does not lose its silky whiteness,
> When whirling wind excites it, it does not raise up waves.
> No one plumbs its depths:
>> It contains the Four Pneumas from the far edge of the sky.
> No one measures its distance:
>> It overshadows the many rivers down among men.
> And when it happens that
>> As the year enters upon Triple Autumn,
>> Its conatus is set straight across a thousand *li*.
> [Then by]
> Measuring dragon cart for easy sweep,
> [Or]
> Framing magpie bridge for zig-zag link,
> [We can see]
> Made up with rosy dawn-light, dimpled with stars:
>> We know that Minx Woman cannot be compared to her.
> With every sort of bangle and shining bauble,
>> She presents the certain likeness of Heng-o.
> Seventh Eve:
>> Devise it with all manner of well-found methods,
>> And, even more, emulate the solidly talented—
>> Still its flow will not be gauged,
>> Its far reach will not be measured.
> None can discern its tip or terminus—
> No more can one know whether it is dense or thin.
> And as to its modesty:
>> When Grand Yang is radiant it does not contest that light.
> Though, as to its virtue:
>> If the huge sea should wither it will in no way be drained.
> It wrests dense massiness from the congealed frost;
> It borrows bright hardness from the white stone.
> It stays aloft with hazard,
> Embedded in empty nothing it is self-preserved.

Its name connects it with the veins of earth,
Its reflection is dispersed among the heavenly signs.
It is just on the clearest nights that it is most radiant,
But with the congealing of faint rain it briefly glimmers.
Its luminous whiteness deserves praise:
   It is as frosty-white as shimmering silk,
So straight and flat that it could have been spun:
   Yet also piled vertically like racked-up clouds.
Hung across the Purple Pole so that it rotates on a slant,
It spans the cyan void dividing it in twain.
It emits the empyreal light in flowing waves,
It encloses faint-glowing hue in swollen mists.
If only I could ask about it of the raft visitor—
How I wish that it could be judged for me by Milord Yen!

I leave the interpretation of most of these lines to the erudition or imagination of my readers, but must comment on a few of them.

1. "Mainstay" (*kang*)—these are the stellar braces of the cosmos. Compare the story of Er the Armenian in Plato's *Republic*.[116] The narrator died, was resurrected, and told the tale of his visit to the other world. He saw a stream of souls stretching from heaven to earth: "this light binds the heavens, holding together all the revolving firmament like the undergirths of a ship of war." The traditional Chinese vocabulary of basic features of the cosmos (and also of the political and social order of this world, which derives from it), is full of images from the language of threads, textiles, weaving, cords, and nets. Too little attention has been paid to this subject by philologists.
2. I have explained the use of English "Minx" as the equivalent of *wu* elsewhere.[117] This starry charmer, sometimes called Woman Star, is, as we have seen in earlier chapters, often the companion or peer of the moon goddess or of the Weaver Maid in poetry.
3. "When the huge sea withers . . ." This verse refers to the putative connection between the Star River and the world ocean. The waters of the divine river are inexhaustible, even if those of Ocean are not.
4. "Its name connects . . ." That is to say, its name is "River," which points up its connection with the waters that pulse under the earth.
5. "Purple Pole" refers to the powerful circumpolar asterisms, within the Wall of Purple Tenuity.
6. "Raft visitor" refers to the tale of a man whose raft carried him up to the Sky River. We shall tell his story soon.
7. "Milord Yen" is Yen Ling, the intimate of a Later Han emperor. His counterpart appeared in the sky as a nova or "visitor star."[118]

SKY RAFTS

His soul proud Science never taught to
   stray
Far as the solar Walk or milky way.

Alexander Pope, *Essay on Man*

Flight Beyond the World

"Han Water has from the first communicated with the flow of Starry Han."
So wrote Li Po, expressing something more than a pretty myth. His great
contemporary Tu Fu, ferrying the River Kiang in Szechwan, imagined with
an abundance of classical support for his fancy that he saw the further end
of the river merging its waters with those of the Cloudy Han.[119] Since sky
river and earth rivers were not merely mirror images of each other, but were
actually joined—the subcelestial water being rarefied by its ascent, and the
heavenly water condensed to a mere compound of hydrogen and oxygen by
its descent, just as a star was petrified when it fell to earth—it followed that
means existed whereby, either through luck or magic, an adventurer might
sometimes ride upwards into the fields of the sky: "The house of the
weaving woman—it may be discovered by an envoy from a Han household;
the ford of the ox-waterer—it is easily observed by a man out on the sea."[120]

The "sky raft," the normal vehicle of these explorers, appears frequently in early medieval literature under a variety of names. It may be a
"star raft" (*hsing ch'a*), shining as brilliantly as the star palaces among
which it floats;[121] it may be a "dipper-invading raft" (*fan tou ch'a*);[122]
frequently it is a "numinous raft" (*ling ch'a*);[123] or it is a "sylph's raft"
(*hsien ch'a*) that bears the immortal spirits to their mysterious harbors;[124] it
is also the "raft of the eighth month" (*pa yüeh ch'a*) that crosses the clear
Ford of the Stars in mid-autumn.[125] This is the season when the sky raft
usually appeared to mortal eyes, and also the time of the sky-girl (Vega)
festival, celebrated on the seventh evening of the seventh month. The legends about the raft may be related to the presence of the asterism Sky Raft
(*t'ien fu*, using another word for a simpler sort of raft) on the celestial
equator, near the Herd Boy. That raft is θ Aquilae, and it does indeed shine
high in the sky on autumn nights, as it floats on the great Sky River.[126]

The prototype of all of these fairy rafts was the one which carried the
Great Yao around the primitive skies above his inchoate realm:

In the thirtieth year of Yao's ascent of the See, a huge raft came floating on the
Western Sea. On the raft was a light which shone by night and was extinguished by
day. Men out on the sea saw this light in the distance, now great, now small, like the
coming and going of a star or the moon. The raft floats around the Four Seas, making
a regular circuit of the sky each twelve years. After one circuit it starts up again. Its
name is "Moon-threaded Raft" (*kuan yüeh ch'a*); it is also called "Star-hung Raft"
(*kua hsing ch'a*). The Plumed People nest and rest on it.[127]

(The twelve-year circuit corresponds, of course, to the Jovian year. By implication Yao is identified as the spirit of the clock-planet Jupiter, cruising
solemnly through the twelve signs of the zodiac. Evidently he permits occasional passage to transcendent beings, dancing their unknowable sarabands
as they visit among the sidereal villas.) The journeys made possible by this
remarkable flow had its appeal for a certain kind of tourist, although it seems

to have been mere chance that would bring a hopeful raftsman into the sky. Sometimes the transit upwards took place from the Pacific Ocean, sometimes from the western source of one of the great rivers of China. The latter route provided access to some ancient and anonymous Speke or Burton, looking for the source of the Yellow River, supposed to be in the highlands of northeastern Tibet:

He saw a woman washing silk gauze. When he asked her about it she said, "This is the Sky Ho." Then she gave him a stone, which he took back with him. He inquired of Yen Chün-p'ing, who said, "This is a stone that propped the loom of the Weaving Woman."[128]

(Yen Chün-p'ing was a celebrated astrologer of the first century A.D., and a connoisseur or appraiser of celestial rarities. The story, of course, is much later than that.) An almost identical version of this tale has survived to our times and was well-known to the men of T'ang. In this, the hardy adventurer was the famous explorer Chang Ch'ien—an obvious enough protagonist since he had speculated about the source of the Yellow River. His informant, however, was not Yen Chün-p'ing but the even more famous Tung-fang Shuo.[129]

Li Shang-yin, whom we tend to think of unjustly as a poet of boudoirs, tinkling lutes, and satin hangings, alludes to a similar stone wedge in his quatrain entitled "Visiting from the Sea." The poet offers such a marvelous stone to a guest from overseas, not only as a valuable souvenir, but to remove from himself the onus of bearing the jealousy of the Weaver Maid's star-lover. The whole thing is immensely flattering to the doubtless delighted foreigner.

> Overseas visitor, riding on your raft up through the purple vapors,
> Did the Star Fairy stop her weaving once she heard you?
>
> You must not quail at Ox-hauler's jealousy;
> I shall use this stone which propped her loom as a gift
>   to you, milord![130]

Curious stone wedges such as this suggest some sort of Neolithic artifact—perhaps a variety of polished stone adze, thought to have a celestial origin, as did the meteoric axes of the thunder gods. A rare gift indeed!

The theme of the divine stone from the shore of the Sky River occurs repeatedly in T'ang poetry under a number of guises—almost always with overtones of rare adventure, or tall tales, or handsome gifts received. The "sylph's stone" (hsien shih) or "curious stone" (ch'i shih) is frequently an enigmatic and perhaps slightly troublesome artifact fallen or sent from among the stars. Think of the slab in the film "2001." It ought to be incised with runes to make its full purpose clear—but there are none.[131] Perhaps not all are props from galactic looms; some may be fragments of the blue firmament; some are only scarily unidentifiable.

The archetypal tour of the sky by a superhuman also furnished a neat pattern for symbolic descriptions of royal journeys through the Chinese realm. Li Chiao had described one in a poem he calls simply "Stars":

A numinous raft turns away from the County of Shu;
The treasure swords are renewed at the City of Feng.
A leader of armies approaches the northern bastions.
A son of heaven enters into Western Ch'in.

No longer to pose as the Three Eminences or as the Stabilizers,
Better that we play the five aging vassals;
Through this night—the Song of Ying River!
Who would recognize in us a gathering of statesmen?[132]

All the imagery here is stellar, and could be satisfactorily treated as the inauguration of an epoch of peace revealed in the sky. But sky and earth interact, and this is the kind of poem that requires a commentary.

1. Possibly this refers to the peace established after prolonged wars with the Tibetans, and the marriage of the Chinese princess Chin-ch'eng kung-chu to King Srong-tsan-sgam-po in 641.[133]
2. The reference is to the famous swords of Lei Huan whose potent purple glow hangs benignly over the south.
3. This line probably refers to the pacification of the Eastern and Western Turks between A.D. 630 and 648, making the northern frontiers safe for the men of T'ang. The stellar commander of the victorious Chinese arms may well be the asterism Heaven's Great Army Leader (*t'ien ta chiang chün*), in Andromeda.
4. The allusion is apparently to T'ai Tsung's defeat of his rival Hsüeh Chü, Prince of Western Ch'in, and his son, early in the seventh century. There may be a simultaneous reference to the Ch'in asterism in the Wall of the Sky Market (δ Serpentis). Through this first quatrain, then, we seem to have an astrological progress, symbolizing the victories of T'ang T'ai Tsung in each of the Four Quarters.
5. Now that peace has been established, and the throne made safe, five great ministers, whose policies have presumably made these successes possible, can celebrate at leisure, like ordinary citizens. The Three Eminences (in the feet of the Great Bear) and the Two Stabilizers (in Draco—not the same as the Stabilizer adjacent to Mizar) are the props and supports of the heavenly palace, and the wise men who direct the activities of the ruling family and its armies.
6. These five old men appear also in ancient legend as the essences of the Five Planets in mortal form. Yao pointed them out to Shun, and they are said to have appeared at the birth of Confucius.[134]
7. How this song is appropriate escapes me.
8. Self-explanatory.

Pacing the Void

In modern times, the celestial Raft of Empire became the flagship of the impressive fleet which Cheng Ho steered into the Indian Ocean early in the fifteenth century to establish a transitory thalassocracy among the states of tropical Asia. The eunuch-admiral represented the holy Yao making a circuit of "All under Heaven," and also the more historical Chang Ch'ien, who took the emblems of Chinese sovereignty to the barbarians of the west. It was appropriate then that one of Cheng Ho's officers entitled his account of the expedition "Surpassing Vista from the Raft of Stars."[135]

On the subtler plane of hope, expectation, and dream—coupled with their opposites of frustration and disappointment—the sky raft provided a convenient image of the journey to a happier world. Like the flight of the swan and the wild goose, it was invoked in poetry in situations of melancholy and despair. Ch'u Kuang-hsi, sailing on the Yellow River, is filled with sadness as he watches the moon set:

If only I might chance to meet that raft-riding stranger
Who would talk endlessly of his voyage on the Starry Han.[136]

A better river, leading to a cleaner world—and that is also the thought of the travelling salesman, oppressed by his separation from his family, his meaningless movements, his sense of failure:

Following the wind, chasing wild waves—parted year after year—
But how I would laugh if I were on time for the Raft of the
Eighth Month![137]

For some writers, though, the sky raft is at once more mysterious and more natural than these examples suggest. It is an element in a sacred landscape. For instance, an anonymous T'ang poet tells of a monastery in a dark, coniferous forest. Imperceptibly the manmade structure blends with its organic setting, and finally merges with the sky and its lights. The poem concludes with this couplet:

The tall pines brush the Starry Han—
Each and every one—a raft for sylphs![138]

Somehow the tall trees have slid into the sky river, each to float off into eternity with its cargo of divine passengers.

A fu-rhapsody of the T'ang period has preserved for us—in the style characteristic of that genre—a richly ornamented summary of the lore of the sky raft. The author is K'o-p'in Yü, a man of the late eighth century.[139] The magic boat floats silently and mysteriously from an unknown embarkation to an unknown destination:

Vehicle fading afar, ah! none knows its movements;
Route through a watery waste, ah! leaving no trace behind.

266

Man with wood, ah! floating together,
Sky and sea, ah! some hue of cyan.
Stage by stage through crystalline flashes of Yellow Road,
Threading the vivid glitter of White Elms.
It is not recorded whence it came,
Nor can one know whither it goes.

(The Yellow Road is the ecliptic—the segmented route of the zodiac. "White Elms" is an ancient trope for "stars," suggesting the spirit-like glimmer of the elms of the sub-arctic forests.) Raft and passenger are indispensible to one another; the raft gives passage, the man gives purpose:

The traveller, without the raft, would only strive and toil,
    while his business could in no way be achieved;
The raft, lacking traveller, might go and come, while no one
    in this world would know of it.

The lone voyager in a celestial boat became a stock image—or rather, a set of stock images—in poetry. By T'ang times he was a poetic persona of considerably greater distinction than his ethereal raft. His story is probably very ancient indeed, but we know it mainly from the Six Dynasties period. Here it is in an early medieval version that probably retains its third-century form in most respects:

An old saying tells that the Sky Ho is connected with the sea. In recent times a certain man lived on a sea isle. Year after year, in the eighth month, a raft floated in, and its arrival and departure never missed the date. The man, possessed by a strange ambition, installed a flying gallery on the raft, which he provided with an abundance of provisions. Then he boarded the raft and went away with it. During some ten days he could still observe the stars, moon, sun and planets, but after that they became obscure and indistinct and he was unaware of either day or night. On it went for ten days or so until it reached a place where there were the semblances of city walls, fortifications, houses, and other buildings—all most imposing. He could see, far off, a woman in a palace, and she was much engaged in weaving. He also saw a male person who was leading a cow and had stopped to water it. The man who led the cow asked him in surprise, "From what place have you come here?" Our man told him of his wish to make the trip, and asked in his own turn what place this might be. "Go back to the county of Shu, milord," was the reply, "and inquire of Yen Chün-p'ing—then you will understand." In the end he did not go ashore at all, and was returned on schedule. Later he went to Shu and enquired of Chün-p'ing who said, "On a certain day of the month that year there was a Stranger Star that trespassed on the constellation Ox-hauler." He calculated the year and the month—and it was exactly the time when this person had arrived on the Sky Ho![140]

The Cowboy and Weaver stars are readily identifiable. The word "trespassed"—an astronomical term used conventionally of occultations and planetary invasions of constellations—has shivery suggestions of an unwanted suitor in another man's mating territory. What concerns us here is the

Stranger Star—that is, the intrepid islander as seen from the surface of the earth. This epithet is the usual one given a new "fixed" star which appears where none had ever been seen before, that is, a nova or supernova.

The image of the brilliant new star in the sky—one that we might speak of as enjoying a meteoric career, although Stranger Star never refers to a meteor—had already been given a fixed identity in a classical tradition. The tale, in which apparently there was more than a germ of truth, tells how Yen Kuang, or Yen [Tzu-]Ling, whose adventure we are about to summarize, became indissolubly fused with the anonymous rafter whose strange story has just been narrated. It is a story of the Later Han period, and tells how Yen Kuang, a crony of Han Kuang Wu Ti, fell asleep with a leg lying across his sovereign's body—a clear case of *lèse majesté*. A busybody of an astrologer saw a relationship with the contemporary appearance of a nova in the asterism *Ti tso* "Seat of the Divine King," an obviously angry sign from on high. The disrespectful Yen should have suffered a nasty fate—but he was forgiven, and retired to a life of idle fishing. *Ti tso* (our α Herculis, a variable double star, "orange-red and green") was evidently a star to be trifled with. In any case, the trope "Stranger Star" became the constant companion, not only of "Star Raft," but of "Theocrat's Seat," and a stereotyped allusion to every kind of intrusion of a common man among his betters—an excellent and sometimes jocular figure for poets who, with assumed modesty, adopted it for their own when inspired to write about their own humble advent among a group of courtiers or men of letters.[141] At the same time, the tale provided the easily exploitable image of a hermit living in a pure mountain forest, away from courts and courtiers. This idea was repeatedly alluded to by such phrases as "fishing in the shoals" (since Yen Kuang dedicated himself to angling in shoal water by Fu Ch'un Mountain in Chekiang), or casting a line from Yen Kuang's "fishing terrace"—a name given to a rock in this region. Here is a simple quatrain by Li Po which treats the traditional subject unambiguously:

> Yen Ling roves no more in the company of a myriad carriages;
> Back he goes to lie in the empty hills and fish the dark blue flow.
> Henceforth the Visitor Star takes leave of the God-King's See:
> He has never been "Grand White" who gets drunk in Yang-chou.[142]

There is undoubtedly some personal reference here to a high impeachment officer named Ts'ui, to whom the poem is dedicated, which evades us now. It appears, however, that the magnate is advised to seek the remote silences, while the poet, figured as Venus (Grand White) takes his exile in the fleshpots of a big commercial city.[143]

The poetic treatment of the transmogrification of men into stars, whether new stars or old, varies widely and can be compared to the various ways in which the lonely but splendid goddess in the moon has been presented by poets. The persona of the shameless courtier can, for instance, be represented as prudent and content in exile, as in this poem written late in the ninth century by Wang Tsun:

Once he fished, bleak and chilled, out in the monotonous gloom—
A mandate flown from a former friend—off into the mountain's door;
At last—after both royal love and disfavor, disdaining lordly
   coach and coronet,
He lies high up among the five clouds—a "stranger star" even there.[144]

(The chill of outer space and the cold stars is, of course, transferred here into the atmosphere of the imperial court. The finale has the banished Yen Ling happily fishing among the five-colored auspicious clouds—a kind of earth-bound supernova. The "flying mandate" is an allusion to the urgent communications sent to the Taoist hermit T'ao Hung-ching by the sovereign of Liang.)

On the other hand, Lu Kuei-meng's treatment of the same subject in his "Early Passage" does not dwell on the quiet pleasures of a mountain retreat; it looks rather towards supernal bliss only to find an awful sense of alienation in the abyss of the stars:

Shall we admit that in the cyan void is limitless pleasure?
Yet the name or nickname of "Stranger Star" is joyless to men.[145]

(The poet, leaving by boat before dawn, sees the constellations reflected in the dark water, and is despondent at the prospect of his lonely journey.)

269

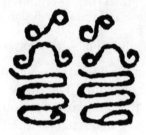

Wh' on sky, clouds, sea, earth, rocks
  doth rays disperse,
Stars, rainbows, pearls, fruits, diamonds
  pierce;
The world's eye, source of light, soul
  of the universe.

Edward Benlowes, "Theophila"

13 A Potpourri of Images

George Santayana, dismissing the notion that we find the starry sky beautiful or sublime because of such "scientific" notions as the plurality of habitable worlds and the vastness of empty space, or that we are struck with awe as we contemplate it as the habitation of the angels of theology or of the more abstract powers of astrology, has given us, in his customary fine style, the plain sensuous facts:

Now, the starry heavens are very happily designed to intensify the sensations on which their beauties must rest. In the first place, the continuum of space is broken into points, numerous enough to give the utmost idea of multiplicity, and yet so distinct and vivid that it is impossible not to remain aware of their individuality . . . In the second place the sensuous contrast of the dark background—blacker the clearer the night and the more stars we can see—with the palpitating fire of the stars themselves, could not be exceeded by any possible device. This material beauty adds incalculably, as we have already pointed out, to the inwardness and sublimity of the effect.

And again:

. . . whereas a star, if alone, because the multiplicity is lacking, makes a wholly different impression. The single star is tender, beautiful and mild; we can compare it to the humblest and sweetest of things.[1]

These observations surely underlie attitudes held by men generally about the stars, however modified by the molds of culture. The serenity of spirit suggested or generated by the clear night sky appears commonly in Chinese poetry, modified in many ways. Some poets, looking out over the great capital city from a high pagoda, are filled with peace and confidence at the sight of the crystalline constellations—a reaction only secondarily conditioned by the belief in the relation between the Northern Dipper and the Imperial Metropolis.[2] Others, for the same reasons in the end, see the soaring vault with its trains and chains of eternal gems as a sign of the tranquillity properly induced by Heaven's benign approval of the behavior of the Son of Heaven and his responsible agents.[3] The explanation is somewhat different; the source in the recesses of the human heart is the same.

The effects produced by the splendid sight of the star-studded blackness is, however, not the same as that suggested by individual and nameable components of the glittering hierarchy. Once a star is conceived as an individual with a unique appellation, or even as representing the class of which it is a member, it is differently conceived. Stars with labels lend themselves readily to metaphorical manipulation, and ultimately—and at worst—to petrifaction as prefabricated images and frozen clichés, whose employment seldom requires much imagination on the part of a writer, and whose lost luster can be restored only by the most ingenious of pens.

In what follows I shall present a mixed bag of common and commonplace stellar images, none of them really rare, without regard to such niceties as the distinction, say, between gems as metaphors for stars and stars as metaphors for gems. Moreover, in the nature of things, I shall be selective in the interests of simplicity. The reader of this study will have noticed that most of my examples from the field of poetry belong to "late T'ang"—that is, to writings of the late eighth, the ninth, and the early tenth centuries. (I use the expression "late T'ang" purely as a temporal label. I hope one day to dedicate myself to finishing off the traditional pejorative associations of such pigeon-holes as "waning T'ang," especially as a mode of describing the writing of the ninth and early tenth centuries.) One important implication of this treatment is that a significant feature of much of the poetry of this era must be partly concealed. This is the poetic device I shall style "image reflection." It might just as well be styled "microcosmic reiteration." It is characterized by the use, all at once, of a gamut or spectrum of images. If the voice of the deity is to be heard, it is echoed by gibbon, by chukar, and, mysteriously, out of the storm. If a goddess is present, she appears as a rainbow, as a wisp of pink sunrise cloud, as a wild azalea. The whole of an alchemical poem gleams with secret fires, molten minerals, and enigmatic smoke. The sense of moving from beginning to conclusion is more difficult to detect under this rich patina.

Let us look at an example. In the poem translated below there is no single moon metaphor; rather, we are in a linguistic hall of mirrors. Not only is there a built-in affinity among white clouds, magical asterisms along the ecliptic, and drifting disconsolate ghosts, but the moon itself is imaged in the precious toad, and again in the eye of the spring, and again in the hyaline lunar soul, and in the dewdrops, and in the disembodied "cool light" as well. The poem was written in the second half of the ninth century by Fang Kan:

> In cool empyrean, beyond haze and piled clouds,
> At Triple Five—the jade toad in autumn!
> The starry chronograms are correct in their serried fields;
> Set in the void, the ghosts and trolls are gloomy.
>
> Wellhead limpid—cold protopsyche crystalline;
> Dew dripping—chilly light floating.
> Until the snapping off of the green cinnamon
> I chant as I watch; I could not bear to rest.[4]

A few notes are in order: "Triple Five" is the fifteenth day of the lunar month, when the moon is full; in autumn it appears as a Harvest Moon or Hunter's Moon. The toad shows forth the cool translucent white of fine jade. The "starry chronograms," the ominous star patterns along the ecliptic through which the sun, moon, and planets pass, shine down on corresponding "fields" of influence below. The goblins seem out of place up in the void; although they are *yin* like the moon, they are creatures of darkness, but doubtless they are steering clear of that orb, confounded by the brightness of the lunar toad. The gushing spring is an earthly counterpart of the full moon, whose "cold protopsyche" is neither dark nor crescent, but fully realized. Similarly the drops of dew are miniature moons, whose great mother above is now represented only as a floating light. Until moonset, and the extinction of the cinnamon tree in the moon, the poet will sit raptly and sing. In short, we have here a medley of moon images: moon as toad, as embodied soul, as pure light with a corresponding series of sublunar reflections.

In the imagistic miscellany which follows, I shall try to avoid this sort of thing, and rather to compress the illustrative data into reasonable dimensions, shying away from multiple referents, and devoting myself to the presentation of one image at a time, denuded of pieces of its costume and divorced from its intimate associates.

Among the lights of the sky the moon is a prime favorite, and well noticed in the poetries of the world. Metaphors for the moon show a remarkable similarity across the boundaries of cultures—to no one's surprise. Indeed, "the history of poetry in all ages is the attempt to find new images for the moon."[5] In China there is a perennial linkage between moon, pearl, dewdrop, teardrop, and less perfect analogies between the moon and crystal balls, crystal or silver dishes, jade wheels, and whatnot. To say of the glittering costume of a dancing girl—in our tradition it would be studded with sequins—that

> The brocade gathers dew from on the flowers—
> These pearls attract the lymph from within the moon—[6]

makes a pleasant combination of unembarrassing stereotypes and suggests that the beads of her dress are miniature moons, and that she herself is an avatar of the Moon Goddess. But we cannot consider now every little moon image, or we shall never get on to the stars proper.[7]

Still, we cannot leave the moon without a look at the hallucinations induced by its physiognomy. The Chinese saw a human face there only rarely, and the face of the beloved—comparable to the full moon—is notably lacking in Chinese literature. Yet it was possible to identify one's beloved—like Omar-Fitzgerald's "moon of my delight"—with the moon, but it is an inconstant moon, not one "that knows no wane":

> I think of you, lord as the full moon—
> Night after night the splendor of your
> countenance diminishes.[8]

But this refers to no inconstancy of the soul, but to the gradual and inevitable erosion of the memory of a beloved face. On a less romantic level, Po Chü-i personified the moon as the constant companion of a traveller, each of them experiencing similar vicissitudes:

> Moving at dawn in the wake of a shrinking moon,
> Lodging at evening in the company of a new moon,
> Who can say that the moon lacks feelings—
> It has pursued me over the distance of a thousand miles![9]

So much now for the full moon. The crescent moon tends to be restricted to more material associations: it is a curtain-hook, or it is a nomad's compound bow:

> Many nights with the moon by the head of the wall:
> Curved and bent—like a drawn bow.[10]

273

This is a threatening moon, very different from the familiar, friendly full moon which reminds us constantly of home and loved ones.

The stars, whether fixed or restless, have a wider range of metaphorical availability than does the moon, since they have the general sensuous appeal described by Santayana, but usually lack the attributes of an individual personality, so uniquely displayed by the sun and moon. Perhaps the most obvious of the available tropes was the visual one of "star" and "fiery speck"—and therefore "spark." "Fire to cinders—a lonely star extinguished," wrote Yüan Chen on the death of the light in his brazier,[11] but this last spark, like the unstable spark of the firefly, also symbolized the light of the soul, and "fire stars" were one way in which the souls of the dead might show themselves. The new pyrotechnical arts also suggested the creation of stars from chemical reagents, and Lu Kuei-meng wrote of "starry bullets and swords of frost" along with performing elephants and tame rhinoceroses in the general context of "Miscellaneous Arts."[12] These were the magical creations of men, almost equal to those of the heroes of antiquity or of the gods themselves.

It is but a short remove from the fiery spark to the glittering flake. "Flying stars" are treated in poetry as flying snowflakes, as flower petals flying before the wind, as drifting fireflies, and as the frost crystals flying from the sky to the ground. They may even be the white tips of spume on wind-blown waves: "Heaven seems an empty river—the stars seem waves."[13] And indeed "shooting stars" regularly suggest flowing water (liu hsing and liu shui). All of these pretty thoughts derive not only from vision, but from the ontological affinity of stars with water and ice.

We are familiar with "starry eyes" and find them most appropriate in sparkling women, while for the Chinese they characterized divine women like the jade-faced goddess of Mount Hua, the West Marchmount, whose "star eyes" opened wide as she laughed.[14] But for the Chinese stars can be the flashing eyes of a bird of prey—by no means "starry-eyed" in our sense of that phrase. So the tenth-century poet Kao Yüeh combined two kinds of divine whiteness in his picture of a magnificent hunting goshawk: "Snow talons star pupils—rare in our age."[15]

Even glistening little fruits could be likened to stars. P'i Jih-hsiu wrote of the sourpeel tangerine that its skin was comparable to the surface of the sun and its "marrow" like a star.[16] Hsü Yin somewhat more aptly, we might think, thought it proper to describe the pearly fruit of the lichee enclosed in its scarlet husk as "Star pill in vermilion bullet—splendent as with sunlight."[17] This imagery is almost more lunar than stellar, more pearly than fiery, and contains an intellectual or imaginary element, suggesting a concept of the quality of star as we might see it close up as a crystal sphere, rather than as we actually see it—a dazzling point of light.

A Potpourri of Images

Sky-born and air-born are similar modes of being, and birds and stars had much in common—an affinity enhanced when the bird is as white as a snowflake in the distance:

> A myriad miles—lake and sky deep blue;
> A single star—flying egret white.
> At such a time I indulge all cherished hopes,
> Not averse to being a stranger adrift.[18]

Such a solitary sky-being could be a consolation, but the sight of a whole flock of winged stars tended to induce a sense of the cooperative action of creatures of light—simultaneously useful and decorative. So it seems in Li Tung's poem about a great Ch'ang-an monastery, where a flock of roosting cranes parallels a procession of saffron-robed monks:

> The cranes' lodging—they star a thousand trees;
> The monks' return—they fire the entire slope.[19]

The reader will have to judge for himself how compatible these lofty similes are with the use of the word "star" to describe (even in poetry) polished white objects such as polo balls. It was natural that the flying pellet should be compared with a specific breed of star, the shooting star: "hard and round, clean and smooth, acting the single star!"[20] This line was included in an impression of a polo match by the Ch'ang-an courtesan Yü Hsüan-chi. More specific are the ornamental lines of Hsü Yin, which transport a valiant polo-pony to the game-fields of the sky, to carry the luminous ball through the goal:

> Chasing to catch the white sun, to course up the Han in the blue;
> Taking a drifting star in his mouth, to pass into the
>   painted gate.[21]

Meteors, then, are often triumphant globules, with an aura of divinity. They suggest brilliance, urgency, and exalted station, all simultaneously. The image could be applied even to the splashing oars of a boat bearing a learned young gentleman to a destination of undoubted dignity:

> From western Kiang, the eastward gush is urgent;
> His solitary oar is like a gliding star.[22]

But the "gliding stars" are ready currency in many markets. If they fall to earth they suggest spiritual things turned to stone:

> Who could say that a heart—far parted—will not change:
> A star of heaven—fallen to earth—can become a stone.[23]

This metaphor is one of many that classify transient, impermanent, fleeting, changeable, restless things. The Chinese soldiers looking blindly into the

275

Gobi from their stone towers by the Jade Gate, hearing the whimpers of the restless spirits of dead predecessors, and seeing the flickering fires of the Hunnish chief on the horizon, naturally thought of will-o'-the-wisps and all such ghost-lights, and saw themselves transformed at no great date.

> Our bodies seem fleeting as stars—our relics like tumbleweeds.[24]

The intertwinings of impermanent location—at the mercy of stronger powers, whether physical or spiritual—allied shooting stars with all uses of the word *liu* "follow a current; drift with the tide; be carried willy-nilly, and so on." *Liu* was the word for penal exile, a fearful penalty for high crimes. The identity of meteors was as ambiguous as that of Chuang Chou's butterfly.[25] They are as undependable as the passions of men and their female objects are fickle: "First like the moon, steadily crossing the sky; in the end dashing off like a floating star."[26] Unsettled minds with uncertain prospects twist through the world like the Cloudy Han, or are carried away like "fire stars" by the unseen currents of the cosmos.[27] Sheer speed is itself meteoric: "The post-rider, like a star, comes by way of the gorge."[28] But government messengers have a special status, and so we have treated the matter of star messengers earlier in this book.

Stars may also function as elegant ornaments in the setting of an aristocratic feast—in the imagery adopted by the poet, a carouse of the immortals in a kind of crystal paradise:

> Through the fine night the lordly youths feast in the orchid hall;
> Their persons censed with viscous musk, beasts belching perfumes.
> Netted in cloud, golden dragons bite the painted candles,
> Netted in stars, a silver phoenix drains a rose-gem liquor.
>
> A tree of pearls that fills the screen opens on the spectacle of spring;
> The sounds of singing through a ballad whirl round turquoise beams;
> They on the mat—they still do not know of the dawn beyond
>   the curtained awnings:
> A nymph in blue, with lowered voice, points to the eastern quarter.[29]

A line-by-line commentary on this scene of elegant debauch, titled "Drinking at Night" and written in the ninth century by Ts'ao Sung, follows:

1. A night-long party in luxurious surroundings.
2. Braziers in the shape of animals emit gusts of incense, to add to the musk-perfumed clothes of the guests.
3. There are also sconces or candlesticks in the shape of dragons.
4. "Star nets" are common in poetry for any patterned array of stars. Apparently the expression refers to an arrangement of sparkling gemstones on a ewer shaped like a divine bird (examples from T'ang are known).
5. A large decorated screen dominated by a painted paradisical tree, gives the illusions of a changeless springtime setting.

A Potpourri of Images

6. The revellers, inspired, shout traditional ballads.
7. Their world is an artificial one, an illusory spring in paradise—but day is already breaking outside. They are unaware of the passage of the night.
8. A maid-servant whispers to the host that the sun is lighting up the eastern sky.

In set groups—i.e., as constellations—the stars were bereft of part of their natural sensuous imagery, and assumed the lower role of marking the outlines of a kind of Platonic idea realized in the sky. The Dipper, quite naturally, was for pouring wine: "I have not heard that the Northern Dipper was drained of Yao's wine."[30] This line, from a set of verses on ritual, in which the south wind vibrates the zither of Shun (an obvious precursor of the aeolian harp), is a little concerned with the beauty or virtue of the component stars themselves. We are even further from the diamond-lights in the sky when the eyebrows of Lao-tzu are described as like "the Northern Dipper."[31] The simile suggests only the odd but clearly supernatural shape of the godlike person's superciliary prominences. Similarly, the Bow and Arrow constellation (in Argo Navis) aimed at the Wolf (i.e., Sirius) does not glitter or beam when it is reduced to a mere symbol made to stand for a more mundane foe and fiend.[32] We have then to do with ordinary conceits. Similarly, even the grand conception of heavenly and earthly correspondents could be neatly turned to the uses of flattery, as in the following quatrain by Po Chü-i, which refers to the proposal of Li Ch'en (T'ang Hsüan Tsung) that two willow branches be planted in the imperial park, in the hope that they would take root there. The poet suggests that the sovereign's virtue would immediately have its reaction in the sky; two new stars could be looked for in the lunar asterism Willow:

A single tree, failing, fading, neglected in the muddy ground:
A pair of branches, burgeoning bright, planted in the heavenly court;
We know as certainty that, up among the mystic replicas, after
  this spring,
Within the light of Willow Lodging—an added pair of stars.[33]

Inventions such as these hardly fall even within the domain of wit, and had best be left to the historian of manners. In any case, they are far removed from images derived from the visual quality of stars. At the opposite extreme are the tropes which trade on stars collectively, as a field of multiple glittering specks. This was done in China through the use of the reduplicated form "star [on/by/with] star" (hsing hsing)—commonly with reference to the frosty spangles of white in the hair of aging men, most often in their earlocks or sideburns. It is but a step from "Starry frost—my earlocks already invaded,"[34] to "From today onwards my earlocks (show) star on star."[35] The prevalence of figures of speech based on similar analogies led a tenth-century

277

lexicographer to the expression "metallic powder" (*chin hsieh*) as a synonym of "star."[36] Such also were the phrases used by P'i Jih-hsiu in his description of the natural ornamentation of a peacock: "Filigreed down, golden fibers—entirely star on star."[37] But the figure is considerably faded in a "Pacing the Void" poem by Liu Yü-hsi, which refers back to the story of Ting Ling-wei, a Han Taoist who turned into a sacred crane, flew back to the gate of his hometown, only to be frightened away forever when a boy tried to shoot him down:

> In a showy coat—after a thousand years—come home
>   as solitary crane;
> Congealed cinnabar makes him a crown—snow makes him a dress.
> But (like) star with star is transcendent's speech—and men
>   are through with listening,
> And off towards the five clouds with flapping wings he flies.[38]

The standard usage employed here is "sparse, scanty (of words), as far-dispersed minute points of light in infinite space," but we may also like to see a reference, not only to the infrequency of the divine being's words—greater prolixity might have spared him an ignominious assault—but their mystery, like the mystery of the stars.

Appendix A

T'ang Star Icons

Artistic representations of the forms of the star deities existed, apparently in abundance, in T'ang times. Other than possible examples among tomb figurines, hardly any have survived to the present day. A notable exception to this generalization is the painting of the Tejaprabha Buddha and the Five Planets, found at Tun-huang, and dated A.D. 897.[1] Occasionally we find references to depictions of the stars and planets in volumes of history or in the writings of the poets. An example is a notice of the discovery of "depictions of the true shapes" of twenty-seven of the twenty-eight inhabitants of the Lunar Lodgings. (The missing one is "Base" in Libra.) These images, carved from jade, were found in "Jade Flower Grotto"[2] in Lo-ch'uan *hsien* in Ning-chou, northwest of Ch'ang-an and below the Ordos.[3] They are thought to represent the forms of the Perfected Beings of Taoism.

Aside from such casual encounters, however, most known examples of T'ang star art—chiefly a genre of Taoist religious art—are known only by name, as listings in a catalogue of paintings in the famous collection of Sung Hui Tsung. In this book, "Catalogue of Paintings of [the reign of] Extended Harmony" (*Hsüan ho hua p'u*), virtually all references to paintings of stars, planets and constellations—or rather to the humanized forms of the beings concealed in them—are listed among the works of religious painters, many of whom were quite as ready to paint pictures of Maitreya as portraits of the cosmic lords of the Dipper. These icons must have been true cult objects, either actual objects of worship or at least intended for the edification of the faithful or the conversion of undirected visionaries. Here is a tabulation of those of T'ang date—or reputed to be of that period—taken from the category of Taoist and Buddhist works (*tao shih*, i.e., "Tao and Shakya") distinguished in that catalogue. (In my translation of their titles, "effigy" represents *hsiang*—perhaps "icon" would have served better; "illustration" translates *t'u*.)

Yen Li-te, "Effigies of the Seven Planetoids," 2.
Yen Li-pen, "Effigies of the Five Stars [i.e., Planets]," 2.
Yen Li-pen, "Effigy of Grand White [Venus]," 1.
Yen Li-pen, "Effigy of the [Lunar] Lodging 'Chamber,' " 1.
Yen Li-pen, "Effigy of the Great Theocrat of Northern Culmen in Purple Tenuity," 1.

Ho Ch'ang-shou, "Effigy of Chronographic Star [Mercury]," 1.
Wu Tao-hsüan, "Effigy of the Theocratic Lord of Grand Yang," 1.
Wu Tao-hsüan, "Effigy of Chronographic Star," 1.
Wu Tao-hsüan, "Effigy of Grand White," 1.
Wu Tao-hsüan, "Effigy of Sparkling Deluder [Mars]," 1.
Wu Tao-hsüan, "Effigies of Rahu," 2.
Wu Tao-hsüan, "Effigy of Ketu," 1.
Wu Tao-hsüan, "Effigies of the Five Stars," 5.
Wu Tao-hsüan, "Illustrations of the Five Stars," 1.
Wu Tao-hsüan, "Effigies of the Twenty-eight Lodgings," 1.
Yang T'ing-kuang, "Effigies of the Five Stars," 1.
Yang T'ing-kuang, "Effigy of a Star Magistrate," 1.
Fan Ch'iung, "Effigy of the Star Lord of the Southern Dipper," 1.
Ch'ang Ts'an, "Effigy of a Star Magistrate," 1.
Sun Wei, "Illustration of a Star Magistrate," 1.
Chang Su-ch'ing, "Effigy of a Sky Magistrate," 1.
Chang Su-ch'ing, "Effigies of the Nine Planetoids," 1.
Chang Su-ch'ing, "Effigy of Longevity Star," 1.[4]

The famous Chou Fang is a special case in that one of his stellar pictures is listed in a separate source—the "Register of Renowned Painters from the Levee of T'ang" (*T'ang ch'ao ming hua lu*). Here, among other subjects, both supernatural and secular, said to have been painted by him, there is an "Illustration of the Five Stars."[5] The "Extended Harmony" catalogue does not neglect Chou Fang; it attributes the following items to him:

Chou Fang, "Illustrations of the True Shapes of the Five Stars," 1.
Chou Fang, "Illustrations of the Five Stars," 1.
Chou Fang, "Illustrations of the Five Planetoids," 1.
Chou Fang, "Effigy of a Star Magistrate," 1.
Chou Fang, "Peerless Effigy of the Great Theocrat of Northern Culmen."[6]

However these representations are not listed among the works of painters of the "Tao and Shakya" category, but in the "Human Being" (*jen wu*) category. The reason for this different classification is not clear. It might be argued that while the artists listed in the earlier group were *primarily* known as religious painters, Chou Fang painted religious subjects only occasionally. This supposition must be rejected, however, because—for example—of the conspicuous case of Yen Li-pen, best known for his many secular and courtly works.

Appendix B

Planetary Conjunctions
According to *T'ang Shu*, 33

Date	Planets	Location
30 July 618	Saturn, Venus, Mercury	Well (Gemini, not including Castor or Pollux)
16 May 619	Saturn, Venus, Mercury	Well
21 July 626	Jupiter, Mercury	Well [omen: "revolutionary plots"]
[between 10 June and 9 July] 644	Venus, Mercury	Well [omen: "plots of armed activity"]
7 18 July 645	Venus, Mercury	Well [omen: irruption of barbarians; refers to Li Shih-min's Korean campaigns]
26 August 650	Jupiter, Venus	Willow (stars in Hydra)
31 October 707	Venus, Mars	Barrens and Roof (stars in Aquarius, Equuleus and Pegasus)
[between 19 August and 16 September] 711	Saturn, Venus	Spread (stars in Hydra)
[between 10 May and 8 June] 712	Mars, Venus	Well
[between 6 September and 4 October] 750	The Five Planets (Mars arrived and left first)	Tail (stars in Scorpius) [omen: ambiguous; good if there is virtue in high places]

281

Pacing the Void

Date	Planets	Location
[between 17 March and 15 April] 755	Mars and Venus "fighting" among the asterisms	Net (Hyades) and Mane (Pleiades and Aldebaran); Well and Ghost (Gemini and Cancer)
[between 3 June and 1 July] 756	Mars, Saturn	Barrens and Roof
17 May 757	Jupiter, Mars, Venus, Mercury (Mars left first; Jupiter remained)	Quail Head (Well and Ghost)
[between 18 September and 16 October] 757	Venus occluded Jupiter; visible by day	Quail Fire (Willow, Star, Spread)
[between 12 May and 10 June] 758	Mars, Saturn, Venus	House (stars in Pegasus)
17 August 768	The Five Planets	"in the east" [omen: good for the Middle Kingdom]
19 December 773	Venus, Mercury	Roof
24 February 775	Jupiter, Mars	[Southern] Dipper (stars in Sagittarius)
[wrong date, apparently in August] 775	Venus, Mercury	Willow
[between 26 June and 25 July] 781	Mars, Venus	Well
[between 4 July and 2 August] 783	Mars, Venus	"fighting" in Well
spring [27 January to 23 April] 784	Mars, Jupiter	Horn and Gullet (stars in Virgo)
6 July 788	Jupiter, Mars, Saturn	House
[wrong date, apparently in late spring] 790	Venus, Mercury	Well

Appendix B

Date	Planets	Location
[same] 790	Mars, Saturn	Straddler (stars in Andromeda and Pisces)
13 December 814	Mars, Saturn, Venus	Woman (stars in Aquarius)
10 August [but garbled] 815	Jupiter, Mars, Venus, Mercury	Well
1 June 816	Jupiter, Mercury	Well
23 July 816	Jupiter, Mercury	Well
19 December 816	Saturn, Mars	Barrens and Roof
[between 23 December 816 and 20 July 817]	Saturn, Venus, Mercury	Roof
24 September [but garbled] 819	Jupiter, Venus, Mercury	Axletree (stars in Corvus)
[between 17 April and 15 May] 820	Saturn, Venus	Straddler
[between 8 January and 5 February] 821	Mars, Saturn	Straddler
9 March 822	Jupiter, Mars	[Southern] Dipper
28 August 822	Mars, Saturn	Mane and Net
31 August 824	Mars, Saturn	Well: successive attacks by Mars on Saturn [omen: "internal disorders"]
17 September 826	Mars, Saturn	between Well and Ghost
[between 13 October and 10 November] 828	Jupiter, Mars, Saturn	Star (stars in Hydra)
29 May 829	Jupiter, Saturn	[not given]
27 June [but garbled] 830	Jupiter, Venus	Well

Pacing the Void

Date	Planets	Location
[between 6 February and 5 March] 832	Venus, Mars	Feather Forest (Aquarius)
[between 28 October and 25 November]832	Venus, Mars, Saturn	Axletree
wrong date, [possibly summer] 834	Venus, Mars	Wing (stars in Crater)
26 June 838	Venus, Mars	Spread
22 January 839	Mars, Venus, Mercury	[Southern] Dipper
winter [between 10 November 839 and 6 February 840]	Jupiter, Mars (both in retrograde)	Well
13 July 842	Mars, Jupiter	Wing
17 November 844	Venus, Mars	[Southern] Dipper
[860–873]	Mars, Saturn, Venus, Mercury	Net and Mane
[between 10 September and 8 October] 888	Jupiter, Saturn, Venus	Spread
[between 26 October and 24 November] 900	Venus, Saturn	[Southern] Dipper

T'ang Sons of Heaven

Name		Posthumous Title		Years of Reign
Li Yüan	李淵	Kao Tsu	高祖	618–626
Li Shih-min	李世民	T'ai Tsung	太宗	626–649
Li Chih	李治	Kao Tsung	高宗	649–683
Li Hsien	李顯	Chung Tsung	中宗	683–684
Li Tan	李旦	Jui Tsung	睿宗	684–690
		CHOU		
Wu Chao	武曌	Tse t'ien huang-hou 則天皇后		690–705
		T'ANG (restored)		
Li Hsien	李顯	Chung Tsung	中宗	705–710
Li Chung-mao	李重茂	Shang Ti	殤帝	710
Li Tan	李旦	Jui Tsung	睿宗	710–712
Li Lung-chi	李隆基	Hsüan Tsung	玄宗	712–756
Li Heng[d]	李亨	Su Tsung	肅宗	756–762
Li Yü	李豫	Tai Tsung	代宗	762–779
Li Kua	李适	Te Tsung	德宗	779–805
Li Sung	李誦	Shun Tsung	順宗	805
Li Ch'un	李純	Hsien Tsung	憲宗	805–820
Li Heng[h]	李恒	Mu Tsung	穆宗	820–824
Li Chan	李湛	Ching Tsung	敬宗	824–826
Li Ang	李昂	Wen Tsung	文宗	826–840
Li Yen	李炎	Wu Tsung	武宗	840–846
Li Ch'en	李忱	Hsüan Tsung	宣宗	846–859
Li Ts'ui	李漼	I Tsung	懿宗	859–873
Li Hsüan	李儇	Hsi Tsung	僖宗	873–888
Li Yeh	李曄	Chao Tsung	昭宗	888–904
Li Chu	李柷	Ching Tsung	景宗	904–907

Note: Dates are from year of actual accession to year of death or abdication; not, as customarily, from proclamation of one new regnal era to the next.

285

Notes

CHAPTER 1. INTRODUCTION

1. "Yo conozco distritos en que los jóvenes se prosternan ante los libros y besan con barbarie las páginas, pero no saben descifrar una sola letra." My translation. The original appears in Jorge Luis Borges, "La Bibliotheca de Babel," in *Ficciones* (Emecé, Buenos Aires, 1956).

2. George Steiner, *After Babel: Aspects of Language and Translation* (Oxford University Press, 1975), p. 74.

CHAPTER 2. THE T'ANG ASTRONOMERS

1. Venice, 1655; as quoted in Mario Praz, *The Flaming Heart*.

2. See Dubs, 1958, 295.

3. See Nakayama and Sivin, 1973, xxv, for a clear statement of this point.

4. See Sivin, 1969, 4.

5. These discoveries generally belong to the Han dynasty, except that precession was a discovery of the fourth century.

6. Needham and Wang, 1958, 197–198, 264, 276.

7. Bezold, 1920, *passim*.

8. Bezold, 1920, 45.

9. TS, 33, 2a.

10. Yabuuti, 1954, 585; Nakayama, 1966, 450.

11. Yabuuti, 1954, 585.

12. In the *Ta chi ching*. Nakayama, 1966, 450. Cf. ———, 1975b.

13. Nakayama, 1966, 450.

14. See Yabuuti, 1961, and Yabuuti, 1963c.

15. Nakayama, 1966, 450; Yabuuti, 1963c.

16. Yabuuti, 1961; Yabuuti, 1963c; Nakayama, 1966, 450.

17. Yabuuti, 1961; Yabuuti, 1963c.

18. Huber, 1906, 40–41; Schafer, 1963, 276. See the latter reference also for other information about imported astrological and calendrical writings.

19. Nakayama, 1966, 450–452.

20. TS, 59, 12a.

21. Yabuuti, 1954, 586–589.

22. Sivin, 1969, 7.

23. TLSI, 9, 82. See also Nakayama and Sivin, 1973, xxvi. Such laws were in effect in other periods as well, e.g., in the state of Later Chao in the fourth century where "the private study of stellar prognostics was not allowed." CS, 106, 1360d. The elaborate text of an edict of 767 justifying this principle is preserved in CTS, 11, 11a–11b. See also the tenth century

edict of Chou T'ai Tsu along the same lines. CTW, 124, 4a–4b. On the other hand some rulers were strongly opposed to astrology, e.g. Chin Wu Ti and Sui Yang Ti. See Ho, 1966, 22.

24. The limit of information provided by the standard sources. Doubtless the perusal of late T'ang texts would yield some information for the troubled decades beyond the reign of Hsüan Tsung.

25. TS, 47, 6a.

26. Eg. in TS, 5, 4a.

27. TS, 47, 5b–6a.

28. LWHWC, 2b–3a.

29. CS, 106, 1360d.

30. TLT, 10, 23a–24a; TS, 47, 5b–6a.

31. Needham and Wang, 1958, 298, fig. 118. The observatory is represented on a number of Korean postage stamps. It was called *Chan hsing t'ai* (Korean *Ch'ōmsōngdae*).

32. TS, 38, 1b–2a.

33. Needham and Wang, 1958, 297.

34. Darwin, 1825.

35. CTS, 35, 1a.

36. Needham and Wang, 1958, 197–198; Ho, 1966, 13.

37. The *Fa hsiang chih*, CTS, 35, 1a. See also Needham, 1958, 544.

38. Among other sources see, for example, Schafer, 1963b, 276.

39. CTS, 35, 2b–3b; TCTC, 212, 9b.

40. Needham and Wang, 1958, 360.

41. Wheatley, 1971, 385–386.

42. Sivin, 1969, 7–9.

43. TS, 25, 1b. This page lists only eight new calendars, omitting the "Utmost Virtue" almanac of Su Tsung, which is referred to further along in the section. See also Yabuuti, 1963a, 455–456, which provides the names of the astronomers responsible for each one, along with other useful information.

44. Nakayama, 1966, 442.

45. Po Chü-i. "Szu t'ien t'ai," CTS, han 7, ts'e 1, ch. 3, 7a. There is a complete translation, with an outline of the actual events which inspired the poem, in South, 1972.

46. Soothill, 1952.

47. LWHWC, 3b.

48. HCWYSC, b, 4a.

49. TCTC, 204, 3a.

50. In what had been Wan-nien hsien. The name was retained until 702. For a sketch of the T'ang history of the Hall of Light, see Waley, 1950, 1–4.

51. TS, 37, 2b; THY, 11, 276.

52. Schafer, 1963b, 238; TTHY, 132.

53. TS, 107, 5b.

54. TCTC, 204, 3a; THY, 11, 277; TS, 4, 3b; CTS, 6, 2b. The sources do not agree on whether work was begun in 687 or 688. I follow THY.

55. TS, 4, 4a.

56. TS, 4, 4a.

57. THY, 11, 277; TTHY, 132; Schafer, 1963b, 238.

58. Schafer, *loc. cit.*

59. See CTW, 164.

60. Liu Yün-chi, "Ming t'ang fu," CTW, 164, 13a.

61. TCTC, 204, 7a–7b.

62. TCTC, 211, 18b.

63. THY, 11, 281.

64. THY, 11, 281; TCTC, 214, 18a.

65. TLCCFK, 5, 3a.

CHAPTER 3. COSMOGONY

1. Sivin, 1969, 1, 3, 67.

2. LH, "Tzu jan."

3. *pu chih jan erh jan i.*

4. *yin-yün.*

5. Wu Yün, "Hsing shen k'o ku lun," CTW, 926, 1a–2a.

6. Schafer, 1973a, 29–32, 48–51, 72–73.

7. Li Po, "Shang yün yüeh," CTS, han 3, ts'e 4, ch. 8, 9a. The relevant passage is translated in Schafer, 1973a, 72.

8. P'i Jih-hsiu, "Ou shu," CTS, han 9, ts'e 9, ch. 1, 12a.

9. CT, "Ta tsung."

10. Schafer, 1965a, 156.

11. Shirley, 1974, 109.

12. LH, "Wu shih."

13. The image of the Fashioner's furnace occurs much later in Tu Fu, "K'u T'ai chou Cheng szu hu Su shao chien," CTS, han 4, ts'e 4, ch. 19, 8b.

14. I have translated another of his poems, in a rather different kind of moralizing tone, in Schafer, 1973a, 76.

15. Su Cheng, "Shih mi," CTS, han 11, ts'e 2, 5b.

16. Lü Yen, CTS, han 12, ts'e 6, ch. 2, 4b and 5a.

17. Li Hao, "Ta tan shih, szu shou," no. 2, CTS, han 12, ts'e 6, 10b.

18. Wu Yün, "Pu hsü tz'u," CTS, han 12, ts'e 6, 8b.

19. LT, "T'ien jui."

20. SW, article ti "earth." For Taoist descriptions of this separation see Kaltenmark, 1960, 560.

21. San wu li chi in TPYL, 1, 1a. This is the first cosmological entry in the anthology.

22. HHS, 2, 0653c.

23. Ch'en Tzu-ang, "Ch'ing Yün chang," CTS, han 2, ts'e 3, ch. 1, 1b.

24. Meng Chiao, "Teng Hua-yen szu lou wang Chung-nan shan tseng Lin chiao shu hsiung ti," CTS, han 6, ts'e 5, ch. 4, 6a.

25. Li Shang-yin, "Han pei," CTS, han 8, ts'e 9, ch. 1, 9b–10b.

26. Ti wang shih chi in TPYL, 1, 3a.

27. Kuang ya in TPYL, 1, 4a.

28. T'ai shang Lao chün k'ai t'ien ching, quoted in YCCC, hence the terminal date given. Also in TT, 1059, 1b.

29. Lao chün t'ai shang hsü wu tzu jan pen ch'i ching, TT, 678, 1a.

30. Ti wang shih chi, TPYL, 1, 4b.

31. I ch'ien tsao tu, TPYL, 1, 4b.

32. Kuang ya in IWLC, 1, 2.

33. T'ai shang Lao chün k'ai t'ien ching, TT, 1059, 2b.

34. Kuang ya, TPYL, 1, 5b.

35. T'ai shang Lao chün k'ai t'ien ching, TT, 1059, 3a.

36. Wang Wei, "Tseng tung yüeh Chiao lien shih," CTS, han 2, ts'e 8, ch. 3, 4a–4b.

37. Ch'ang Kun, "Chung shu men hsiao ho hsüeh piao," CTW, 415, 13a–13b.

38. See Ho, 1966, 52. Ho translates T'ai chi by "Great Absolute," which hardly brings out the connotation of the expression as a final climax.

39. Li Shan, commentary on Su Ch'o, "Yu T'ien-t'ai shan fu," WH, 11, 5a.

40. Liu Tsung-yüan, "T'ien tui," CTW, 585, 15a. For "white-souled," see my discussion of the "soul" of the moon below.

41. Meng Hao-jan, "P'eng li hu chung wang Lu shan," CTS, han 3, ts'e 3, 8b.

42. Li Ch'ün-yü, "Sung T'ao shao fu fu hsüan," CTS, han 9, ts'e 3, ch. 2, 12a.

43. Tu Kuang-t'ing, "T'i Pei p'ing chao," CTS, han 12, ts'e 6, 2b.

44. See TzT, 1, 1376.

45. CT, "Ta tsung." Several crucial cosmogonic terms occur in CT, "Chih pei yu," but are used as illustrative examples rather than in a formal demonstration.

46. HNT, "Chüan yen."

47. LH, "T'an t'ien."

48. LT, "T'ien jui."

49. T'ai shang Lao chün k'ai t'ien ching, 3a.

50. Santillana and Dechend, 1969, 59.

51. Needham and Wang, 1958, 408.

52. This "Great Year" was styled T'ai chi shang yüan "Superior Epoch of Grand Culmination." See Sivin, 1969, 18.

53. Sivin, 1969, 10, gives other examples of such cycles, especially those used to mark the beginning of Han imperial epochs.

54. Sivin, 1969, 65–67.

55. SHC, biography of Ts'ai Ching.

56. Needham and Wang, 1958, 600–601.

57. Yen Chen-ch'ing, "Fu-chou Nan-

ch'eng hsien Ma-ku shan hsien t'an chi," CTW, 338, 7a.

58. Lü Yen, untitled poem, CTS, han 12, ts'e 6, ch. 2, 1a–1b.

59. Lü Yen, untitled poem, CTS, han 12, ts'e 6, ch. 2, 4b. Cf. poems by Shen Pin of the end of T'ang, in which the mulberry fields are a topic frequently on the lips of divine beings, sometimes seriously sometimes flippantly, as in his "Ma-ku shan," and "I hsien yao," CTS, han 11, ts'e 4, 3b and 4a.

60. Lan Ts'ai-ho, "T'a ko," CTS, han 12, ts'e 7, 11b.

61. Po Chü-i, "Tu shih, wu shou," no. 3, CTS, han 7, ts'e 1, ch. 2, 9a–9b.

CHAPTER 4.   THE SKY

1. E.g. *Li t'ung* in IWLC, 1, 2.

2. *Wu li lun* in CHC, 1, 1. Evidently some persons favored the view that "pneuma" was the support rather than the substance of heaven and earth, although this opinion is readily reconcilable with those I have mentioned. For this, see Ho, 1966, 53–55.

3. LH, "T'an t'ien."

4. LH, "Shuo jih."

5. *Ku pien i* in TPYL, 7, 1a.

6. Ts'ai Yung, *T'ien wen chih*, in TPYL, 2, 4a.

7. For further details see Maspero, 1929, 333–356; Needham and Wang, 1958, 210 ff.

8. ATL, 2a–2b.

9. PKLT, 1, 5a.

10. CCYMP, 19a–19b; *Hun t'ien i* in IWLC, 1, 1. There are some discrepancies in these two versions. I have tried to reconcile them.

11. ATL, 2b.

12. ATL, 1b.

13. Yang Chiung, "Hun t'ien fu," CTW, 190, 1a.

14. CT, "Shao-yao yu."

15. HNT, "T'ien wen."

16. LH, "T'an t'ien."

17. YYTT, 2, 9.

18. CCKIY, 1a; *Lo shu chen yao tu* in TPYL, 2.

19. LH, "Shuo jih"; CCYMP, 21a, 22a; PKLT, 1, 2a. The second of these sources refers to a terrestrial rotation opposite to that of the sky. See Needham and Wang, 1958, 213, for more details on this matter. N. Sivin (p.c. of 13 June 1976) has drawn my attention to Vitruvius' use of the millstone image.

20. LH, "Shuo jih."

21. Liu Yü-hsi, "T'ien lun," CTW, 607, 1a–6a. Lamont, 1974, has given us a translation and commentary on the contributions of both Liu Tsung-yüan and Liu Yü-hsi, which, though admirable in many respects, is marred by semantic imprecisions that often tend to distort the arguments of the two men. An example (p. 79) is the expression *shen ming*, which he renders, rather arbitrarily, as "divinely brilliant," as if it were an enthusiastic comment on a well-faceted diamond, with the additional comment that this phrase ". . . does not imply any kind of real divinity." I do not know what Mr. Lamont understands by "real divinity," but I have seen this expression in hundreds of contexts in all of which it plainly refers to what we mean by "supernatural beings," who express themselves through miracles and powerful concentrations of mana. The word *ming* functions as in *ming ch'i*[a] (sometimes *ming ch'i*[b]). The result of this misinterpretation is the falsification of Liu Yü-hsi's argument.

22. YYTT, 2, 9.

23. Hua-jui fu-jen, "Kung tz'u," CTS, han 11, ts'e 10, 4a.

24. The word *hsüan*, which I have tentatively translated "unrestricted," has been glossed as "luminous" or "illuminated" (*ming*) without supporting linguistic evidence. Perhaps it was meant to convey something like "irradiated."

25. Ts'ai Yung, "T'ien wen chih," TPYL, 2, 4a; ATL, 1a; PPT in TPYL, 2, 2a.

Yang Chiung, "Hun t'ien fu," CTW, 190, 2a, states that the doctrine was lost with the state of Ch'u in Chou times.

26. PKLT, 1, 2b, following earlier sources. Cf. Yang Chiung, "Hun t'ien fu," CTW, 190, 1b.

27. PPT in TPYL, 2, 2a. Cf. Needham and Wang, 1958, 220, for Yü Hsi's statement of this position.

28. Yang Chiung, "Hun t'ien fu," CTW, 190, 1b.

29. "It was entrusted to the stones to toughen the earth, but soon they took wing." Pablo Neruda, *Las piedras del cielo* (my translation).

30. Schafer, 1973a, 13–14.

31. LH, "T'an t'ien." Wang Ch'ung vouches for the general acceptance of this view in ths first century, and goes to some pains to refute it.

32. Chang Chiu-ling, "Ch'iu wang teng lou wang nan chiang ju Shih-hsing chün lu," CTS, han 1, ts'e 9, ch. 1, 6a.

33. Schafer, 1961, 89–90.

34. Schafer, 1963b, 230–234. Lapis lazuli was apparently known in late T'ang times as *se-se*.

35. Su Sung (fl. 1061), quoted in PTKM, 10, 2a.

36. PTKM, 20, 5a.

37. *Huang shih kung chi*, quoted in CHC, 1, 12. This book is listed among others on military arts, including divination and board games, in SS, 34, 2450b.

38. Meng Chiao, "Han ch'i," CTS, han 6, ts'e 5, ch. 5, 8a.

39. Ku K'uang, "Ling shuang chih hua i chang," CTS, han 4, ts'e 9, ch. 1, 4a.

40. Hsü Yin, "Ho liu," CTS, han 11, ts'e 1, ch. 1, 2a.

41. Feng Yen-chi, "Ts'ai sang tzu," CTS, han 12, ts'e 10, ch. 10, 8a.

CHAPTER 5. THE STARS

1. As in LH, "Ting kuei."

2. *San wu li chi*, TPYL, 5, 1a; Yang Ch'üan, *Wu li lun*, TPYL, 7, 3a.

3. Chang Heng, *Ling hsien*, TPYL, 7, 1b.

4. LH, "Shuo jih"; SW, in TPYL, 5, 1a.

5. HNT, "T'ien wen."

6. CCWSTT, 4b.

7. Paraphrasing LT, "T'ien jui."

8. SC, 27, 0111b.

9. This subject will be treated more fully in the next chapter.

10. LH, "Shuo jih."

11. *Lou t'an ching* in FYCL, 5, 24a.

12. Ho, 1966, 67; SS, 19, 2402b.

13. Ho, 1966, 67–68; Allen, 1963, 458.

14. CCWWYK, 1a.

15. CCWYMP, 30a; CCWWYK, 1b.

16. *Hsüan men pao hai* in YCCC, 24, 2a.

17. CCYMP, 31a.

18. Wheatley, 1971, 44a.

19. Allen, 1963, 453.

20. See *inter alia* Schlegel, 1967, 523; Ho, 1966, 69.

21. CCHCT, 2a–2b, commentary of Sung Chün.

22. Translation of Ho, 1966, 69.

23. SCHCSPYC, 1a–1b.

24. CCWHCT, 4b.

25. Li Shang-yin, "Ch'en hou kung," CTS, han 8, ts'e 9, ch. 2, 28a.

26. LH, "Lei hsü."

27. CCWHCT, 3a.

28. SC, 25, 0104d.

29. The Greeks sometimes saw two wagons (*hamaxai*) as well as two bears in the sky. See Mair, 1969, 208. See also Allen, 1963, 420 for further information.

30. CCWYMP, 31a.

31. Po Hsing-chien, "Tou wei ti ch'e fu," CTW, 692, 26a–26b.

32. YYTT, 1, 7.

33. Li Po, "Ni ku, shih erh shou," no. 6, CTS, han 3, ts'e 6, ch. 23, 2b.

34. TS, 6, 4b; 32, 6a; CTS, 10, 2a; THY, 43, 767.

35. JYT, 3a.

36. PPT, "Tsa ying."

37. *Hou sheng lieh chi*, TPYL, 662, 1a.

38. Tuan Ch'eng-shih has preserved a curious list of unknown origin. The

names he gives may be approximated in English as follows: 1. "Held-to *Yin*"; 2. "Assonant and Agreeable"; 3. "Looked-at Gold"; 4. "Anti-System"; 5. "Wary of Rival"; 6. "Opened Treasure"; 7. "Far-flight." YYTT, 14, 104.

39. YCCC, 24, 7a–7b.
40. CCWYTS, 4b–5b.
41. CSCN, a, 3a.
42. TLCCFK, 5, 4b.
43. Cheng Yü, "Chin yang men shih," CTS, han 9, ts'e 3, 2b–3a.
44. HC, 4b.
45. Ch'u Kuang-hsi, "Shu hua ch'ing kung," CTS, han 2, ts'e 10, ch. 1, 4a.
46. TS, 33, 2a, 5a.
47. CCWYMP, 71a.
48. In his "Bacchus and Ariadne," in the Ducal Palace, Venice.
49. Allen, 1963, 178.
50. Schlegel, 1967, 516–518.
51. CS, 11, 1103b.
52. CTS, 36, 2b.

CHAPTER 6. ASTROLOGY

1. "I know now, Lord, what the stars of your sky say; what their diamond-points keep writing for me."
2. TCTC, 214, 17a.
3. See TS, 34, 3712c–d; CTS, 37, 3203a–b. These sources tell of the heavy reliance of these kinds of prognosis on the *Hung fan* traditions, especially as exploited by Liu Hsiang and incorporated by Pan Ku in the "Five Activities" section of the "Book of Han."
4. The arts of geomancy too were involved with the stars. See Needham, Wang and Robinson, 1962, 282, for an account of geomancers in the eleventh century, which ". . . laid much more emphasis on *kua*, compass-points and constellations, making particular use of the magnetic compass." See also Wheatley, 1971, 419 for the use of astrology in city planning in Asia.
5. TS, 27b, 3b. Exploded stars were

comets not novae. See below in my treatment of comets.
6. Shumaker, 1972, 11.
7. Nakayama, 1966, 447.
8. Not "symbol" as in modern usage.
9. Yang Chiung, "Hun t'ien fu," CTW, 190, 2b; cf. *I ching* in TPYL, 1, 8a.
10. Chang Ch'ien (T'ang), *Ling kuai chi*, quoted in Schlegel, 1967, 74.
11. This term, and various near synonyms which I have adopted freely, has been given wide currency by the efforts of Needham, Sivin, Nakayama, Wheatley, Porkert and others.
12. Schafer, 1967, 6; Wheatley, 1971, 414, 417–8; Nakayama and Sivin, 1973, xxii.
13. The astrological (*T'ien wen*) section of the *Chin shu* gives a systematic account of the meaning of the asterisms, which is abridged in the *T'ang shu*. The latter, except in describing specific events, restricts itself to the twelve regions of the Jupiter cycle. See Ho, 1966, *passim*. For the titles of astrological books by T'ang authorities, most of them now lost, see TS, 59, 12a. The writings of Li Ch'un-feng are particularly numerous.
14. For much more on portent astrology in China (with Babylonian parallels), see Ho, 1966, 21; Nakayama, 1966, 443–445; Yabuuti, 1973, 94. Ho, 1966, 14, notes that the influence of Liu Hsiang, in the first century B.C., made "the belief in astral influence on state events" a firm (rather than random) part of Chinese astrological doctrine.
15. SC, 12, 0043c.
16. LH, "Pien hsü."
17. TS, 108, 6a.
18. Cf. Schlegel, 1967, 422.
19. *Na yin*, referring to a correlation of the Five Activity cycle with the pentatonic scale. Metal corresponds to the tone *shang*. See Porkert, 1974, 71.
20. TS, 33, 2b–3a; TS, 204, 5b.
21. LH, "Ming i." Cf. Needham, Wang and Robinson, 1962, 356.

22. CTS, han 6, ts'e 3, ch. 12, 22a.

23. LH, "Ming i" seems to complicate the situation. *Ming*, says Wang Ch'ung, is "wealth and nobility, poverty and low station." *Lu*, on the other hand, is "ascendancy and decay, rising and being eliminated." Perhaps the terms have exchanged their definitions in the transmission of the text.

24. Nakayama, 1966, 449. Cf. Chao, 1946, *passim*.

25. Chao, 1946, 283; Nakayama, 1966, 448, n. 23.

26. Han Yü, "Tien chung shih yü shih Li chün mu chih ming," CTW, 564, 13b–15a.

27. Lü Ts'ai, "Hsü lu ming," CTW, 160, 11a–13a. See also his biography in CTS, 79, 7b–8a.

28. SC, 127, esp. p. 0272c. Needham, Wang and Robinson, 1962, 357; Nakayama, 1966, 449.

29. Nakayama, 1966, 448, 450.

30. Han Yü, "San hsing hsing," CTS, han 5, ts'e 10, ch. 4, 4b.

31. Chang Kuo's astrology is best preserved in a composite volume called *Hsing ming su yüan*. The Szu-ku editors of 1781, whose analysis of the book appears as a preface, describe its contents as follows (some notes of my own are included): (1) *T'ung hsüan i shu* relates the doctrines of Chang Kuo (i.e. those of T'ung hsüan hsien sheng, the title awarded him by Li Lung-chi). The section begins with a simplified cosmology (the primordial universe was like an egg) and goes on to describe the origin of man and his special talents. The author (presumably Chang Kuo himself) observes that even a sage familiar equally with the arcana of the cosmic diagrams that emerged from the Ho and Lo Rivers in antiquity must, for mastery of affairs, know also "the bright glitter of the Five Planets and the sparkling shine of the assembled sky-lights, and to distinguish the chronograms and stations with respect to both Energetic and Re-ceptive, and to classify human bodies with respect to Heaven and Earth." (2) *Kuo ch'eng wen ta* purports to be verbal instruction received from Chang Kuo by Li Ch'eng. (3) *Hsüan miao ching chieh* is ascribed by Chang Kuo, with commentary by Cheng Hsi-ch'eng of the Yüan period. (4) *Kuan hsing yao chüeh* is attributed to Cheng Hsi-ch'eng. (5) *Kuan hsing hsin ch'uan k'ou chüeh* is not given an author, but the title page suggests that it is the work of Cheng Hsi-ch'eng with additions by others. The *Hsing tsung*, in ts'e 465–470, *chüan* 567–585 of the *T'u shu chi ch'eng*, has been ascribed to Chang Kuo. It includes horoscope charts. Needham and Wang, 1956, 352, takes this to be a work of the fourteenth century. This evaluation, however, is vitiated in Needham, Wang and Robinson, 1962, 356, where it is assumed that *Hsing tsung* is a T'ang work and the whole *Hsing ming su yüan* is said to be a "book" by Chang Kuo.

32. *T'ung hsüan i shu*, 8a.

33. LH, "Lan shih." In the astrology of the Jupiter cycle, the determination of fate was actually made by an invisible, counterrotating correlate of the planet, called *t'ai sui*. Sivin, 1969, 10.

34. Tu Mu, "Tzu chuan mu ming," CTW, 754, 20b.

35. Soymié, 1972, 664.

36. YYTT, 3, 30.

37. Nakayama, 1966, 453; Shumaker, 1972, 11; Yabuuti, 1973, 94. Some variety of horoscope astrology was transmitted by the Khitans in the tenth century. Yeh-lü Ch'un, author of "Inclusive Summation of Fate in the Stars" (*Hsing ming tsung kua*), visited Korea in 984, transmitting the science of the "Zen Masters of his country." The language suggests the possibility of Indian influence on the art. Huang, 1937, 6b.

38. Allen, 1963, 284.

39. Clerke, 1905, 135.

40. Clerke, 1905, 136. Nowadays astronomers assign a color index to stars,

as follows: the index of blue-white stars is negative; that of white ones is 0; yellow stars like the sun have a color index of +1, while deep red ones are +2. Wallenquist, 1966, 227.

41. William Cullen Bryant, "The Constellations."

42. Allen, 1963, 314.

43. Clerke, 1905, 136. Clerke adds, "Intense tints result from strong selective absorption in the atmospheres of the stars they distinguish, and strong absorption implies large loss of light. Stars shine with the rays that have survived transmission through the glowing vapours in their neighborhood . . ."

44. Clerke, 1905, 143.

45. Clerke, 1905, 140. When characterizing the colors of stars I follow Allen, 1963, who in turn adopts the descriptions in Henry Smyth, *Cycle of Celestial Objects*.

46. LWCMC, 3a.

47. LWTWI, 2a.

48. SC, 27, 0110c; cf. Ho, 1966, 125.

49. CCWKCF, 38, 11b and 55, 20b.

50. CCWKCF, 3b.

51. Thomas Lovell Beddoes, *Death's Jest-book*, Act IV, scene 4.

52. CS, 12, 1106a. Occurrences that do not fit this simple pattern have sometimes been recorded. Thus the scholiast Meng K'ang wrote of a passage in HS that a certain bizarre apparition, a star with a blue interior and red exterior, below which there were three comets, was an emanation of Saturn. Cf. Needham and Wang, 1958, 433.

53. SC, 27, 0109d–0110a.

54. TS, 32, 6a.

55. So in the reports of THY.

56. CCWKCF, 10a.

57. CCWYTS, 29a.

58. *Rhododendron Simsii*, a synonym of *R. indicum v. ignescens*.

59. Schafer, 1965b, 111–113.

60. Meng Chiao, "Ch'ou Cheng P'i chih-chu yung," CTS, han 6, ts'e 5, ch. 9, 2a–2b.

61. Li Po, "Chiu p'u ko, shih ch'i shou," (No. 14), CTS, han 3, ts'e 4, ch. 7, 1b.

62. Lo Pin-wang, "Shang Li shao ch'ang ch'i," CTW, 198, 5b–6a.

63. It is No. 80 in Flamsteed's catalogue. For the color and other data see Allen, 1963, 440. There the name is said to be derived from Arabic *al-Khawwāra* "the feeblest," but this traditional etymology is very doubtful. See Benhamouda, 1951, 84.

64. CS, 11, 1102c. There were other "stabilizers" in the T'ang sky, e.g. the "Four Stabilizers" in Camelopardalis, clustered close to the pole star.

65. SIC, 1, 3a.

66. *Wei chih* cited in IWLC, 1, 12. This omen is referred to also in a poem of Wang Jung, "Yung ming chiu nien ts'e hsiu ts'ai wen," WNSC, 12b.

67. YYTT, 8, 61.

68. Allen, 1963, 68 lists the other Egyptian associations.

69. Ishii, 1968, 144.

70. Wan San-szu, "Ho lao jen hsing chien piao," CTW, 239, 1b–2a.

71. Li Shang-yin, "Wei Ying-yang kung ho lao jen hsing chien piao," CTW, 772, 1a–1b. Other surviving memorials of congratulation on appearances of Canopus are Yang P'ing, "Ho lao jen hsing chien piao," CTW, 478, 7b–8a; Ling-hu Ch'u, "Ho lao jen hsing chien piao," CTW, 539, 13b.

72. Ch'ih Mao, "Lao jen hsing fu," CTW, 361, 10–11a.

73. Yang Chiung, "Lao jen hsing fu," CTW, 190, 15b–16b.

74. WHTK, 294, 1a.

75. CCWWYK in TPYL, 5, 4b.

76. JYT, 2b.

77. Lu Wo, "Fu te shou hsing chien," CTS, han 9, ts'e 3, 6b.

78. Ho Ning, "Kung tz'u, wu shou," CTS, han 11, ts'e 4, 1b.

79. For this name, see Schafer, 1967, *passim*.

80. Schafer, 1967, 123–124.

81. Chang Chi, "Sung Cheng shang shu fu Kuang-chou," CTS, han 6, ts'e 6, ch. 4, 9b.

82. Schafer, 1967, 123.

83. TLSKC, 2a and 3a.

84. SC, 27, 0110c. Cf. Ho, 1966, 21, for comments on the relative significance of various kinds of stellar changes, in color, movement and the like.

85. But the name was also given to an asterism in Capricornus.

86. HC, 7b.

87. HC, 2a.

88. CCWYTS, 33a–33b.

89. CCWCTP, 1b.

90. HC, 7b.

91. CCWYMP, 72b.

92. TS, 33, 8a.

93. HC, 4b.

94. SC, 27, 0110b; JYT, 3a.

95. JYT, 3a. Cf. HS in TPYL, 5, 6b: "yellow is benign."

96. See LTFH, 5, 1b–2a.

97. CCWYMP, 24b.

98. Ho, 1966, 125.

99. Clerke, 1905, 3–5; Wallenquist, 1966, 211.

100. HC, 5a.

101. TS, 33, 7b. This apparition is reported for the winter of 901 and again for the summer of 904, during the final fateful years of the T'ang dynasty.

102. CS, 11, 1103a.

103. Darwin, 1825.

104. Abbreviated from ra's al-ghūl "head of the ogre," doubtless referring to Medusa's head. See Benhamouda, 1951, 100.

105. CS, 11, 1103b, 1103d.

106. HC, 3a.

107. HC, 3a.

108. Li P'in, "Sung yu jen wang Chen-wu," CTS, han 9, ts'e 6, ch. 1, 9a.

109. Ku Fei-hsiung, "Wu Tsung wan ko, erh shou," (No. 1), CTS, han 8, ts'e 4, 7b.

110. Hsüeh Neng, "Shou-an shui kuan," CTS, han 9, ts'e 2, ch. 4, 6b.

111. Li Shang-yin, "Pi ch'eng, san shou," (No. 1), han 8, ts'e 9, ch. 1, 24a.

112. Tu Kuang-t'ing, "Tseng jen," CTS, han 12, ts'e 6, 3b.

113. Examples from Schafer, 1974, 406–407; the final quotation is from a poem written on command at a court banquet by Tu Shen-yen.

114. Nakayama and Sivin have used the expression "field allocation." See Nakayama and Sivin, 1973, xxii.

115. See a good statement in CCWKCF, 2b. The system has Babylonian parallels; see Nakayama, 1966, 447.

116. See, for instance, CL, Ch'un kuan, tsung po, "Pao chang shih."

117. HNT, "T'ien wen," with commentary of Hsü Shen.

118. Santillana and Dechend, 1969, 123.

119. Yabuuti, 1973, 92, traces the system back to late Chou times. Cf. Ho, 1966, 113ff., which is more conservative, fixing the system securely in early Han times. CTS, 36, 1a, assigns it to classical astronomers Kan Te and Shih Shen. These are by no means identical with the twelve signs of the Western zodiac which, however, appear on the ceiling of a tomb of A.D. 1116 recently excavated in Hopei. The foreign scheme seems to have been known in China at least by T'ang times. See ———, 1975b, 42.

120. TS, 31, 7b–12b maps the stations for us.

121. Wu Yüan-heng, "Sung Feng chien i fu Ho pei hsüan wei," CTS, han 5, ts'e 7, ch. 1, 9b.

122. TS, 31, 7b–8a; CTS, 35, 2b; 36, 1a.

123. An early account of these correspondences is given in HS quoted in TPYL, 5, 6b–7a.

124. Henry Wadsworth Longfellow, "Moonlight."

125. Wheatley, 1971, 385, describes this theory.

126. Ho, 1966, 17–18. Ho gives a useful chart of the Chinese-Indian correspondences. For the history and theory

of the *hsiu* see also Maspero, 1929, 282–285; Needham and Wang, 1958, 231–259.

127. Chatley, 1940, 394.

128. ———, 1973, 19.

129. A book on the omens of the twelve Jupiter stations and the twenty-eight lunar lodgings is named in the *Book of T'ang*. It is Shih Ch'ung, *Shih erh tz'u erh shih pa hsiu hsing chan*, listed in TS, 59, 12a. Unfortunately it does not survive. A surviving fragment of the CCWKCF (6b–7b), however, does list the significance of eclipses in each of the lodgings. Unsystematic notes on the auspices of the lunar asterisms may, however, be found in many T'ang sources.

130. FPNTTWYC, 5b.

131. CCWKCF, 6b–7b.

132. *Ibid.*, 10b.

133. Chang Hsiao-p'iao, "Shang Che-tung yüan hsiang," CTS, han 8, ts'e 4, 1a. For the royal platform, see Schafer, 1967, 29.

134. Ch'en T'ao, "Shang chien ch'i," CTS, han 11, ts'e 4, ch. 2, 3b. For other examples see Schafer, 1967, 124–125.

135. Cf. Clerke, 1905, 223 for ". . . wrapped and entangled in an immense cosmical cloud."

136. Hastings, 1962, XII, 64.

137. Schlegel, 1967, 354.

138. PKLY, 1, 26b.

139. Schlegel, 1967, 368.

140. SW. This asterism which, in Chou times, presided over the destiny of the state of Chin, was controlled by the god Shih-ch'en, who subsequently gave his name to one of the twelve Jupiter stations. See *Tso chuan*, Chao, 1.

141. CCWYMP, 72b.

142. Hou, 1979, 194.

143. Tu Kuang-t'ing, "Chao kuo t'ai fu jen mou shih chiao chiao tz'u," KuCC, 8, 12a.

144. TS, 31, 12a.

145. CS, 11, 1103c.

146. Schlegel, 1967, 113; HC, 4b.

147. Li Ho, "Ma shih, erh shih san shou," (No. 4), CTS, han 6, ts'e 7, ch. 2, 2a.

148. *Erh shih szu chih*, 16b–18b (YCCC, 28). This book, attributed to "Heavenly Master Chang," predates the eleventh century, and may well be much older.

149. See FYCL, ch. 5, "Section on the Devas."

150. FYCL, 6, 23a.

151. Needham and Wang, 1958, 253; Allen, 1963, 392–393. My sources are YYTT, 3, 29, and FYCL, 5, 18a ff., which agree very well.

152. CCWKIY, 6a, 7a.

153. HNT, "T'ien wen." SC, 27, 0111d, gives a somewhat different explanation, attributing, for instance, lightning storms and "nocturnal lights" to "movements of the *yang* pneuma."

154. CCWYMP, 4a–4b.

155. LH, "Shuo jih."

156. CIL, a, 1a.

157. SC, 27, 0111c. Cf. commentary of Cheng K'ang-ch'eng on CL, Ch'un kuan, "Pao-ch'ang shih." Ho, 1966, 144, gives examples of color interpretations. For "iridescent clouds" see Minnaert, 1959, 226–227.

158. LH, "Pien hsü."

159. Schafer, 1973a, 13–14.

160. CCWYMP, 32b–33a.

161. Ho, 1966, 139, based on CS; CCWYMP, 41b. Cf. Schafer, 1973a, 22.

162. HNT, "T'ien wen."

163. As for instance in CCWCTP, 13b.

164. Ho, 1966, 144; Schafer, 1973a, 18.

165. Bernhard, Bennett, Rice, 1948, 216; Minnaert, 1959, 182–183.

166. Yang Chiung, "Hun t'ien fu," CTW, 190, 5b. "Crouching turtle" seems to refer to "Dusky Warrior," guardian of the northern skies, who is sometimes styled *pieh* "[softshell] turtle."

167. Ho, 1966, 66.

168. Botley, 1957, 188–189.

169. Botley, 1956, 32.

170. For the record of auroras in early China, the reader is invited to consult Schove, 1951.

171. SC quoted in TPYL, 7, 5a.

172. WHTK, 294, 2b.

173. TS, 34, 10a; CTS, 7, 7a.

174. TS, 34, 10b.

175. The dates enclosed in square brackets represent the first and last days of the lunar month, when the text gives no specific day within that month. (This convention is followed in all lists of dated phenomena in this book.)

176. TS, 34, 10b.

177. TS, 32, 10b.

178. Ho, 1966, 138, based on CS, 12, 1106b; JYT, 3b.

179. CTS, 7, 11a, 12a.

180. TS, 125, 5b.

181. Chang Yüeh, "Ching yün wu," CTS, han 2, ts'e 4, ch. 1, 5b.

182. See Wallenquist, 1966, 164.

183. T'ai shang fei hsing yü ching, in TPYL, 675, 8b. (Cf. TT, 1033).

184. Bernhard, Bennett, Rice, 1948, 216. For the Chinese data see also Ho and Needham, 1959, 124–134.

185. Ho, 1966, 139–144.

186. HC, 3a.

187. TS, 32, 5a.

188. Needham and Wang, 1958, 433.

189. Meadows, 1969, 58.

190. TS reports apparitions of meteors in the Wu hsing chih, ch. 22, 8a ff.; as a sub-heading of "stellar malfunctions" (hsing pien). They are also recorded in the "basic chronicles" of the dynastic histories; CTS however does not put them there (the great brontide of 744 is an important exception).

191. The Chinese word for a comet's tail, mang, is a metaphorical borrowing. Mang is the "awn" or tip of a head of grain. Needham and Wang, 1958, especially the caption to Plate LXII, is mistaken in taking this word to stand for the uncommon phenomenon of a short spike protruding "forward" from a comet's head. Mang, in fact, occurs regularly with tailed comets (hui hsing), just as wei "tail" does with meteors.

192. These refinements are carefully registered in Ho, 1966, 136–138.

193. TS, 166, 10b.

194. E.g. TS, 109, 1a–1b.

195. The bibliography of the Sui imperial library contained a book in one scroll devoted to the nomenclature of meteors and other ominous stars: "Forms, Names and Omens of Uncanny Stars and Streaming Stars" (Yao hsing liu hsing hsing ming chan), SS, 34, 2450c.

196. HC, 2b.

197. CCWKCF, 30b.

198. CCWKCF, 30a.

199. CCWYMP, 70b.

200. CCWYMP, 70b.

201. CCWKCF, 30a.

202. P'ei Yüeh, "Ch'i," CTS, han 11, ts'e 2, 2a.

203. A dictionary reference to Hsiao Kai, Han shu yin i, whose original I have not been able to trace.

204. Tu Fu, "Chung hsiao," CTS, han 4, ts'e 4, ch. 15, 17b.

205. See EY, "Shih t'ien."

206. Hsü Shen commentary on HNT, "T'ien wen."

207. Ho, 1966, 137; descriptions in CCWHCT, 17a; SM, "Shih t'ien"; HS, quoted in TPYL, 7, 5b.

208. CCWYTC, 30a.

209. HC, 4b–5a.

210. This couplet, from a poem entitled "Ch'ang-an chi hou shou ch'iu yeh chi shih," has been attributed both to Ch'en Yü (CTS, han 6, ts'e 1, 3a) and to Lu Lun (CTS, han 5, ts'e 2, ch. 5, 4a).

211. SC, 27, 0111b–c.

212. See for instance Ho, 1966, 138.

213. Schlegel, 1967, 153–154, 433.

214. TS, 35, 10a.

215. WHTK, 291, 17a–17b.

216. CTS, 36, 7a; THY, 43, 774-775.
217. TS, 32, 8a; THY, 43, 775; WHTK, 291, 17b.
218. TS, 32, 8a; THY, 43, 775; WHTK, 291, 18a.
219. TS, 32, 8a; TCTC, 199, 15a-15b.
220. TS, 32, 8a; THY, 43, 775; WHTK, 291, 18b.
221. Schafer, 1963a, 206.
222. THY, 43, 775.
223. THY, 43, 775.
224. THY, 43, 775.
225. TS, 32, 8b.
226. TCTC, 210, 10b.
227. TS, 32, 8b; WHTK, 291, 18b.
228. TS, 32, 8b; WHTK, 291, 18b-19a.
229. TS, 5, 12a; CTS, 9, 6a; THY, 43, 775.
230. TS, 32, 8b.
231. TS, 32, 8b. It was the eleventh month, but the day given did not occur in that month.
232. TS, 6, 9b; TS, 32, 8b; CTS, 11, 12b.
233. TS, 32, 8b; WHTK, 281, 19b.
234. TS, 7, 8a.
235. TS, 32, 9a; WHTK, 291, 20a.
236. TS, 32, 9a; WHTK, 291, 20a.
237. THY, 43, 775.
238. THY, 43, 775.
239. THY, 43, 775.
240. THY, 43, 775.
241. TS, 32, 9a; THY, 43, 775; WHTK, 291, 20b.
242. THY, 43, 776.
243. THY, 43, 776.
244. THY, 43, 776. The date as given is garbled.
245. TS, 32, 9b; CTS, 16, 14b-15a.
246. THY, 43, 776.
247. THY, 43, 776; Allen, 1963, 36.
248. TS, 32, 9b; WHTK, 291, 21b; WHTK seems to have the wrong date.
249. THY, 43, 776.
250. TS, 32, 9b; WHTK, 291, 21b.
251. THY, 43, 776.
252. TS, 32, 9b; WHTK, 291, 21b.
253. THY, 43, 773.
254. TS, 32, 9b; THY, 43, 776.

255. TS, 32, 9b-10a; CTS, 17b, 2b; THY, 43, 776; WHTK, 291, 22a.
256. TS, 32, 10a.
257. THY, 43, 776.
258. THY, 43, 776.
259. TS, 8, 7b.
260. THY, 43, 776.
261. THY, 43, 776.
262. THY, 43, 776.
263. THY, 43, 777.
264. THY, 43, 777.
265. TS, 32, 10a; CTS, 17b, 25b.
266. THY, 43, 777.
267. CT, 18a, 3b.
268. TS, 32, 10a; THY, 43, 777; WHTK, 291, 22b.
269. THY, 43, 777.
270. CTS, 18a, 4a; THY, 43, 777.
271. CTS, 18a, 3b.
272. CTS, 18a, 16a; THY, 43, 777. (THY gives a date corresponding to 26 March.)
273. TS, 32, 10a-10b.
274. TS, 32, 10b; WHTK, 291, 23b.
275. TS, 32, 10b.
276. TS, 32, 10b; WHTK, 291, 24a (WHTK has the wrong year [876]); TCTC, 257, 6a.
277. WHTK, 291, 24a.
278. TS, 32, 10b; WHTK, 291, 24a.
279. TS, 32, 11a; WHTK, 291, 24b.
280. TS, 32, 11a; WHTK, 291, 24b.
281. TS, 32, 11a; WHTK, 291, 24b.
282. TS, 32, 11a.
283. Hastings, 1962, IV, 579.
284. John Milton, *Paradise Lost*, 1, 741-5.
285. Yang Chiung, "Hun t'ien fu," CTW, 190, 4b.
286. CCWKCF, 10a.
287. Ho, 1966, 136-138, classifies divine messengers as "streaming" (my translation) when they come from above, "flying" when they rise from below, and "running", a larger variety. He adds: "the magnitude of a meteor indicates its relative importance as a messenger."
288. Chapin, 1940, 46.

289. Chang Chi, "Chi T'ui-chih," CTS, han 6, ts'e 6, ch. 2, 8b.

290. CCWWYK, 17b.

291. Ch'en tzu-ang, "Chia yü Chi shih yü kuei Ts'ui szu i t'ai chih erh shih," CTS, han 2, ts'e 3, ch. 2, 5a.

292. Ch'üan Te-yü, "Sung Yüan chung ch'eng ch'ieh chieh ts'e Nan chao, wu yün," CTS, han 5, ts'e 8, ch. 4, 2a.

293. Lu Wo, "T'i chia hsiang i," CTS, han 9, ts'e 3, 6b. Cf. the almost identical usage in Lu Shih-cheng, "Fen P'ei we hsiang kung hsi t'ing yeh yen Lu lang chung," CTS, han 5, ts'e 7, 8b.

294. Sun T'i, "T'ung Hsing p'an kuan hsün Lung ch'uan kuan kuei hu chung," CTS, han 2, ts'e 7, 6a.

295. Liu Yü-hsi, "T'ang shih yü . . . Yao yüan wai, so ho chien cheng chi tso," CTS, han 6, ts'e 2, ch. 3, 11b.

296. Sun T'i, "Sung Chou p'an kuan wang T'ai chou," CTS, han 2, ts'e 7, 4b.

297. Hsüan miao nei p'ien, STCN, 8, 4a. The quoted source must be T'ang or earlier.

298. See Schafer, 1973a, 24.

299. Ch'u tz'u, "Chiu huai."

300. FYCL, 5, 24a.

301. LH, "Ju tseng." See also LH, "Shuo jih," in which chapter Wang Ch'ung opposes a view that only the fixed stars (as well call them), unlike the five planets, have stoniness latent in them. The "five stones" of Sung do sound uncommonly like petrifactions of the Five Planets.

302. CCYMP, 74a.

303. FYCL, 6, 24b–25a.

304. Ts'ui Mien, "Lo hsing shih fu," CTW, 273, 1a–1b.

305. HNT, "T'ien wen."

306. Li Po, "Ni ku," CTS, han 3, ts'e 6, ch. 23, 1b.

307. Not to be confused with "siderite" as the name of a mineral which is chemically ferrous carbonate.

308. For data on Chinese records and descriptions of meteorites see Chang, 1921, 322–332; Needham and Wang, 1958, 433–434.

309. Barnard, 1971, 639–640.

310. Wallenquist, 1966, 38.

311. Ho, 1966, 150.

312. See, inter alia, Schafer, 1967, 106, 153–154, based chiefly on LPLI.

313. LMC.

314. Wei Chuang, "Yu hsüan chi hsü," CTW, 889, 4b–5b.

315. Wallenquist, 1966, 238; Barnes, 1969, 704–708; Futrell, 1972, 106–116. The latter source contains some good color photos of tektites. Barnes' report is based on an earlier study in Chinese by Lee Da-ming published in Scientia Geologica Sinica in 1963.

316. Futrell, loc. cit. One writer (Touché-Skadding, 1947), with the help of poorly understood or quite misunderstood versions of Chinese sources, has tried to demonstrate that these correspond to the old Chinese "fire orbs" (huo chu). His arguments are painfully unpersuasive. For fire orbs, see Schafer, 1963b, 237–239; they were normally spherical burning cyrstals.

317. LPLI, quoted in Schafer, 1967, 106. Chang, 1921, 147–150, mentions them but has not identified them with tektites.

318. LMC.

319. Schafer, 1961, 6–7.

320. Po Chü-i, "Wei ch'i tzu t'ai tzu pin k'o tsai ch'u pi shu chien i ch'ang chü ho erh chien chih," CTS, han 7, ts'e 7, ch. 32, 13a.

321. Han Tsung, "Hsing-p'ing hsien yeh chung te lo hsing shih i chih hsien chai," CTS, han 9, ts'e 3, 1b.

322. Po Chü-i, "Wen chih ch'in shih," CTS, han 7, ts'e 7, ch. 31, 3b.

323. Whitelock, 1961, 140.

324. Meadows, 1969, 58.

325. CCWYMP, 71b.

326. See Ho, 1966, 130–131.

327. TS, 32, 7b.

328. TS, 32, 8b.

329. CTS, 19a, 8b.

330. Ho, 1966, 130. See this modern source and various ancient ones for some highly specialized cometary forms, to which I pay little attention here.

331. Yüan Chen, "Tai ch'ü chiang lao jen, po yün," CTS, han 6, ts'e 9, ch. 10, 2b.

332. See for instance, SM, "Shih t'ien"; T'ien wen lu in TPYL, 7, 8a.

333. Allen, 1963, 123.

334. TS, 32, 5b.

335. TS, 32, 5b.

336. E.g., for T'ang, Chia Kung-yen (seventh century), commentary (i su) on CL, "Pao chang shih." Cf. Ho, 1966, 129–130. Comets were regularly reported in the "basic chronicles" of the T'ang dynastic histories; these are not always the same as those in the "Tractates on the Patterns of Heaven." The reason for this distribution is not always readily apparent.

337. Liu Hsiang, Hung fan chuan, TPYL, 7, 4a.

338. HNT, "T'ien wen"; CCWKIY, 7b.

339. SS, 20, 2406b.

340. CS, 13, 1106a.

341. CCWYTS, 30a.

342. HC, 2a.

343. HC, 3a.

344. EY, "Hsing ming"; CS, 12 1105d; HC, 2a, 6b, 7b–8a, 9a–9b; CCWYTS, 9b, 12b–13a, 30b, 31a; CCWKCF, 26b; CCWPCT, 10a; SWHSW, 7a; Ho, 1966, 84–96, 129–130.

345. HAFC, 68, 22b.

346. Kuan-hsiu, "Ching shih ma chung tso," CTS, han 12, ts'e 4, ch. 8, 3b.

347. Liu Ch'ang-ch'ing, "Chih te san nien ch'un cheng yüeh . . . wu shih yün," CTS, han 3, ts'e 1, ch. 4, 9a.

348. Chiao Tao, "P'ang chiu shih," CTS, han 9, ts'e 4, ch. 3, 1b–2a.

349. Tu Fu, "T'i Heng shan hsien wen hsüan wang miao hsin hsüeh t'ang ch'eng lu lao," CTS, han 4, ts'e 2, ch. 8, 16a.

350. Cheng Ku, "Hui luan," CTS, han 10, ts'e 6, ch. 2, 5a.

351. For a list of Chinese observations of Halley's comet see Needham and Wang, 1958, 432.

352. 26 March 626 (THY, 43, 765); 9 May 639 (TS, 2, 7b; THY, 43, 765); 29 September 663 (TS, 3, 5b); 4 November 675 (TS, 3, 9a; CTS, 5, 6b; THY, 43, 768); 20 April 683 (TS, 3, 11a; CTS, 5, 11a; THY, 43, 766); 16 November 707 (CTS, 7, 6b; THY, 43, 766); 16 September 709 (TS, 4, 14a; CTS, 7, 8a; THY, 43, 767); 1 April 738 (CTS, 9, 2a; THY, 43, 767); 22 January 767 (TS, 6, 9b; CTS, 11, 11a; THY, 43, 767); 17 February 817 (TS, 7, 14a; CTS, 15, 11a; THY, 43, 767); 27 February and 6/7 March 821 (TS, 8, 1b; CTS, 16, 7a; THY, 43, 767); 16 April 825 (CTS, 17a, 5a); 3 September 828 (TS, 8, 4b; CTS, 17a, 14a; THY, 43, 767); 22 September 857 (TS, 8, 12a; CTS, 18b, 16a); between 10 February and 10 March 894 (TS, 32, 7b).

353. TS, 2, 6a; CTS, 3, 3a; THH, 43, 765.

354. TS, 102, 4a; CTS, 72, 2b. Although the number of celestial apparitions reported in the astrological section of CTS are much fewer than those in TS, the former reports give much more circumstantial detail, with actual court reportage appended.

355. CTS, 3, 3a.

356. CTS, 3, 7a; 23, 2b; TCTC, 196, 3a; TFYK, 144, 1747; THY, 43, 765.

357. TS, 33, 1b.

358. TS, 3, 7a; CTS, 5, 2a–2b; 36, 6a; TFYK, 144, 1475; THY, 43, 766.

359. TS, 3, 9a; 32, 5b; CTS, 5, 6b; TFYK, 144, 1749; THY, 43, 766.

360. THY, 43, 766.

361. TS, 3, 10b; 32, 5b; CTS, 5, 9b.

362. TS, 32, 5b; THY, 43, 766.

363. Schove, 1956, 132; Wen, 1934, 196 notes a similar description ("like a circular object without a tail" in a European source).

364. TS, 32, 5b.

365. TS, 32, 5b; THY, 43, 766.

366. TS, 32, 5b–6a. This source has "second month" instead of "seventh month." The latter date, which is correct, is given in THY, 43, 766.

367. TCTC, 205, 3a.

368. TS, 5, 3a; CTS, 8, 2b; THY, 43, 767; TCTC, 210, 11a–11b.

369. TS, 5, 9a; THY, 43, 767.

370. Suggested by D. J. Schove in private communication; cf. Shove, 1975, 402. See Bede, 1954, 366-369.

371. TS, 6, 4b; 32, 6a; CTS, 10, 12a; THY, 43, 767.

372. Shove, 1956, 132; Wen, 1934, 196–197.

373. TS, 6, 10b; 32, 6a; CTS, 11, 17a–17b; THY, 43, 767.

374. Schove, private communication.

375. TS, 6, 10b; 32, 6a; CTS, 11, 19b; THY, 43, 767.

376. TS, 8, 2a.

377. TS, 8, 6b; CTS, 17b, 12b.

378. CTS, 36, 13a.

379. CTS, 36, 13a; THY, 43, 768.

380. Needham has pointed out that the Chinese seem to have been the first to recognize that when a comet appears in the morning sky its tail points west, but when it appears at evening it points east—that is, away from the sun. Needham and Wang, 1958, 432. This knowledge goes back at least to T'ang times. See above p. 108. See also Wen, 1934, 196–197.

381. Wen Tsung, "Huo hsing chien hsiu hsing chao," CTW, 73, 1a–2b; cf. text in CTS, 17b, 19b–20a.

382. TS, 8, 7a; 32, 6b; CTS, 17b, 19b–20a; 36, 13a–13b; THY, 43, 768; TFYK, 145, 1757–1758. No Indic title of the sutra, which is in the prajñā-pāramitā tradition, is known. It was translated by Kumārajīva, and later by Amogha.

383. Reischauer, 1955, 47, n. 205.

384. Schove, 1956, 132.

385. CTS, 17b, 21a.

386. Reischauer, 1955, 46–47.

387. TS, 32, has 7 February, while THY, 43, has 17 February. In view of an error of dating elsewhere in the THY account, I am inclined to prefer the TS version. TS, 8, 7b; 32, 6a–6b; CTS, 17b, 24a–24b; THY, 43, 768; TFYK, 145, 1759–1760.

388. Shove, private communication.

389. TS, 8, 8a.

390. TS, 8, 8a; 32, 6b. The latter reference mentions only that "the eastern quarter is the allotment of Yen."

391. Reischauer, 1955, 310.

392. Wu Tsung, "Hui hsing chien pi cheng tien te yin," CTW, 77, 6a–7a.

393. TS, 8, 8b; 32, 7a; THY, 43, 768. (It is THY that has 16 February.)

394. Li Shang-yin, "Wei Ju-nan kung ho hui hsing pu chien fu cheng tien piao," CTW, 771, 12b–13b.

395. TS, 32, 7a.

396. TCTC, 249.

397. TS, 32, 7a; TCTC, 250, 15a. TS gives a cyclical day in the fifth month; TCTC gives the same day in the third month. I follow the latter authority, whose sequence of dates is more convincing.

398. Schove, private communication.

399. TS, 9, 3a; 32, 7a.

400. CTS, 19a, 14b. This source calls it an aphelial comet.

401. TS, 32, 7a; THY, 43, 768.

402. I have borrowed Schlegel's translation.

403. TS, 32, 7a.

404. Translation of Dorothy Whitelock. Whitelock, 1961, 53.

405. TS, 10, 3a; 32, 7a.

406. TS, 32, 7b.

407. TS, 32, 7b.

408. TS, 32, 7b–8a; THY, 43, 768.

409. THY, 43, 768.

410. Ai Ti, "Hui hsing chien pi cheng tien ch'ih," CTW, 94, 2a; TFYK, 145, 11b–12a.

411. Ai Ti, "Yao hsing pu chien ch'ih," CTW, 94, 2b.

412. TS, 32, 7b–8a. CTS reports the event, but has the wrong cyclical day. CTS, 20b, 5b; THY, 43, 768.

413. Whitelock, 1961, 60; Schove, private communication.

414. Darwin, 1825.

415. Needham and Wang, 1958, 426.

416. Supernovae appear once in every one or two centuries in our galaxy, and at similar intervals in other galaxies. Needham and Wang, 1958, 426.

417. Clerke, 1905, 89.

418. Needham and Wang, 1958, 424.

419. Needham and Wang, 1958, 426.

420. Lundmark, 1921, 227–228.

421. Lundmark, 1921, 226; Ho, 1966, 23, 136; Needham and Wang, 1958, 431. See CS, 12, 1105c, for plain evidence of Chinese awareness of the distinctions.

422. HC, 10a.

423. Needham and Wang, 1958, 427.

424. CCWKCF, 30b–31b; CCWKIP, 28b, 31a–31b; CCWYMP, 71a; HC, 5a, 6b, 8a, 9a; Ho, 1966, 84–89, 93, 109–110.

425. CCWKCF, 30b–31a.

426. CCWYMP, 71a.

427. HC, 8a.

428. Li Po, "Ch'ou Chang ching yeh su nan ling chien tseng," CTS, han 3, ts'e 6, ch. 18, 4b–5a.

429. TS, 32, 6b; WHTK, 294, 11b.

430. Lundmark, 1921, 236.

431. TS, 32, 6b; WHTK, 294, 11b.

432. TS, 32, 6b; WHTK, 294, 11b.

433. TS, 32, 6b; WHTK, 294, 11b.

434. TS, 32, 7b; WHTK, 294, 12a.

435. TS, 32, 7b; WHTK, 294, 12a; Lundmark, 1921, 236.

436. TS, 32, 7b; WHTK, 294, 12a.

CHAPTER 7.
EMBODIED STARS

1. Schlegel, 1967, 461–462.

2. Tou Ch'ün, "Ts'ao t'ang yeh tso," CTS, han 4, ts'e 10, 13a.

3. Tu Hsün-ho, "Chi Tou ch'u shih," CTS, han 10, ts'e 8, ch. 1, 17b.

4. Ch'ien Ch'i, "Feng ho chung shu . . . Hsüeh erh shih yü," CTS, han 4, ts'e 5, ch. 3, 18a.

5. P'i Jih-hsiu and others, "Han yeh wen yen," CTS, han 11, ts'e 9, ch. 6, 2b.

6. P'i Jih-hsiu, "Ju lin wu tsung," CTS, han 9, ts'e 9, ch. 3, 3b. This poem on a sacred grotto is one of twenty he wrote in the summer of 870, on a visit to Grand Lake (T'ai hu).

7. Szu-k'ung T'u, "Mou wu san jih hui," CTS, han 10, ts'e 1, ch. 2, 13a.

8. Hsi-ch'ang, "Tu Ch'i-chih shang jen chi," CTS, han 12, ts'e 5, 9b.

9. E.g. Sun Ho, "K'u Fang-ying hsien sheng," CTS, han 10, ts'e 8, 5a.

10. Fang Kan, "K'u pi shu Yao shao chien," CTS, han 10, ts'e 3, ch. 3, 7a.

11. Liu Yü-hsi, "Chi ho Tung ch'uan . . . yin ch'eng shih shih," CTS, han 6, ts'e 3, ch. 7, 13b.

12. Schafer, 1967, 5, 154.

13. Tu Fu, "Heng-chou sung Li ta fu ch'i chang Mien fu Kuang-chou," CTS, han 4, ts'e 4, ch. 18, 13b.

14. See Schafer, 1967, 35; Li Mien held office there in 769–771.

15. Soymié, 1962, 316, relates his name "Grand White" to the moon rather than to Venus, and detects a dedication to that satellite also in the names of the poet's younger sister and two sons.

16. Allen, 1963, 263.

17. Connoisseurs of the Wine Star will find matters of interest beyond the sphere of Li Po's personal involvement. For instance there is a pair of matching poems titled "Wine Star" by P'i Jih-hsiu and Lu Kuei-meng—playful, but a little mannered—in CTS, han 9, ts'e 9, ch. 4, 6b and ibid., ts'e 10, ch. 4, 6a respectively.

18. Li Po, "Yüeh hsia tu cho, szu shou," No. 2, CTS, han 3, ts'e 6, ch. 22, 2a.

19. Yang Ch'i-hsien, in a scholium on Li Po's poem, "Ch'ou Ts'ui shih yü," in FLPCLTPS, 19, 13b–14a.
20. Frodsham, 1970, 51.
21. Li Ho, "Ch'in wang yin chiu," CTS, han 6, ts'e 7, ch. 1, 9a.
22. P'i Jih-hsiu, "Li han lin," CTS, han 9, ts'e 9, ch. 1, 7a.
23. Cheng Ku, "Tu Li Po shih," CTS, han 10, ts'e 6, ch. 2, 13a.
24. P'ei Yüeh, "Huai-su t'ai ko," CTS, han 11, ts'e 2, 1a–1b.
25. Schafer, 1963c, 97.
26. Li Po, "Teng T'ai po feng," CTS, han 3, ts'e 6, ch. 20, 2b.
27. Santillana and Von Dechend, 1969, 255. The authors' word is a natural derivative of Greek *katasterizein*.
28. Thompson, 1895, xiii.
29. Meadows, 1969, 87.
30. Needham, Wang and Robinson, 1962, 355, referring particularly to CSCN, a, 1a.
31. But see, for instance, Schafer, 1973a, 23–24.
32. SIC, 3, 4b.
33. CSCN, a, 5a.
34. Schlegel, 1967, 629, Bauer, 1956, 143–146; Needham and Wang, 1958, 282.
35. CT, "Yüan yu"; CCWYMP, 32a; Schlegel, 1967, 158, reporting from SC and elsewhere.
36. Chiao-jan, "Wen t'ien," CTS, han 12, ts'e 2, ch. 6, 13a.
37. Solger, 1922, 163–166.
38. TPKC, 6, 4b. This tenth-century version of the tale is a composite one based on two separate sources.
39. FSTI, 2, 14a; Needham and Wang, 1958, 282.
40. Hsien Ch'eng, *Hou Han shu,* in TPYL, 5, 7b.
41. Kuan-hsiu, "Wen Wang Ts'ao ch'ang shih tsu," CTS, han 12, ts'e 3, ch. 6, 1b.
42. PKLT, 1, 22b.
43. Yüan Chen, "Hsien Tsung chang wu hsiao huang ti wan ko tz'u, san shou," CTS, han 6, ts'e 8, ch. 8, 5a.
44. Allen, 1963, 364.
45. Po Chü-i, "Te Tsung huang ti wan ko tz'u, szu shou," No. 2, CTS, han 7, ts'e 4, ch. 18, 13a. Compare Ch'üan Te-yü, "Hui chao huang t'ai tzu wan ko tz'u," CTS, han 5, ts'e 8, ch. 8, 2a.
46. Meng Chiao, "Tiao Lu Yin," CTS, han 6, ts'e 5, ch. 10, 10.
47. Li Tung, "Kuo Chia lang hsien chiu ti," CTS, han 11, ts'e 2, ch. 3, 12b.
48. Ling-i, "K'u Wei shang shu," CTS, han 12, ts'e 1, 6b.
49. Tu Fu, "Wei chiang chün ko," CTS, han 4, ts'e 2, ch. 8, 2a.
50. Chang Yüeh, "Sung Wan Chün tzu yü lin fu Yung-ch'ang ling," CTS, han 2, ts'e 4, ch. 3, 6b.
51. E.g. Ts'en Shen, "Sung Chang lang chung fu Lung yu chin hsing ch'ing kung," CTS, han 3, ts'e 9, ch. 3, 3a.
52. Ch'en Tzu-ang, "Cheng tsung chih Ch'i men ta Sung shih i ts'an chün chih wen," CTS, han 2, ts'e 3, ch. 1, 11a.
53. Ts'ui T'ai-chih, "Feng chih sheng chih sung Chang shang shu hsün pien," CTS, han 2, ts'e 4, 6b.
54. Wang Ch'i-yung, "Chang chien hsing," CTS, han 2, ts'e 7, 4a.
55. Liu Ch'ang-ch'ing, "Shih p'ing hou sung Fan Lun kuei An-chou," CTS, han 3, ts'e 1, ch. 5, 14a.
56. Lu Ts'ung-yüan, "Feng ho sheng chih sung Chang Yüeh hsün pien," CTS, han 2, ts'e 6, 7a.
57. Hsiang Szu, "Ch'ang-an t'ui chiang," CTS, han 9, ts'e 1, 14b.
58. Li Po, "Ming t'ang fu," CTW, 347, 3b.
59. SWHSW, 3b; cf. a T'ang story derived from this tradition in CHL, TPKC, 59, 1a.
60. Li Po, "Hsi yüeh yün t'ai ko sung Tan ch'iu tzu," CTS, han 3, ts'e 4, ch. 6, 3a.
61. YCCC, 43, 13a.
62. YYTT, 11, 84.

63. SKCC, 35, 21b.
64. Maspero, 1950b, 125, citing *San t'ien cheng fa ching* and its commentary.
65. Lü Yen, in CTS, han 12, ts'e 6, ch. 3, 1a.
66. Liu Yü-hsi, "Ho Yen chi shih wen T'ang ch'ang kuan Yü jui hua hsia yu yu hsien erh chüeh," CTS, han 6, ts'e 3, ch. 12, 15a. The prose tale is from K'ang P'ien, *Chü t'an lu*, in TPKC, 69, 1b. It appears also in TLCCFK, 4, 2a, along with the notice of the location of the friary.
67. Wu Yung, "Yü nü miao," CTS, han 10, ts'e 7, ch. 3, 9b.
68. *San chiu su yü*, in TPYL, 674, 1b.
69. *Kuei shan yüan lu*, in TPYL, 674, 7b.
70. YYTT, 2, 17.
71. HWTNC, 1b. Cf. the translation in Schipper, 1965, 68.
72. YYTT, 2, 13.
73. HC, in TPYL, 672 and 673.
74. Schipper, 1965, 37–42.
75. *Chin ken ching*, in TPYL, 672, 2b. "White-silk . . ." is a real book, cited, for instance, in TPYL, 676, 5a.
76. *T'ai i ti chün tung chen yüan ching*, in TPYL, 672, 6a.
77. Schipper, 1965, 58–59.
78. HNT, "T'ien wen."
79. Lo Yin, "Ch'ü," CTS, han 10, ts'e 4, ch. 5, 3b.
80. Li Shang-yin, "Shih i yüeh chung hsün chih Fu-feng chieh mei hua," CTS, han 8, ts'e 9, ch. 1, 292.
81. It has been translated in Graham, 1965, 155. I presume to offer another version, since Graham has missed this matching, having translated the last line". . . amid the frost on the moon," which, to my mind, should read ". . . the frost and the moon," parallel to "The Dark Maid and the White Beauty" of line three, both of whom he places in the moon.
82. Li Shang-yin, "Shuang yüeh," CTS, han 8, ts'e 9, ch. 1, 2b.

83. Mair, 1969, 212; Olivieri, 1897, 5.
84. Olivieri, 1897, 16. Cf. Allen, 1963, 168 ff. for many fine tales about this constellation.
85. Whicher, 1965, 52.
86. CL, "T'ien kuan."
87. TS, 47, 10b.
88. *Ta hsiang lieh hsing t'u*, TPYL, 6, 2a.
89. TS, 47, 9b–10; CTS, 44, 4a. The Chinese is *hou fei szu hsing*, which includes the "Heirgiver" (*hou*), *Hui fei, Li fei*, and *Hua fei*.
90. CCWHCT, 4a.
91. Schlegel, 1967, 459, citing various sources.
92. *Wu chen ti chün chiu yin hung ho tsung ching wan hua yin t'ien ching*, in YCCC, 31, 10b–11a.
93. ESPCC, 1a.
94. Chang, 1973, 68, doubts whether she is to be identified with the moon goddess Ch'ang-o.
95. Yanagisawa, 1967, *passim*.
96. An honorific title bestowed on Confucius during the reign of T'ang T'ai Tsung.
97. The traditional "Confucian classics."
98. *Shen hsien kan yü chuan*, TPKC, 65, 1a–3a.
99. Or "Hoyden," as I have also styled her in Schafer, 1967, 124.
100. HC; SC, 27, 0109c, commentary of (T'ang) Chang Shou-chieh.
101. Mao Wen-hsi, "Huan ch'i sha," CTS, han 12, ts'e 10, ch. 5, 2b. I feel that I may have missed something about the connection between the "little cicada" (a distinct species) and Minx Woman. Also, I have emended *liang* "both" to *yü* "rain" in the last verse, the characters being alike—but doubtless sense could be made out of *liang*.
102. Schafer, 1967, 125, giving examples from Li Shang-yin and Wang Po.
103. Hsü Ching-tsung, "Feng ho ch'i hsi yen hsüan p'u ying, erh shou," No. 2, CTS, han 1, ts'e 9, 4a.

104. A full account of the story is probably best seen in Solger, 1922, 168-194, along with much interesting speculation about the antiquity and origins of various elements in it. He associates the Weaver Maid with the moon and with rain—the Seventh Evening being the symbolic end of the rainy season, and indeed in T'ang poetry poems about her frequently allude to clearing skies, thinning clouds, and brighter stars. He believes that the tale of the two lovers originally was applied to two quite different stars (i.e. not Vega and Altair), and that the episode of Vega crossing the Sky River originally referred either to the moon or to Venus. If this is so, the mythical crossing took place far from where it has in historical times, since the ecliptic crosses the galactic circle in the northern hemisphere between Taurus and Gemini, almost half way around the celestial sphere from Aquila and Lyra. (There is, of course, a southern crossing in Scorpius and Sagittarius.)

105. SC, 27, 0109c.

106. Chang Yüeh, "An-le chün chu hua chu hsing," CTS, han 2, ts'e 4, ch. 2, 15b.

107. Schafer, 1973a, 100-103, gives several T'ang examples.

108. Li Shang-yin, "Sheng nü tz'u," CTS, han 8, ts'e 9, ch. 2, 14b.

109. Li Shang-yin, "Yu ch'un," CTS, han 8, ts'e 9, ch. 3, 12a.

110. Sung Chih-wen, "Ch'i hsi," CTS, han 1, ts'e 10, ch. 2, 11b.

111. Wang Ch'u, "Yin ho," CTS, han 8, ts'e 2, 4b.

112. Benhamouda, 1951, 128, 132.

113. YYTT, 10, 76.

114. SC, 27, 0109c.

115. Chang Heng, Hsi ching fu, in WH, 2.

116. EY, "Shih t'ien," with commentary of Kuo P'u.

117. Schafer, 1970, 28.

118. Chih nü, "Tseng Kuo Han, erh

shou," and "Kuo Han ch'ou chih nü," CTS, han 12, ts'e 7, 7b.

119. Hsü Ching-tsung, "Ch'i hsi fu yung ch'eng p'ien," CTS, han 1, ts'e 8, 6a.

120. Solger, 1922, 172.

121. Schafer, 1973a, 21.

122. CCWTCC, 2a.

123. CCWYMP, 24a.

124. Waley, 1963, 149-151. The MS. is No. 6174 in the Giles catalogue.

125. See for instance Ch'üan Te-yü, "Ch'i hsi chien yü chu sun t'i ch'i ch'iao wen," CTS, han 5, ts'e 8, ch. 10, 5b, for a poetic account of these activities.

126. KYTPIS, 18a.

127. Liu Tsung-yüan, "Ch'i ch'iao wen," CTW, 583, 1b.

128. CCWHCT, 7a. For the relation of this celebration to other celestial banquets and to the peaches of immortality see Schipper, 1965, 51-53.

129. HC, 7b.

130. Allen, 1963, 316-317; Benhamouda, 1951, 173.

131. Allen, 1963, 202.

132. Wei Chuang, "Ho Cheng shih i ch'iu jih kan shih i po yün," CTS, han 10, ts'e 9, ch. 3, 10a-12b.

133. KCC, a, 6.

134. Chang Wei, "Yü-yang chiang chün," CTS, han 11, ts'e 3, 5a.

135. Bernhard, Bennett and Rice, 1948, 122, 130. For the mystery of the two swords at the Last Supper see McCready, 1973, passim.

136. Chapin, 1940, 17, quoting CS, 36.

137. Tu Mu, "Ho Hsüan-chou Shen Ta-fu teng pei lou shu huai," CTS, han 8, ts'e 7, ch. 5, 10b.

138. See Chapin, 1940, passim, for the best account of these wonders. Straughair, 1973, 60-63, also gives a standard version, but her account contains some unfortunate errors: she confuses the Ox constellation with the star Altair (which is associated with the lover of the Weaver Woman)—but this is an ancient confusion. She puts the Southern Dipper

incomprehensibly in Hercules instead of in Sagittarius, and seems not aware of the astrological authority of these asterisms in the Wu-Yüeh region—an important matter in this story. There are other solecisms, for instance when the "fiery star" (p. 112) is unaccountably called the second of the twenty-eight "Chinese zodiacal constellations," but should be either the planet Mars, or (as more rarely) the red star Antares.

139. Hsüeh Ying, "Lo shen chuan," translated in Schafer, 1973a, 134.

140. Li Ch'ün-yü, "Pao chien," CTS, han 9, ts'e 3, ch. 2, 17b.

141. Wei Chuang, "Ho Li hsiu ts'ai shu tsao ch'un yin ch'ü shih yün," CTS, han 10, ts'e 9, ch. 4, 4b.

142. Tu Mu, "Chi Shen pao hsiu ts'ai," CTS, han 8, ts'e 7, ch. 5, 8b.

143. TS, 36, 10a; also translated in Schafer, 1973a, 22. Cf. also Schlegel, 1967, 172, which notes the connection between "Dipper" and dragons.

144. Hu Tseng, "Yen-p'ing chin," CTS, han 10, ts'e 2, 11b.

145. Wang Tsung, "Yen-p'ing chin," CTS, han 9, ts'e 8, 6b.

146. Li Chiao, "Chien," CTS, han 2, ts'e 1, ch. 3, 6b.

147. Translated in Schafer, 1973a, 17, from CS, 97.

148. Ch'in T'ao-yü, "Pien chiang," CTS, han 10, ts'e 5, 3b.

149. Lo Pin-wang, "Tsai chün chung tseng hsien huan chih chi," CTS, han 2, ts'e 3, ch. 3, 4b.

150. Kao Shih, "Sung Hun chiang chün ch'u sai," CTS, han 3, ts'e 10, ch. 3, 5a.

151. Lo Pin-wang, "Ho Li ming fu," CTS, han 2, ts'e 3, ch. 3, 1b.

152. Shao Yeh, "Lan Meng Tung-yeh chi," CTS, han 9, ts'e 8, 2a–2b.

153. Yüan Chen, "Ch'ou Liu Meng chien sung," CTS, han 6, ts'e 8, ch. 8, 2b.

154. Kuan-hsiu, "Sai shang ch'ü, erh shou," (No. 2), CTS, han 12, ts'e 3, ch. 2, 2b.

155. Hsü Yin, "Lung chih, erh shou," (No. 1), CTS, han 11, ts'e 1, ch. 1, 10a.

156. Lo Pin-wang, "Chiu shu pien ch'eng yu huai ching i," CTS, han 2, ts'e 3, ch. 3, 11a. An almost identical couplet appears in Lo Pin-wang, "Sung Cheng shao fu . . . ts'ung jung," CTS, han 2, ts'e 3, ch. 2, 3b.

157. Hsü Hun, "Teng suan shan kuan fa chün," CTS, han 8, ts'e 8, ch. 10, 3a.

158. "Ching chen chien wen," CTS, han 12, ts'e 8, 11a.

159. Li Chiao, "Pao chien p'ien," CTS, han 2, ts'e 1, ch. 1, 4b.

160. Lo Pin-wang, "Ts'ung chün hsing," CTS, han 2, ts'e 3, ch. 2, 1a. The translation of the third line, yeh jih fen ko ying, may be overdone. I suspect a disjunction of the expression fen yeh, the key term of "disastrous geography," which would imply that the sun directs his axe to the place where national destiny is about to be played out.

161. Ch'i-chi, "Ku chien ko," CTS, han 12, ts'e 5, ch. 10, 4b.

162. Wang Wei, "Lao chiang hsing," CTS, han 2, ts'e 8, ch. 1, 24b–25a.

163. Liu Ch'ang-ch'uan, "Pao chien p'ien," CTS, han 5, ts'e 6, 1a.

164. Wei Ying-wu, "Ku chien hsing," CTS, han 3, ts'e 7, ch. 9, 3a–3b.

165. Li Ho, "Kung wu ch'u men," CTS, han 6, ts'e 7, ch. 4, 5b.

166. SC, 66, 0183b; Chapin, 1940, 45–46.

167. Wei Kao, "Tseng Hsia," CTS, han 5, ts'e 7, 5a–5b.

168. Owen Meredith (E. R. B. Lytton), The Wanderer. Cf. Allen, 1963, 391.

169. Schlegel, 1967, 448–449.

170. For further information see Schlegel, 1967, 502.

171. PPT, "Tsa ying."

172. SCHHCCT, 5a–7b.

173. Tu Mu, "Sung Jung-chou T'ang chung ch'eng fu chen," CTS, han 8, ts'e 7, ch. 2, 6a.

174. Wang Wei, "Tseng P'ei Min chiang chün," CTS, han 2, ts'e 8, ch. 4, 11a. Curiously there is a poem by Wei Kao, written about a century later, whose first and last lines are identical with those of Wang Wei's quatrain. I have not solved this puzzle to my satisfaction.

175. TS, 217a, 10a. Cf. similar descriptions of the nomadic Hsiung-nu and Hsien-pei in earlier times.

176. Allen, 1963, 36.

177. ———, 1928, Vol. 4, Pl. 9.

178. Boots, 1934, plates 12 and 13.

179. Bertuccioli, 1974, 412.

180. Forbes, 1950, 401, 415.

181. See p. 103, in the discussion of meteorites. Barnard, 1971, 639–640.

182. CCWYTS, 12b.

183. E.g. CCWYTS, 30a.

184. CS, 27, 0111c.

185. Li Shang-yin, "Ming jih," CTS, han 8, ts'e 9, ch. 1, 8a.

186. CKT, 4, 49b.

187. Many classical commentaries on the CL, *Li chi, et al.*

188. Ch'u Kuang-hsi, "Kuan Fan-yang ti fou," CTS, han 2, ts'e 10, ch. 2, 10b–11a.

189. Liu Ch'ang-ch'ing, "Sung Ch'i lang chung tien Kua-chou," CTS, han 3, ts'e 1, ch. 1, 20a.

190. Wu Yüan-heng, "Sung Hsü yüan wai huan ching," CTS, han 5, ts'e 7, ch. 1, 5b.

191. Nieh I-chung, "Hu wu jen hsing," CTS, han 10, ts'e 1, 2a.

192. YTCS, 124, 9a.

193. YTCS, 124, 9a.

194. HYKC, 3, 33.

195. Lo Pin-wang, "Ch'ou hsi p'ien," CTS, han 2, ts'e 3, ch. 1, 10a.

196. Li Chiao, "Ch'iao," CTS, han 2, ts'e 1, ch. 3, 5a.

197. Li Shang-yin, "Ch'i hsi," CTS, han 8, ts'e 9, ch. 1, 31b.

198. TT, 85, 485b.

199. WHTK, 138, 1227b; CTS, 29, 11b.

200. CFC, 7a.

201. SHC, 16a.

202. NS, 76, 2725a.

CHAPTER 8. THE SUN

1. "The sun has perished from the sky and an evil mist has spread over all."

2. Yang Chiung, "Hun t'ien fu," CTW, 190, 4b. HNT, "T'ien wen," gives other details about the mythological cosmography of the sun's wayfaring.

3. Gk. *ortyx* "quail," hence Ortygia, where the bird's cult was celebrated. Thompson, 1895, 125.

4. Thompson, 1895, 3–10; Allen, 1963, 57.

5. CCWYMP, 2b.

6. LH, "Shuo jih."

7. ChS, 4, 2270d.

8. SS, 3, 2353d.

9. TFYK, 25, 4a; (C)TS, in TPYL, 920, 6a.

10. YYTT, in TPKC, 462, 8a. Cf. Schafer, 1963a, 209. The episode involving Jui Tsung is said by our authority to be recorded in the *T'ang shu*, although I have not yet found it there. In any event, the original *Yu yang tsa tsu* can hardly have quoted from this high source, and we must suspect an additional note by the Sung editors of TPKC which has become merged and made continuous with the original text of YYTT. The passage appears in TPYL, 920, 6a, as from (C)TS without reference to YYTT.

11. MT, "Fei kung." Cf. Schafer, 1963a, 212–213.

12. PPT, in TPYL, 920, 6b.

13. Schafer, 1963a, 213, quoting SKC, 47, 1036b.

14. As translated in Schafer, 1963a, 213, from the text of *Chüan san kuo wen*, 66, 6b.

15. Schafer, 1963a, 212, quoting CTS, 37, 13b.

16. Schafer, 1963a, 213.

17. PPT, in TPYL, 920, 4a.

18. Yüan Chen, "Ch'iu hsi yüan huai," CTS, han 6, ts'e 8, ch. 5, 2b.

19. *Tsuan yao,* in CHC, 1, 5.

20. *Tsuan yao,* in CHC, 1, 5.

21. *Kuang ya,* in CHC, 1, 5.

22. So, for instance, *Fan tzu chi jan,* in TPYL, 3, 8a—presumably an authentic Chou source; HNT, "T'ien wen."

23. LH, "Shuo jih."

24. [*Chou*] *I ts'an t'ung ch'i,* in TPYL, 3, 3a.

25. Ho, 1966, 121.

26. CCWCTP, 1b.

27. Sivin, 1969, 6. The greatest number of eclipses possible in any one year is seven, taking solar and lunar together; the smallest number is two, both of which must be solar.

28. LH, "Shuo jih"; Needham and Wang, 1958, 411–414. There was also an ancient belief that the sun-crow was a kind of eclipse-demon who devoured the orb he inhabited.

29. He is quoted to this effect in PKLT, 1, 13b.

30. CCWKCF, 6a.

31. CCWCTP, 2a–5a. This source was much quoted by the T'ang astronomer Gautama Siddharta in his *K'ai yüan chan ching.*

32. See also, for example, the eclipse of 22 May 635, in the asterism Net. TS, 32, 1a. See also those of 646 and 665; and compare with other ominous phenomena in the same regions of the sky.

33. TS, 32, 1b. The eclipse of the third month of 665 was the same in all respects.

34. CTS, 12, 4a.

35. Bernhard, Bennett, and Rice, 1948, 108.

36. Needham and Wang, 1958, 436; Ho, 1966, 23.

37. TS, 32, 4b. Confusion about the nature of sunspots has been world-wide. In early medieval Islam and in Europe they were sometimes taken to be transits of Mercury and Venus. Shirley, 1974, 131.

CHAPTER 9.   THE MOON

1. Cf. Hawkes, 1962, 46–48. Hawkes regards these "questions" as sheer entertainment, more in the nature of riddles than serious problems. I am not quite convinced.

2. LC, quoted in TPYL, 4, 6b; HNT, quoted in TPYL, 4, 8a.

3. Needham interprets a classic statement of the irradiation of the moon by the sun to mean that the Chinese, like Parmenides before them (presumably) understood early that the moon shines by reflected light. But the text he cites says merely "the light of the moon then comes forth"—nothing about reflection. Needham and Wang, 1958, 227.

4. CCWYMP, 60a.

5. "Tung chen huang ch'i yang ching ching," quoted in WSPY, 3b.

6. *Ibid.,* 3b–4a.

7. Whicher, 1965, 30.

8. *Paradise Lost,* I, 63.

9. Cf. Needham and Wang, 1958, 415, for the mutual interaction and interdependence of sun and moon.

10. HNT, "T'ien wen"; LH, "Shuo jih"; Schafer, 1973a, 7–8.

11. PPT, quoted in TPYL, 4, 8b; cf. Needham and Wang, 1958, 485.

12. See examples in Needham and Wang, 1958, 390 note c.

13. Fertility of women also had lunar connections. Eberhard, 1968, 246, 382.

14. Eberhard, 1968, 239; Needham and Wang, 1958, 252. Both give possible origins of this myth.

15. Schafer, 1963b, 227. Needham has the *ching* of the moon radiating "sem-

308

inal virtue." See Needham and Wang, 1958, 492.

16. LC in TPYL, 4, 6b; FTCJ in TPYL, 4, 8b; CCWYMP, 22b.

17. Bernhard, Bennett and Rice, 1948, 84–85. Many of these Latin names are apparently owed to Johann Hevelius, the seventeenth-century lunar topographer.

18. See Schafer, 1965a, *passim*.

19. Hsü Yin, "Yüeh," CTS, han 11, ts'e 1, ch. 3, 14b.

20. *Romeo and Juliet*, II, 2.

21. SM, in TPYL, 4, 6a. Perhaps fortunately, good rhymes for *ngywăt were not common in Middle Chinese, so we are spared witless jingles comparable to our "June" and "moon." The only common ones are *k'ywăt (1) "imperial watchtower," (2) "defective"; *mywăt "socks"; *Ywăt "Viet[-nam]"; *ywăt "quoth"; *ywăt "stone axe"; *pywăt "shoot, emit"; *pywăt "hair of head"; *bywăt "invade"; all of these are listed in Karlgren's dialect dictionary in his *Phonologie Chinoise*, p. 871. *Kuang yün* lists 119 rhymes, an impressive number until one observes that most are alternate, rare, or obsolete forms. But compare note 80 below.

22. E.g. *Mathews' Chinese-English Dictionary*, and one of the definitions in Morohashi, *Dai Kanwa jiten*. See also Needham and Wang, 1958, 414.

23. SS, 20, 2405b.

24. These, all from PKLT, 1, 15b, correspond to *yao t'u, ch'an kuang, ch'an hui, su yüeh, su o,* and *su kuang*.

25. Ch'un-t'ai Hsien, "Yu ch'un t'ai shih," CTS, han 12, ts'e 7, 1b.

26. PKLT, 1, 15b, corresponding to *ch'an p'o, chin p'o,* and *yüan p'o*.

27. HCWYSC, a, 10a–10b.

28. YTFY, "Wu po."

29. Li Hua, "Tsa shih," CTS, han 3, ts'e 2, 1b.

30. TC, Chao kung, 7.

31. Cf. *Ts'an t'ung ch'i* (TSCC), 19–20, and SW, under *p'o* written as if

for "war-chief" meaning "new-born moon."

32. HCWYSC, a, 5a. Cf. LC, "Hsiang yin chiu i."

33. For readers who have not recently watched the monthly evolution of the moon perhaps it will serve some small purpose here to recapitulate: (1) at "new," the dark moon is closest to the sun; this is its position during solar eclipses; (2) when first visible it is just east of the sun, and shows a thin bright crescent on its western edge, which faces the sun; the faint moon follows the sun across the daytime sky, to become prominent as it sets in the west; (3) as the month goes on, the moon waxes through the gibbous stage towards "full," meanwhile falling further behind the sun in the east; (4) finally it is reduced once more to crescent size, but the crescent is now on its eastern edge, and the moon is just west of the sun, rising in the east before sunrise and coming closer to the sun until it disappears at the next conjunction ("new moon") at the end of the lunar month. About 29½ days are involved.

34. CYHYMC, 4a.

35. T'ai Tsung, "Liao-ch'eng wang yüch," CTS, han 1, ts'e 2, 6a.

36. Ch'ing Hua and Ts'ui Tzu-hsiang, "Yü Ts'ui Tzu-hsiang fan chou . . . ," CTS, han 11, ts'e 9, ch. 7, 4b. Liu Ch'ang-ch'ing, "Hu shang yü Cheng T'ien," CTS, han 3, ts'e 1, ch. 3, 16a. A similar construction occurs in a poem of Li Shang-yin: "Pi ch'eng," CTS, han 8, ts'e 9, ch. 1, 24a. I would translate it "first bears its new soul (*p'o*)," but A. C. Graham, whose version of this has made it widely known, has been misled by the dictionaries, and gives us "the dark begins to grow"—and this is a poem about "Seventh Night," when the moon is waxing, not waning!

37. Ma Tai, "Chung ch'iu yüeh," CTS, han 9, ts'e 2, ch. 2, 2b.

38. Compare the moon in eclipse, pictured by Lu T'ung as a pearl deep in a dragon's brain, or concealed in an oyster's womb. "Yüeh shih shih," CTS, han 6, ts'e 7, ch. 1, 1b. See Graham's translation, 1965, 84.

39. Hsüeh T'ao, "Yüeh," CTS, han 11, ts'e 10, 1b.

40. Yüan Chen, "Han feng hsi," CTS, han 6, ts'e 8, ch. 5, 3a.

41. Li Ching-fang, "T'ien-t'ai ching wang," CTS, han 8, ts'e 4, 6b.

42. Yüan Chen, "Fu te chiu yüeh chin," CTS, han 6, ts'e 9, ch. 14, 2a.

43. Szu-k'ung Shu, "Sai hsia ch'ü," CTS, han 5, ts'e 4, ch. 2, 11a.

44. SC, 27, 0111c.

45. PHT, "Feng shan"; JYT, 1b–2a, quoted in various places, as in KYCC, 77; SS, 20, 2406, and other sources. Wang Ch'ung reports that when once in Wang Mang's time Venus crossed the sky "its essence resembled a half moon." LH, "Shih ying." This has given me pause. Is it possible that some Chinese astronomers observed the phases of Venus? If so, many of these reports would be accounted for. But supposing that the crescent of Venus was actually discernible, why would such a regular phenomenon be regarded as a startling and unusual event?

46. Kühnert, 1901, *passim*. Cf. Needham and Wang, 1958, 422–423; Ho, 1966, 129.

47. So described in Wallenquist, 1966, 16.

48. I would emend the description of SC, 27, 0111c, by substituting *ching* "clear skies" for *ching* "germ; essence." Instead of a sentence that has troubled some commentators, we then get "a phosphor star appears when the sky is clear."

49. Kühnert, 1901, 634–637. Kühnert gives the necessary mathematics in full. I disagree with his interpretation of the phrase *chung k'ung*, which he makes "right in the space (left by the cres-

cent)," a highly improbable usage of *k'ung*. The "star" is not centered in the dark area of the moon, but is "centered in space," like Ezekiel's wheel "way up in the middle of the air."

50. LH, "Shih ying"; SC, 27, 0111c, commentary of Meng K'ang; Sung Chün, commentary on CCWYTC, 3a; SuS, 27, 1498b; SS, 20, 2406a.

51. TLT, 4, 18a–20a; TPYL, 7, 3a–4a. Needham and Wang point out that a true conjunction of the planets never occurs, since this requires that they all agree in longitude and latitude. Needham and Wang, 1958, 408. Nonetheless this does not seem to preclude the occasional occurrence of the splendid sight of the "string of beads" type of conjunction, as suggested in Sivin, 1969, 17, n. 2. A T'ang occurrence of the event is celebrated in a *fu* by K'o-p'in Yü, entitled "Assembly of Stars of Virtue," for which the rhyme scheme is derived from the couplet, "When estimable men gather below // stars of virtue assemble above." CTW, 525, 12b–13b. Here the writer seems to have extended the term "star of virtue" to include another type of auspicious occurrence—that of the conjunction of the five planets like a string of gems. The word I have translated "assemble" (*chü*) is the same one that is normally used of planetary conjunctions.

52. HCWSSC, a, 6b. Cf. similar accounts in CCWKCF, 10b; JYT, 1b–2a; PHT, "Feng shan." The latter two sources specify that the kingly virtue must be beneficial to the people generally.

53. SC, 12, 0044d.

54. TS, 204, 4b.

55. For instance, P'ei Tu, "Erh ch'i ho ching hsing fu," CTW, 537, 8a–9b; Hsia-fang Ch'ing, "T'ien ching ching hsing chien fu," CTW, 615, 3a–4a; T'ao Kung, "T'ien ching ching hsing chien fu," CTW, 594, 9a–9b.

56. P'ei Tu, *loc. cit.*

57. Po Chü'i, "Chi hsien pei tu liu shou P'ei ling kung," CTS, han 7, ts'e 7, ch. 34, 1b.

58. Tu Fu, "Hsing tz'u Yen-t'ing hsien . . . k'un chi," CTS, han 4, ts'e 3, ch. 13, 7b.

59. PKLT, 1, 18b–19a.

60. See, for instance, Eberhard, 1968, 91, where, among other things, the author notes that the cinnamon tree, being evergreen, is a symbol of immortality.

61. Ch'en, 1957, 335.

62. YYTT, 1, 6–7.

63. Yü Hsi, "An t'ien lun," 2b, both in CHC, 1, and TPYL, 4.

64. Wu Yung, "Ch'iu hsi lou chü," CTS, han 10, ts'e 7, ch. 2, 14b.

65. Probably in the Indic sense of the seven holy and precious substances, the *saptaratna*, readily confounded in China with many-colored cloisonné insets of minerals and glass.

66. YYTT, 1, 8.

67. Either *$k'w\bar{e}t$ (1) with the graph "to emerge" as determinative, or (2) with the "cave" determinative. The latter form appears in late Chou literature in the sense of "underground room; subterranean dwelling." See TzT, p. 2382 for alternates, e.g. *$ngyw\bar{a}t\ yw\bar{e}k$ "precinct of moon."

68. For this opposition see, for example, Liang Chien-wen ti, "Ta fa sung," CLW, 13, 4b.

69. Modern Chinese *yüeh-chih*.

70. Tu Fu, "Yu shih yü nan chiao fu," CTW, 359, 7b.

71. P'i Jih-hsiu "Wu chung k'u yü yin shu i po yün chi Lu Wang," CTS, han 9, ts'e 9, ch. 2, 5a.

72. CS, 51, 1220c.

73. Wu Yung, "T'i hua po," CTS, han 10, ts'e 7, ch. 4, 13a.

74. Li Po, "Su Wu," CTS, han 3, ts'e 6, ch. 21, 6b.

75. Liu Yü-hsi, "Sung kung pu Chang shih lang ju fan tiao chi," CTS, han 6, ts'e 3, ch. 9, 1a.

76. Eberhard, 1968, 91, provides the standard corpus of information about the subject. Unfortunately he accepts the unfounded tradition that there are no hares in Kwangtung as evidence that the moon-hare motif, which can be traced back to the late Chou or the beginning of Han, cannot be of southern origin. In fact, China, including south China, is well furnished with hares, including the "South China Hare" (*Lepus sinensis*), the "South Eurasian Field Hare" (*L. europaeus*), the "Hainan Hare" (*L. Hainanus*), and even the "Woolly Himalayan Hare" (*L. oiostalus*). There are no true rabbits (*Oryctolagus*) in China. Tate, 1947, 205–206. The Kalmuks profess to see a hare—which they regard as an incarnation of the Buddha—in the moon; the ancient Mexicans and the Hottentots also place him there. Hastings, 1962, I, 518. I have not been able to verify the idea that it is because the hare (unlike the rabbit) is born with open eyes and is accordingly (like the moon) an open-eyed sky-watcher, that it is identified with the moon. See G. S. Hawkins, *Stonehenge Decoded*, p. 181.

77. CCWYMP, 3b (HHTTS ed., 60–61).

78. Yao Ho, "Tui yüeh," CTS, han 8, ts'e 3, ch. 3, 13b.

79. Compare the similar illusion of flying roof-beams, an image exploited by Wang Po in his famous "Gallery of the Prince of T'eng."

80. The word translated "hop" (*$kyw\bar{a}t$) is one of the few perfect rhymes for "moon."

81. Szu-k'ung T'u, "Tsa yen," CTS, han 10, ts'e 1, ch. 1, 5a.

82. Marlowe, *Doctor Faustus*.

83. TCTC, 204, 2b.

84. Soymié, 1962, 296, based on HNT.

85. CCWYKT, 10b; TPYL, 4, 7b.

86. CCWYMP, 27.

87. The last two epithets are *hsien ch'an* and *yao ch'an*.

88. Li Chung, "Szu chü yang ch'un yu kan chiu chi Ch'ai szu t'u," CTS, han 10, ts'e 5, ch. 4, 5a.

89. "Frog" is *wa* or *ha-ma*.

90. PKLT, 1, 18a, quoting Li Po.

91. Li Chung, "Su ch'ing ch'i mi ch'u shih yu ch'un," CTS, han 11, ts'e 5, ch. 3, 4b.

92. Lu T'ung, "Yüeh shih shih," CTS, han 6, ts'e 7, ch. 1, 2a.

93. Li Po, "Ku feng," CTS, han 3, ts'e 4, ch. 1, 2a.

94. Schafer, 1967, 256; Eberhard, 1968, 203–204. Eberhard finds that the frog has sexual overtones.

95. Eberhard, 1968, 205.

96. *Wa*. The form *Kua* seems to be related to *kua* "snail." Schafer, 1967, 256.

97. Solger, 1922, 167. "Toad" is M.C. *chem-zhyo*. The moon goddess is *gheng-nga* or *zhang-nga*.

98. Meng Chiao, "Ch'iu huai," CTS, han 6, ts'e 5, ch. 3, 1a. A. C. Graham, 1965, 68, renders this "The face of the autumn moon freezes." But the final word in the line is *ping* in the "departing tone"—a verbalized noun "to become ice," not *tung*, the everyday word "to freeze."

99. Soymié, 1962, 305–306.

100. See Eberhard, 1968, 90, for the basics of his legend; Eberhard thinks he may be a debased moon god. See also Soymié, 1962, 304–305.

101. Li Ho, "Li P'ing k'ung-hou yin," translated in Schafer, 1972, 986.

102. Ernest Bramah, "The Great Sky Lantern," in *Kai Lung Unrolls his Mat*.

103. Shu Hsi, "Hsüan chü shih," CS, 51, 1221b.

104. Schafer, 1973a, 39–40.

105. Meng Chiao, "Chi yüan chung chu kung," CTS, han 6, ts'e 5, ch. 7, 2a.

106. Meng Chiao, "K'an hua," CTS, han 6, ts'e 5, ch. 5, 2b.

107. Po Chü-i, "Tsui hou t'i Li Ma erh chi," CTS, han 7, ts'e 3, ch. 15, 19b.

108. For the traditional coldness of the moon palace discovered by Hsüan Tsung, see Soymié, 1962, 300–301.

109. *Ch'i shih ching*, a sutra translated into Chinese in Sui times, in FYCL, 7, 2a.

110. Po Chü-i, "Tung ch'eng kuei, san shou," No. 3, CTS, han 7, ts'e 5, ch. 24, 6a.

111. Mao Wen-hsi, "Yüeh kung ch'un," CTS, han 12, ts'e 10, ch. 5, 4b.

112. Cheng Yü, "Chin yang men shih," CTS, han 9, ts'e 3, 3a–3b.

113. Ku K'uang, "Wang shih Kuang-ling san chi," CTW, 529, 9b.

114. CFC, 7a.

115. CFC, 5a.

116. Ch'eng Yen-hsiung, "Chung ch'iu yüeh," CTS, han 11, ts'e 6, 1a. Ch'eng Yen-hsiung (fl. 960), lived at the end of the Five Dynasties and at the beginning of the Sung period. He took his *chin shih* degree under Southern T'ang.

117. Li Hsien-yung, "Ch'un feng," CTS, han 10, ts'e 2, ch. 2, 1a.

118. TS, 223b, 1a–3a; CTS, 135, 1a–4a.

119. *I shih*, in TPKC, 64, 4a–5a.

120. Duyvendak, 1947, *passim*.

121. KTCHC, 7a–7b. Another version of this tale appears in *SHKYC*, 4, 6b–7a. There the "Song of the Purple Clouds" is transmitted to Hsüan Tsung by ten transcendents who appear out of the clouds; the sovereign plays the tune for Kao Li-shih on his jade flute. But there is no journey to the moon. On the face of it, the tale translated here is a fusion of this Taoist anecdote with a separate story about a trip to the moon.

122. Frankel, 1954, 149.

123. I have used the text of TTTS, 6b–7a.

124. For this much reported-on observance, see, *inter alia*, Soymié, 1962, 314–319.

125. For the traditional versions of the tale, see Soymié, 1962, 301–302, 308–314. For a more credible account of the origin of this dance and its music, see (*inter alia*) Schafer, 1963b, 114–115.

126. Wei Chuang, "Kuei kung tzu," CTS, han 10, ts'e 9, ch. 1, 5b.

127. The forms Ch'ang-hsi, Ch'ang-i (in two versions) and other variants are recorded in early texts. They show a strong phonetic resemblance to each other, exemplified by the constant presence of a graphic element (Modern Chinese *wo* "I"), which yields both a version with front vowel *ngiĕ* in T'ang times, and one with back vowel *nga*. We are confounded, however, by the fact that the first part of the goddess's name (M.C. *zhang*) appears to be a substitute for the tabooed personal name of a Han emperor (M.C. *gheng*). *Gheng* and *zhang*, in the sense of "regular, constant," are close synonyms. Is it possible that they are also related phonetically? For the basic story of Ch'ang-o's flight to the moon, with some admixed nonsense, see Soymié, 1962, 292–294 and Eberhard, 1968, 87–89. Much of interest about her early history can be found in TCTL, 13.

128. Eberhard, 1968, 47–48.

129. Schafer, 1954, 95.

130. Schafer, 1967, 125.

131. Liu Yü-hsi, "Huai chi," CTS, han 6, ts'e 3, ch. 8, 8a.

132. Graham, 1965, 107, makes her, or at least her sky home, one of "Li Ho's favorite symbols of escape." This follows from the myth of the goddess's flight which established her there. It is the other side of the coin of loss and separation.

133. Cf. Schafer, 1967, 125, and many other authorities.

134. Lo Ch'iu, "Pi hung erh shih," CTS, han 10, ts'e 5, 5b. For the story of Lo Ch'iu and Hung-erh see, for instance, Schafer, 1973a, 86–87.

135. Tu Fu, "Yüeh," CTS, han 4, ts'e 4, ch. 15, 22b.

136. Anon., "Yen ko," CTS, han 11, ts'e 9, 3a. The title means something like "song of amorous beauty," or even "voluptuous/erotic song." Whether the author or an editor gave it this title is impossible to say. Perhaps we should not allow it to infect our interpretation too much, but let the verses speak for themselves.

137. Meng Chiao, "Shan-chüan p'ien," CTS, han 6, ts'e 5, ch. 1, 9b.

138. Ts'ao T'ang, "Hsiao yu hsien shih, chiu shih pa shou," No. 38, CTS, han 10, ts'e 2, ch. 2, 4b.

139. Schafer, 1973a, 73.

140. Mao Wen-hsi, "Wu shan i tuan yün," CTS, han 12, ts'e 10, ch. 5, 3a.

141. LT, "Chou Mu wang."

142. Persons of different surname, i.e. relations in the female line.

143. Yang Chiung, "Hun t'ien fu," CTW, 190, 4b.

144. Ho, 1966, 121–122.

145. Unicorns fight during eclipses of the sun and moon. HNT, "T'ien wen."

146. TS, 33, 1a. Sivin (p.c. of 13 June 1976) rightly urges me to note that "trespass" (*fan*) in such places means to occult, or nearly so, the determinative star of the asterism.

147. TS, 33, 1b.

148. Allen, 1963, 236.

149. TS, 33, 2b.

150. TS, 33, 3b.

151. TS, 33, 3b.

152. TS, 33, 4a.

153. TS, 33, 6a. This was 25 October 822.

154. TS, 33, 1b–2a.

155. Cf. Lu T'ung's eclipse poem, in which the moon symbolizes the Son of Heaven. "Yüeh shih shih," CTS, han 6, ts'e 7, ch. 1, 1a–5a. This long and fascinating poem (translated by A. C. Graham in his 1965, 84) deserves careful study, especially in connection with the related one by Han Yü, "Yüeh shih shih hsiao yü ch'uan tzu," CTS, han 5, ts'e 10, ch. 5, 13b–15a.

156. Sivin, 1969, 57, 62–63, notes that Chinese eclipse prediction never reached perfection because it never took

purely empirical considerations sufficiently into account, remaining always
fettered by formal beliefs.

157. Berhard, Bennett and Rice, 1948,
104.

158. HNT, "T'ien wen."

159. KYTPIS, 28b.

160. Needham and Wang, 1958, 228,
as corrected by Sivin (p.c. of 13 June
1976), following Yabuuti, who has Rahu
representing both lunar nodes, and Ketu
standing for the precession of the lunar
perigee.

161. TS, 33, 4b–5a.

CHAPTER 10. THE PLANETS

1. Bernhard, Bennett, and Rice, 1948,
147, 154.

2. *Shang sku k'ao ling yao*, in TPYL,
5, 3b; PPT, in TPYL, 6, 6a; Needham and
Wang, 1958, 398.

3. *Tung chen pa su chen ching*, in
WSPY, 3, 7a.

4. Maspero, 1950a, 22; Needham and
Wang, 1958, 398–408.

5. For a list of the various ominous
possibilities see Ho, 1966, 21.

6. TS, 33, 2a.

7. TS, 33, 4a.

8. E.g. Chang Shu-liang (eighth century), "Wu hsing t'ung se fu," CTW,
441, 1a–2a.

9. The translations are those of Ho,
1966, 126.

10. SHC, 10, 7b.

11. Ho, 1966, 124.

12. E.g. TS, 33, 2b, for 16 May 685,
when Mercury appeared in Well; the
planet is described as a sort of police
agent, and Well as a symbol of statutes
and decrees. The intrusion by the planet
means "the Way is misapplied."

13. CCWYMP, 73b.

14. Other, very uncommon names
appear in *Kuang ya*, TPYL, 5, 4b.

15. Omens summarized in Ho, 1966,

124. See especially CCWYMP, 73a:
"When the Grand White star is luminous, it is neither good nor evil . . ."

16. TS, 33, 2a.

17. TS, 33, 3a.

18. TS, 33, 4b.

19. TS, 33, 3a.

20. TS, 33, 2b.

21. TS, 33, 1b.

22. So in 626 and 680; TS, 33, 1a–1b,
2b.

23. TS, 33, 1a.

24. Wang Wei, "Lung t'ou yin," CTS,
han 2, ts'e 8, ch. 1, 24a.

25. Liu Ch'ang-ch'ing, "Lü tz'u Tan-
yang . . . ," CTS, han 3, ts'e 1, ch. 4,
2a–2b.

26. *Kuang ya*, TPYL, 5, 4b.

27. TS, 33, 1b.

28. TS, 33, 4a.

29. See, for instance, SC, 27, 0110a;
CCWYMP, 73a; Ho, 1966, 122–123.

30. TS, 33, 5a.

31. TS, 33, 1b, 2a, 2b.

32. TS, 33, 3a.

33. TS, 33, 1a.

34. TS, 33, 1b.

35. TS, 33, 2a.

36. TS, 33, 3b.

37. TS, 33, 3b.

38. TS, 33, 7b.

39. TS, 146, 6a. See also the case of
Hsiao Yü, who tried to resign his minis-
terial position when Mars entered
"Holder to the Law." TS, 101, 7b.

40. SKC, Wu chih, 3, 1040b, commentary, quoting from *Sou shen chi*;
SuS, in TPYL, 7, 6a–6b.

41. Dubs, 1958, 299. The monk I-
hsing played an important part in this
determination.

42. Ho, 1966, 122.

43. TS, 33, 1b.

44. TS, 33, 2a.

45. TS, 33, 3b.

46. TS, 33, 3b, 4a.

47. TS, 33, 3b.

48. TS, 33, 2b.

49. TS, 33, 1b.

50. E.g. in 624 (TS, 33, 1a) and in 767 (TS, 33, 3b).

51. E.g. the drought of 648 (TS, 33, 1b); battle and death in 777 (TS, 33, 4b).

52. TS, 33, 1a. For other interpretations, good and evil, of Saturn's movements, see Ho, 1966, 123–124.

53. TS, 33, 8b.

54. TS, 33, 9a.

55. Nakayama, 1966, 447, points out that planetary conjunctions remained ominous despite their predictability on the basis of the Grand Conjunction period, in that they differed from such random apparitions as comets and novae in that "astral influence was simply considered to operate in regular sequences" in such cases.

56. TS, 33, 8a.

57. TS, 33, 8a.

58. Summarized, for instance, in WHTK, 293, 2a–2b.

59. Examples in Ho, 1966, 126.

60. TS, 33, 8a.

61. Santillana and Von Dechend, 1969, 177.

62. TS, 33, 8a–8b. The same fight between "Metal Fire" (Venus) and "Penalty Star" (Mars) took place in Well in the summer of 783. TS, 33, 8b.

63. Ho, 1966, 127. Ho also observes the notion that although such a congregation "in the east" favored China, it stood for the triumph of foreigners when in the west. I have seen no T'ang examples of the latter omen. Similarly, the statement in CCWYMP, 9a, to the effect that "with an assembly of the Five Stars, the Son of Heaven is at the end of his resources," is not borne out, in my experience, by actual examples.

64. TS, 33, 8a.

65. TS, 33, 4a.

66. TS, 33, 2a.

67. Ho, 1966, 127, gives some exact particulars: when it is Jupiter that is occluded, the state shall be undermined by famine; when Mars, by revolution; when Saturn, by assassination; when Venus, by defeat in battle; when Mercury, by women.

68. TS, 33, 1b, 2b, 5a, 7a.

69. TS, 33, 5a.

70. TS, 33, 3a.

CHAPTER 11.
ASTRAL CULTS

1. Santillana and Dechend, 1969, 177, referring chiefly to western star lore, write "... the constellations were seen as the setting, or the dominating influences or even only the garments at the appointed time by the Powers in various disguises on their way through their heavenly adventures." Assuming that this characterization is a reasonable one, it is not entirely applicable to the Chinese stellar hierarchy, which was far from being an adventurous one.

2. See, for instance, Li Ho, "Shen hsien," CTS, han 6, ts'e 7, ch. 4, 10a–10b.

3. Examples of early T'ang tombs continuing the practice of the preceding northern dynasties, which show the potent celestial asterisms on their ceilings, have been revealed in modern times, although their existence has been alluded to in the accumulated literature of centuries. See Anon., 1961, 97–98.

4. HHS, 17, 068d.

5. LH, "Chi i."

6. HS, 25b, 0397d.

7. KYTPIS, 27b.

8. TS, 11, 4b–5a; TS, 12, 4a; Wu ching t'ung i, quoted in CTC, 1, 2; Yang Chiung, "Hun t'ien fu," CTW, 190, 3a; CS, 11, 1102b.

9. TS, 12, 3b.

10. The texts of some T'ang ceremonial chants survive, for instance one used during the great rain-making ritual, which invokes the "Dragon Asterism." See TS, 30, 10b.

11. THY, 22, 426.

12. THY, 22, 427.

13. The birthday celebration, under this name, had been inaugurated in 729.

14. THY, 22, 427; TT, 44, 257a; CTS, 8, 19b. HHS, 15, 0685c, notes a national sacrifice to "Old Man Star" in the "mid-autumn month"—obviously one of the precedents for the present revival.

15. This fundamental fact has been commented on frequently by students of Taoism. For a good reference, see Maspero, 1950b, 144.

16. See Ishii, 1969, *passim*, for the names and some of the attributes of the *ling pao* gods, and their hierarchical status.

17. See, for instance, Strickmann, 1975, 1048.

18. SHC, 12a–12b.

19. Wei Chuang, "Nü kuan tzu," CTS, han 12, ts'e 10, ch. 4, 9b.

20. Li Hsün, "Wu shan i tuan yün," CTS, han 12, ts'e 10, ch. 8, 8b.

21. Schafer, 1963b, 206–207.

22. FHKYLT, 3b.

23. Quotations in TPYL, 675, 1b, 2a, 3a, from *Shang ch'ing ching, Shang ch'ing pien hua ching,* and *Shang ch'ing yüan lu* respectively.

24. Li Ch'i, "Wang Mu ko," CTS, han 2, ts'e 9, ch. 2, 2b.

25. Examples in Schafer, 1963b, 112–115.

26. Tai Shu-lun, "Han kung jen ju tao," CTS, han 5, ts'e 1, ch. 1, 26a.

27. Han Wo, "Ch'ao t'ui shu huai," CTS, han 10, ts'e 7, ch. 3, 9b.

28. Chiao Tao, "T'i tai sheng," CTS, han 9, ts'e 4, ch. 4, 12b.

29. *Chen ching*, in TPYL, 676, 7b.

30. *Wu ch'eng fu shang ching*, in TPYL, 676, 9b.

31. Meng Chiao, "Yu Hua-shan Yün t'ai kuan," CTS, han 6, ts'e 5, ch. 4, 6b.

32. Hsiang Szu, "Ku kuan," CTS, han 9, ts'e 1, 1b.

33. Wu Yung, "Ho P'i po shih fu hsiang ching . . . . chien chien chi," CTS, han 10, ts'e 7, ch. 4, 7a.

34. Yü Yen, in CTS, han 12, ts'e 6, ch. 1, 4a.

35. Li Chao-hsiang, "Hsüeh hsien tz'u chi Ku Yün," CTS, han 10, ts'e 8, 5b.

36. Li Chung, "Fan Tung shen kung Shao tao che pu yü," CTS, han 11, ts'e 5, ch. 1, 1b.

37. Ch'en Yü, "Pu hsü tz'u, erh shou," No. 1, CTS, han 6, ts'e 1, 9b.

38. TLSKC, 8a.

39. PSC, 23b–24a.

40. HHHP, 2, 89.

41. FPNTTWYC, 6b.

42. Ma Tai, "Yeh hsien kuan, erh shou," No. 2, CTS, han 9, ts'e 2, ch. 2, 6b.

43. Ma Tai, "Tseng tao che," CTS, han 9, ts'e 2, ch. 2, 11b.

44. CCSHSC, 8b.

45. *Wei fu jen nei chuan*, in STCN, 8, 22b.

CHAPTER 12.
FLIGHT BEYOND THE WORLD

1. "Word-hoard" is blessed by Webster, although it represents a fairly modern restoration of Old English *word-hord*. The anthology's most recent translator—into English at any rate—David Hawkes has paraphrased it "Songs of the South." My version shows more resemblance to the one suggested by Arthur Waley some years ago.

2. See, for instance, Waley, 1956, and Hawkes, 1962, *passim* , for translations and comments, especially Hawkes, 1962, 29, 81–87. See also Laufer, 1928, 17.

3. See Waley, 1956, 42, 45; Hawkes, 1962, 41–42, 85.

4. Laufer, 1928, 18.

5. Laufer, 1928, 26–27.

6. Laufer, 1928, 15–16.

7. Kaltenmark, 1953, 14.

8. Kaltenmark, 1953, 10ff., Wang, 1962, 6.

9. Thomas Moore, "Paradise and the Peri," *Lalla Rookh.*

10. WTS, 65, 4469d.
11. Hsü Hsüan, "Kuan chi wang ts'ung ch'ien hua chu," CTS, han 11, ts'e 5, ch. 6, 9a.
12. Cf. such expressions as "flame flowers" in T'ang literature. Examples in Wang, 1947, 164.
13. Laufer, 1928, 28.
14. Li Po, "Ch'ao Lu ju," CTS, han 3, ts'e 6, ch. 24, 3b.
15. Li Po, "Chiang sheng sun nü tao shih Ch'u san ch'ing yu nan yüeh," CTS, han 3, ts'e 5, ch. 17, 1b.
16. Li Shang-yin, "Wa," CTS, han 8, ts'e 9, ch. 1, 34a, translated in Schafer, 1973a, 89–90.
17. Laufer, 1928, 23.
18. Laufer, 1928, 29–30.
19. CT, "Chiu pien." Conventional Confucian scholiasts take the magic out of this by interpreting it as "walk under the stars until dawn," like an insomniac.
20. Yü pu and wu pu.
21. Granet, 1925, 147–148; Granet, 1926, 551, 561; Eberhard, 1968, 74–75.
22. Eberhard, 1968, 74–75.
23. Granet, 1926, 549–551, thought that the left foot was always trailing, never passing the right; Ware, 1966, 198, allows occasional priority to the left. The passage is in PPT, "Hsien yao."
24. PPT, "Hsien yao"; "Tsa ying"; and "Teng she." All are englished in Ware, 1966, 180, 260 and 285 respectively.
25. Granet, 1926, 574.
26. PSC, 24a.
27. CHL, in TPKC, 56, 2a.
28. Ta tung yü ching, in TPYL, 660, 2a; for the "Grand Supreme Perfected Men" see also T'ai chen k'o, in TPYL, 660, 2b, and other standard Taoist sources.
29. CTPC, 16, 10b–11a.
30. Huang lao ching, in YCCC, 24, 7a–7b; Pei chi ch'i yüan tzu t'ing pi chüeh, in ibid.
31. Hsüan men pao hai ching, in YCCC, 24, 3a.

32. WHCYKCC has even more complex diagrams.
33. TLSKC, 4a–5b.
34. Soymié, 1972, 662, based on CSYTSC and other early texts in the Taoist canon.
35. CCYTSC, 1c.
36. CCYTSC, 1d.
37. Hsüan men pao hai, in YCCC, 24, 1b.
38. YYTT, 2, 9.
39. TS, 31, 9a.
40. CCYTSC, 1a.
41. Huang lao ching, in YCCC, 24, 7a–7b. The name I have rendered "Solar Luminosity" is an important one in the Chinese doctrine of cosmic correspondences. Porkert has shown us its position in the cycle at the overlap of the positions of Major Yang (t'ai yang) and Minor Yang (shao yang), where these two reinforce each other. Porkert, 1974, 35, 39.
42. Yüan Chen, "K'ai yüan kuan hsien chü ch'ou Wu shih chü shih yü, san shih yün," CTS, han 6, ts'e 9, ch. 10, 3a–4a.
43. For a useful study of the historical development of the major types of yu hsien poetry, see Chu, 1948, passim. See also, if possible, Wang, 1962, 8–9.
44. CTS, 192, 8a. Wu Yün's official biographies may be found in TS, 196, 6a–6b; CTS, 192, 8a–8b.
45. Wu Yün, "Pu hsü tz'u," CTS, han 12, ts'e 6, 7b.
46. Wu Yün, "Pu hsü tz'u," CTS, han 12, ts'e 6, 7a.
47. Wu Yün, "Pu hsü tz'u," CTS, han 12, ts'e 6, 7a.
48. Wu Yün, "Yu Hsien," CTS, han 12, ts'e 6, 2b.
49. Po Chü-i, "Ch'ang hen ko," CTS, han 7, ts'e 3, ch. 12, 13a.
50. Lü Yen, in CTS, han 12, ts'e 6, ch. 1, 3a. My "potent and latent" translate ch'ien and k'un.
51. Lü Yen, in CTS, han 12, ts'e 6, ch. 1, 1b.

52. PS, 6, 2760c.

53. Wei Ch'ü-mou, "Pu hsü tz'u, shih chiu shou," No. 3, CTS, han 5, ts'e 7, 1b.

54. Tu Kuang-t'ing, "Tung t'ien fu ti yüeh tu ming shan chi hsü," CTW, 932, 4b–5a.

55. Minnaert, 1959, 52.

56. See Soymié, 1954, *passim*, for more on this subject.

57. Ch'en T'ao, "Wu shan kao," CTS, han 11, ts'e 4, ch. 1, 12a.

58. Stein, 1942, 43. For a full account, see Soymié, 1954, especially pp. 89–93.

59. Tu Kuang-t'ing, "Tung t'ien fu ti yüeh tu ming shan chi hsü," CTW, 932, 5a. Cf. Schafer, 1967, 145.

60. *T'ien ti kung fu t'u*, in YCCC, 27, 3a, therefore a text datable to before the year A.D. 1000.

61. Lü Yen, in CTS, han 12, ts'e 6, ch. 2, 7b.

62. *Ibid.*, 9b. "White horse's tooth" is the arcane name of a variety of cinnabar. Sivin, p.c. of 13 June 1976.

63. *Ibid.*, 2a.

64. Lü Yen, "Tseng Lo-fou tao shih," CTS, han 12, ts'e 6, ch. 3, 9b.

65. Tu Kuang-t'ing, "Tung t'ien fu ti yüeh tu ming shan chi hsü," CTW, 932, 4b–5b.

66. Fu Tai, "T'i Li Pao-po tung," CTS, han 7, ts'e 9, 2a.

67. YYTT, 2, 9. M. Strickmann has pointed out to me that the ultimate source of this information is CK, 15, 1a ff., in which the underground caverns are described as the homes of the dead, presided over by Taoist superbeings.

68. See characters in glossary. TzT, p. 0043.

69. For the list, see Schafer, 1973a, 13–14.

70. Chang Chih-ho, "K'ung-tung ko," CTS, han 5, ts'e 6, 1a.

71. Ku K'uang, "Pei ko," CTS, han 4, ts'e 9, ch. 2, 4a–4b.

72. Kaltenmark, 1960, 571.

73. Stein, 1942, 45.

74. An example from the New World comes from southwestern Costa Rica, where the powerful shamans, called *usékar*, regularly entered caves to learn from spiritual beings. This seems to have been a common practice throughout aboriginal America. Aguilar, 1971, 31.

75. Aldred, 1971, 8–9.

76. *Chen kao*, in TPYL, 663. 2b.

77. Described in the excellent story of Ts'ui Wei, preserved in TPKC, 34. A summary may be found in Schafer, 1967, 97.

78. Ts'ao T'ang, "Hsiao yu hsien shih, chih shih pa shou," No. 17, CTS, han 10, ts'e 2, ch. 2, 2b.

79. Hsü Hsüan, "Pu hsü tz'u," CTS, han 11, ts'e 5, ch. 5, 3b.

80. Frye, 1966, 96–97.

81. *T'ai chen k'o*, in TPYL, 659, 3a.

82. *T'ai shang ching*, in TPYL, 660, 7a.

83. Hsü Ching-tsung, "Feng ho hsi hsüeh ying chih," CTS, han 1, ts'e 8, 3b.

84. Tu Fu, "Ta yün szu tsan kung fang, szu shou," No. 1, CTS, han 4, ts'e 1, ch. 1, 21a.

85. Meng Chiao, "Ho Huang-fu p'an kuan yu Lang-yeh ch'i," CTS, han 6, ts'e 5, ch. 4, 10b. The *liu-li* of the poem is cognate to our word "beryl," but in China was most often used in the sense of "colored glass."

86. SIC, 10, 4a. "Turquoise" is a conjectural lost meaning of the archaic word *yao*; see Schafer, 1963c, 96. The quotation is from a description of Ying-chou, the fairy island in the eastern sea.

87. Strontium sulphate.

88. Schafer, 1963c, 83–85, 93–94.

89. Schafer, 1963c, 86.

90. Schafer, 1963c, 77, 85, 99. I owe the idea underlying the last clause of this sentence to Nathan Sivin. See also Schafer, 1963b, 85, for more on coral trees and dendritic minerals.

91. Translated in Schafer, 1963b, 83.
92. Tu Kuang-t'ing, "Tung t'ien fu ti yüeh tu ming shan chi hsü," CTW, 932, 4b–5a.
93. Schafer, 1963b, 91–93.
94. Lo Pin-wang, "Tai nü tao shih Wang ling fei tseng tao shih Li Jung," CTS, han 2, ts'e 3, ch. 1, 13a.
95. Schafer, 1963b, 101.
96. Lou t'an ching, in FYCL, 5, 24a.
97. Lu Chüan, "Chi tung sung hu pu lang chung shih Ch'ien fu hsüan pu," CTS, han 2, ts'e 6, 7a.
98. FHKYLT, 5a.
99. Wang Po, "Shang-ming yüan wai ch'i," CTW, 180, 6a–8a. For yao as turquoise see note 86 above.
100. As, for instance, in Ch'u Tsai, "Yüeh," CTS, han 10, ts'e 8, 7b–8a.
101. Olivieri, 1897, 52.
102. For instance in Schafer, 1974, 401–403.
103. See summary in Schafer, 1974, 403, based on SC, 27, 0111b; Chang Heng, ling hsien, in TPYL, 8, 10b; HTKTH, 1b; Yang Ch'üan, Wu li lun, in TPYL, 8, 11a, and other sources.
104. Schafer, 1974, 403.
105. In his poem "Sky Ho," translated in Schafer, 1974, 404.
106. CS, 11, 1103a; Schafer, 1974, 404.
107. Meng Chiao, "Lo chiao wan wang," CTS, han 6, ts'e 5, ch. 5, 5b.
108. SC, 27, 0111b; Schafer, 1974, 403.
109. Schafer, 1974, 406.
110. Schafer, 1974, 406.
111. The phrase is from Yüan Chi-ch'uan, "Shan chung hsiao," CTS, han 4, ts'e 8, 7b.
112. Ma Tai, "T'ung Chuang hsiu ts'ai su chen hsing kuan," CTS, han 9, ts'e 2, ch. 1, 8a–8b.
113. PPT, "Wai i wen," and in TPYL, 68, 5b; Schafer, 1974, 404–405. Cf. Needham's treatment in Needham and Wang, 1958, 489.
114. San fu huang t'u in TPYL, 8, 11a.

115. CTW, 156, 8b–9a.
116. I have consulted the translation of F. M. Cornford (Oxford University Press, 1972 reprint), p. 353.
117. Schafer, 1967, 124.
118. HHS, 13, 0892c; Li Po, "Ch'ou Ts'ui shih yü," discussed elsewhere in my treatment of novae.
119. Schafer, 1974, 405, from Li Po, "Shan huang hsi hsün nan ching ko, CTS, han 3, ts'e 4, ch. 7, 4a–4b; Tu Fu, "Po sha tu," CTS, han 4, ts'e 1, ch. 3, 15a.
120. Yang Chiung, "Hun t'ien fu," CTW, 190, 4b.
121. E.g. in Sung Chih-wen, "Yen An-le kung-chu tse te K'ung tzu," CTS, han 1, ts'e 10, ch. 3, 2b.
122. Sung Chih-wen, "Lu chung wang wan tz'u," CTS, han 1, ts'e 10, ch. 2, 9a.
123. Li Shang-yin, "Jen shen ch'i hsi," CTS, han 8, ts'e 9, ch. 1, 25b.
124. Chang Cheng-chien (6 cent.), "Ch'iu ho shu keng keng," CSCC, 23b.
125. Lu Kuei-meng, "Ho Hsi-mei han jih shu chai chi shih," CTS, han 9, ts'e 10, ch. 10, 4b.
126. I owe this observation to a graduate student at the University of California, Richard Kunst.
127. SIC, 1, 9a.
128. Translated in Schafer, 1974, 405, from Chi lin, in TPYL, 8, 11a.
129. CCSSC, 13b.
130. Li Shang-yin, "Hai k'o," CTS, han 8, ts'e 9, ch. 2, 16a.
131. Two examples out of many, both by the same writer, are Li Te-yü, "Tiao t'ai," CTS, han 7, ts'e 10, 21b; and "Lin-hai t'ai shou hui tzu ch'ih ch'eng shih pao i shih shih," CTS, han 7, ts'e 10, 26b.
132. Li Chiao, "Hsing," CTS, han 2, ts'e 1, ch. 3, 1a–1b.
133. TCTC, 196, 1a.
134. CSCN, a, 5a; SIC, 3, 4b. For more on these personified stars see p. 126.
135. HCSLCC.

136. Ch'u Kuang-hsi, "Yeh tao Lo k'ou ju Huang Ho," CTS, han 2, ts'e 10, ch. 1, 7b.

137. Wu Jung, "Shang jen," CTS, han 10, ts'e 7, ch. 1, 6a.

138. "San hsüeh shan p'an t'o shih shang k'o shih," CTS, han 11, ts'e 8, 5b.

139. K'o-p'in Yü, "Ch'a k'o chih tou niu fu," CTW, 525, 13b–14b.

140. This is a slightly modified version of my translation from PWC, p. 19, as it appears in Schafer, 1974, 404–405.

141. Exemplified, for instance, in Po Chü-i, "K'an Meng-te t'i ta Li shih lang shih, shih chung yu wen hsing chih chü yi hsi ho chih," CTS, han 7, ts'e 7, ch. 34, 4a; Hsü Hun, "Ch'ou Hsing Tu erh yüan wai," CTS, han 8, ts'e 8, ch. 8, 8a.

142. Li Po, "Ch'ou Ts'ui shih yü," CTS, han 3, ts'e 6, ch. 18, 6a.

143. A few titles out of many of Yen Kuang as a supernova: Tu Fu, "Chi Yüeh-chou Chia szu ma liu chang Pa-chou Yen pa shih chün liang ko lao," CTS, han 4, ts'e 3, ch. 10, 22a–22b; Chang Chi, "T'i Yen Ling tiao t'ai," CTS, han 4, ts'e 6, 1b–2a; Li P'in, "Sung Shou-ch'ang Ts'ao ming fu," CTS, han 9, ts'e 6, ch. 3, 3b; Lin K'uan, "Sung Li yüan wai P'in chih Chien-chou," CTS, han 9, ts'e 8, 1a; Szu-k'ung T'u, "K'uang t'i," CTS, han 10, ts'e 1, ch. 3, 1a; Wu Yün, "Yen Tzu-ling," CTS, han 12, ts'e 6, 6a.

144. Wang Tsun, "Yen Ling t'ai," CTS, han 9, ts'e 8, 4b.

145. Lu Kuei-meng, "Tsao hsing," CTS, han 9, ts'e 10, ch. 12, 4b–5a.

CHAPTER 13.
A POTPOURRI OF IMAGES

1. Santayana, 1896, 104–105.

2. Chieh Yen-yung, "Yen t'a," CTS, han 11, ts'e 7, 7a.

3. Han Yü, "Nan-hai shen miao pei," CTW, 561, 5a–7b.

4. Fang Kan, "Chung ch'iu yüeh," CTS, han 10, ts'e 3, ch. 2, 9b.

5. J. Isaacs, The Background of Modern Poetry.

6. Wu Yung, "Ho Han Chih-kuang shih lang wu t'i, san shou shih szu yün," CTS, han 10, ts'e 7, ch. 2, 6b.

7. The image-clusters seem to trouble some translators. I note, for example, that a couplet in a poem by Li Shang-yin, which suggests that if the moon were only stuck to the sky, we might enjoy its beauty forever, is called by A. C. Graham (1965, 167) ". . . an extraordinary example of a constellation of images which holds its irrational spell a thousand years after its meaning has been lost." (The poem is "Pi ch'eng, erh shou," No. 1, CTS, han 8, ts'e 9, ch. 1, 24a.) But the meaning of Li Shang-yin's lines are by no means opaque; they say simply that if "the pearl of dawn" should be fixed in place, one might look at the "plate of water-crystal" forever. We have only a case of the moon called by two different names in as many lines. Perhaps readers are predisposed to find Chinese poetry simple, and are dazzled and blinded by a combination that would be simplicity itself in an Elizabethan conceit.

8. Hsin Hung-chih, "Tzu chün chih ch'u i," CTS, han 11, ts'e 7, 6b.

9. Po Chü-i, "K'o chung yüeh," CTS, han 7, ts'e 3, ch. 12, 5b.

10. Liu Wei, "Sai shang tso," CTS, han 9, ts'e 3, 2a.

11. Yüan Chen, "Huai tseng ch'eng Chang shih yü," CTS, han 6, ts'e 8, ch. 9, 5a–5b.

12. Lu Kuei-meng, "Tsa ch'i," CTS, han 9, ts'e 10, ch. 13, 11b.

13. Han Wo, "Yeh tso," CTS, han 10, ts'e 7, ch. 3, 4a.

14. Wang Shao-chih (380–435), "T'ai ch'ing chi," Ku chin shuo pu ts'ung shu, vol. 9.

15. Kao Yüeh, "Yung ying," CTS, han 11, ts'e 4, 3a.

16. P'i Jih-hsiu, "Tsao ch'un i chü tzu chi Lu Wang," CTS, han 9, ts'e 9, ch. 6, 7b.

17. Hsü Yin, "Li-chih, erh shou," CTS, han 11, ts'e 1, ch. 1, 12b.

18. P'i Jih-hsiu, "Ch'iu chiang hsiao wang," CTS, han 9, ts'e 9, ch. 8, 1a.

19. Li Tung, "T'i Tz'u en yu jen fang," CTS, han 11, ts'e 2, ch. 2, 3a.

20. Yü Hsüan-chi, "Ta ch'iu tso," CTS, han 11, ts'e 10, 3a.

21. Hsü Yin, "Shan shu ta ch'iu hsiao ts'ung pu tsou tsui ch'i yin yu so chien," CTS, han 11, ts'e 11, ch. 1, 13b–14a.

22. Ch'i-chi, "Sung Hsü hsiu ts'ai yu Wu kuo," CTS, han 12, ts'e 4, ch. 4, 8b.

23. Chang Chi, "Yüan pien li," CTS, han 6, ts'e 6, ch. 1, 11a.

24. Wu Shang-hao, "Sai shang chi shih," CTS, han 11, ts'e 7, 2a.

25. Lo Pin-wang, "Yüan shih hai ch'ü ch'un yeh to huai," CTS, han 2, ts'e 3, ch. 3, 2b–3a.

26. Ch'iao Chih-chih, "Ting ch'ing p'ien," CTS, han 2, ts'e 3, 3a.

27. Sun T'i, "Yen tao Jun-chou," CTS, han 2, ts'e 7, 7a.

28. Tu Mu, "Wen k'ai Chiang hsiang kuo Sung hsia sung, erh shou," CTS, han 8, ts'e 7, 1b. Virtually the same poem is attributed to Hsü Hun as "Wen k'ai Chiang Sung hsiang kung shen hsi hsia shih, erh shou," CTS, han 8, ts'e 8, ch. 9, 10a.

29. Ts'ao Sung, "Yeh yin," CTS, han 11, ts'e 2, ch. 2, 9b.

30. Ts'ao T'ang, "San nien tung ta li, wu shou," No. 2, CTS, han 10, ts'e 2, ch. 1, 5b.

31. YYTT, 2, 14.

32. Tu Mu, "Feng ho Po hsiang kung . . . ch'ang chü szu yün," CTS, han 8, ts'e 7, ch. 2, 5a.

33. Po Chü-i, "Chao ch'ü Yung-feng liu chih chin yüan kan fu," CTS, han 7, ts'e 8, ch. 37, 5a.

34. Keng Wei, "Yü chung su I-hsing szu," CTS, han 4, ts'e 10, ch. 1, 9a. Cf. also his "T'ang tsao ch'an ko," CTS, han 4, ts'e 10, ch. 1, 1b.

35. Liu Yü-hsi, "Wen Han pin . . . Han mu Yung-chou," CTS, han 6, ts'e 3, ch. 6, 8a.

36. CIL, a, 3a. Taken from the section on "Heaven" of a book called Po hsüeh chi, current in the Later Chou period.

37. P'i Jih-hsiu, "Ping k'ung ch'üeh," CTS, han 9, ts'e 9, ch. 6, 8a.

38. Liu Yü-hsi, "Pu hsü tz'u, erh shou," No. 2, CTS, han 6, ts'e 3, ch. 12, 2b.

APPENDIX A

1. Stein Collection, The British Museum, Waley Catalogue No. 31.

2. Unfortunately the same name was given to a holy grotto in a mountain of that name in Chekiang; it was the place where the ancient sage called "Red Pine" achieved enlightenment.

3. The discovery is described in TS, 37, 6a (the geographical section) and in TS, 59, 5a (encomiums in the Taoist bibliography).

4. HHHP, 1, 55, 59, 60, 63, 71, 72, 75, 79, 81, 83, 88, 89.

5. TCMHL, 18.

6. HHHP, 6, 169.

Bibliographies

PRIMARY SOURCES

Parenthetical abbreviations stand for books listed under "Collectanea, Encyclopedias and Anthologies," or "Secondary Sources." They represent the editor or source of the edition used in footnote documentation, unless otherwise stated in the note. Works reprinted in CTS, CTW, and TPYL are not listed separately in the bibliography; their titles appear only in the relevant footnotes in romanized form. Especially important among these writings are Hsieh Yen 謝偃, "Ming Ho fu" 明河賦, CTW, 156; Liu Yün-chi 劉允濟, "Ming t'ang fu" 明堂賦, CTW, 164; Yang Chiung 楊炯 "Hun t'ien fu" 渾天賦, CTW, 190; Tu Kuang-t'ing 杜光庭 "Tung t'ien fu ti yüeh tu ming shan chi hsü" 洞天福地嶽瀆名山記序, CTW, 932. Taoist books cited from TT, however, are listed separately in this bibliography, but those quoted from the Taoist anthologies STCN, WSPY, and YCCC are not; they are handled like poems from CTS, for instance. Many Taoist books are known under alternate titles, commonly a longer form and a shorter form; I have made arbitrary decisions about which form to choose for bibliographic purposes.

ATL   Yü Hsi 虞喜, *An t'ien lun* 安天論(YHSFCIS)

CCSHSC   (*Tung chen t'ai shang fei hsing yü ching*) *chiu chen sheng hsüan shang chi* (洞眞太上飛行羽經) 九眞昇玄上記 (TT, 1033)

CCSSC   Tsung Lin 宗懍, *Ching Ch'u shui shih chi* 荊楚歲時記(LSCSTS)

CCWCTP   *Ch'un ch'iu wei ch'ien t'an pa* 春秋緯潛潭巴(YHSFCIS, 56)

CCWHCT   *Ch'un ch'iu wei ho ch'eng t'u* 春秋緯合誠圖 (HWTS, 38; YHSFCIS, 55)

CCWKCF   *Ch'un ch'iu wei kan ching fu* 春秋緯感精符 (HWTS, 37; YHSFCIS, 54)

CCWKIY   *Ch'un ch'iu wei k'ao i yu* 春秋緯考異郵 (YHSFCIS, 55)

CCWSTT   *Ch'un ch'iu wei shuo t'i tz'u* 春秋緯說題辭(YHSFCIS, 56)

CCWTCC   *Ch'un ch'iu wei tso chu ch'i* 春秋緯佐助期 (YHSFCIS, 56)

CCWWYK   *Ch'un ch'iu wei wen yao kou* 春秋緯文耀鉤 (YHSFCIS, 55)

CCWYKT   *Ch'un ch'iu wei yen k'ung t'u* 春秋緯演孔圖 (YHSFCIS, 56)

CCWYMP   *Ch'un ch'iu wei yüan ming pao* 春秋緯元命苞 (HWTS, 35; YHSFCIS, 57)

323

Bibliographies

CCWYTS	*Ch'un ch'iu wei yün tou shu* 春秋緯運斗樞 (YHSFCIS, 55)
CCYTSC	*Chin chien yü tzu shang ching* 金簡玉字上經 (TT, 1027)
CFC	Ts'ui Ling-ch'in 崔令欽, *Chiao fang chi* 教坊記 (TTTS)
CHL	Tu Kuang-t'ing 杜光庭, (*Yung ch'eng*) *chi hsien lu* (鏞城)集仙錄
ChS	*Chou shu* 周書 (KM)
CIL	T'ao Ku 陶穀, *Ch'ing i lu* 清異錄 (1920 ed.)
CK	T'ao Hung-ching 陶弘景, *Chen kao* 真誥 (TT, 637–640)
CKT	*Chan kuo ts'e chiao chu* 戰國策校注 (SPTK)
CL	*Chou li* 周禮
CS	*Chin shu* 晉書 (KM)
CSCC	Chang Cheng-chien 張正見, *Chang San-ch'i chi* 張散騎集 (HWLCPSCC)
CSCN	*Chu shu chi nien* 竹書紀年 (PSNIC)
CSYTSC	*Chin shu yü tzu shang ching* 金書玉字上經 (TT, 581)
CT	*Chuang tzu* 莊子
CTPC	Hung Hsing-tsu 洪興祖 *Ch'u tz'u pu chu* 楚辭補註 (1846 ed.)
CTS	*Chiu T'ang shu* 舊唐書 (SPPY)
CYHYMC	(*T'ai shang*) *chiu yao hsin yin miao ching* (太上)九要心印妙經 (TT, 112)
ESPCC	(*T'ai shang pei tou*) *erh shih pa chang ching* (太上北斗) 二十八章經 (TT, 341)
EY	*Erh ya* 爾雅
FHKYLT	(*Tai shang san yüan*) *fei hsing kuan* (*chin chin shu*) *yü lu t'u* (太上三元)飛星冠(繁金書)玉籙圖 (TT, 534)
FLPCLTPS	*Fen lei pu chu Li T'ai-po shih* 分類補註李太白詩 (SPTK)
FPNTTWYC	(*T'ai shang*) *fei pu nan tou t'ai wei yü ching* (太上)飛步南斗太微玉經 (TT, 341)
FSTI	Ying Shao 應邵, *Feng su t'ung i* 風俗通義 (SPTK)
HAFC	*Hsi-an fu chih* 西安府志 (Taipei, 1970)
HCSLCC	Fei Hsin 費信, *Hsing ch'a sheng lan chiao chu* 星槎勝覽校注 (Shanghai, 1954)
HCWNST	*Hsiao ching wei nei shih t'u* 孝經緯內事圖 (YHSFCIS, 58)
HCWYSC	*Hsiao ching wei yüan shen ch'i* 孝經緯援神契 (YHSFCIS, 58)
HHHP	*Hsüan ho hua p'u* 宣和畫譜 (TSCC)
HHS	*Hou Han shu* 後漢書 (KM)
HNT	*Huai nan tzu* 淮南子
HS	*Han shu* 漢書 (KM)
HTKTH	*Ho t'u kua ti hsiang* 河圖括地象 (HHTTS), 1b
HWTNC	*Han wu ti nei chuan* 漢武帝內傳 (SSKTS)
HYKC	*Hua yang kuo chih* 華陽國志 (TSCC)
JYT	Sun Jou-chih 孫柔之, *Jui ying t'u* 瑞應圖 (KKTSCS)
KCC	Ts'ui Pao 崔豹, *Ku chin chu* 古今注 (TSCC)
KTCHC	Cheng Ch'i 鄭綮, *K'ai T'ien ch'uan hsin chi* 開天傳信記 (HCTY)
KuCC	Tu Kuang-t'ing 杜光庭 *Kuang ch'eng chi* 廣成集 (TT, 339)
KYTPIS	*K'ai yüan T'ien pao i shih* 開元天寶遺事 (TSCC)
LC	*Li chi* 禮記
LCL	*Lung ch'eng lu* 龍城錄 (TTTS)
LH	Wang Ch'ung 王充, *Lun heng* 論衡
LKC	Chang Chien 張薦, *Ling kuai chi* 靈怪集

Bibliographies

LMC	Shen Chi-chi 沈既濟, *Lei min chuan* 雷民傳 (TTTS)
LPLI	Liu Hsün 劉恂, *Ling piao lu i* 嶺表錄異
LT	*Lieh tzu* 列子
LWCMC	*Li wei chi ming cheng* 禮緯稽命徵 (YHSFCIS, 54)
LWHWC	*Li wei han wen chia* 禮緯含文嘉 (YHSFCIS, 54)
LWTWI	*Li wei tou wei i* 禮緯斗威儀 (YHSFCIS, 54)
LY	*Lun yü* 論語
MCPT	Shen Kua 沈括, *Meng ch'i pi t'an* 夢溪筆談 (TSCC)
MT	*Mo tzu* 墨子
NS	*Nan shih* 南史
NTTWYC	*(T'ai shang fei pu) nan tou t'ai wei yü ching* (大上飛步) 南斗太微玉經 (TT, 341)
PHT	*P'o hu t'ung* 白虎通
PPT	Ko Hung 葛洪, *Pao p'u tzu* 抱朴子 (SPTK)
PS	*Pei shih* 北史 (KM)
PSC	*(T'ai shang ch'u san shih chiu ch'ung) pao sheng ching* (太上除三尸九蟲) 保生經 (TT, 580)
PTKM	Li Shih-chen 李時珍, *Pen ts'ao kang mu* 本草綱目 (Hong Kong, 1972)
PWC	*Po wu chih* 博物志 (TSCC)
SC	*Shih chi* 史記 (KM)
SCHCSPYC	*Shang ch'ing hua ch'en san pen yü chüeh* 上清華晨三奔玉訣 (TT, 191)
SCHHCCT	Szu-ma Ch'eng-chen 司馬承禎, *Shang ch'ing han hsiang chien chien t'u* 上清含象劍鑑圖 (TT, 196)
SHC	Ko Hung 葛洪, *Shen hsien chuan* 神仙傳 (HWTS)
SHKYC	Tu Kuang-t'ing 杜光庭, *Shen hsien kan yü chuan* 神仙感遇傳 (TT, 328)
SIC	Wang Chia 王嘉, *Shih i chi* 拾遺記 (PSNIC)
SKC	*San kuo chih* 三國志 (KM)
SKCC	Wu Jen-ch'en 吳任臣, *Shih kuo ch'un ch'iu* 十國春秋 (1672 ed.)
SM	*Shih ming* 釋名
SS	*Sui shu* 隋書 (KM)
STCCW	*San t'ung ch'i cheng wen* 參同契正文 (TSCC)
SuS	*Sung shu* 宋書 (KM)
SW	*Shuo wen* 説文
SWHSW	*Shih wei han shen wu* 詩緯含神霧 (YHSFCIS, 54)
TC	*Tso chuan* 左傳
TCMHL	Chu Ching-hsüan 朱景玄, *T'ang ch'ao ming hua lu* 唐朝名畫錄 (MSTS)
TCTC	Szu-ma Kuang 司馬光, *Tzu chih t'ung chien* 資治通鑑 (Tokyo, 1892)
TCTL	Yang Shen 楊慎 *Tan ch'ien tsung lu* 丹鉛總錄
THIS	Chang Kuo 張果, *T'ung hsüan i shu* 通玄貴書 (HMSY)
THY	*T'ang hui yao* 唐會要 (TSCC)
TLCCFK	Hsü Sung 徐松, *T'ang liang ching ch'eng fang k'ao* 唐兩京城坊考 (CARY)
TLSH	Yen Yü 嚴羽, *Ts'ang lang shih hua* 滄浪詩話 (TSCC)
TLSI	Chang-sun Wu-chi 長孫無忌, *T'ang lü su i* 唐律疏義 (KHCPTT)

325

Bibliographies

TLSKC   Li Ching 李靖, *T'ien lao shen kuang ching* 天老神光經 (TT, 578)

TLT   *T'ang liu tien* 唐六典 (Kyoto, 1935)

TS   *T'ang shu* 唐書 (SPPY)

TTHY   Liu Su 劉肅, *Ta T'ang hsin yü* 大唐新語 (TTTS)

TuT   Tu Yu 杜佑, *T'ung tien* 通典 (KM)

TzT   Chu Ch'i-feng 朱起鳳, *Tz'u t'ung* 辭通 (Shanghai, 1934)

WHCYKCC   (*T'ai shang*) *wu hsing ch'i yüan k'ung ch'ang chüeh* (太上) 五星七元空常訣 (TT, 581)

WHTK   Ma Tuan-lin 馬端臨, *Wen hsien t'ung k'ao* 文獻通考 (KM)

WNSC   Wang Yung 王融, *Wang Ning-shuo chi* 王寧朔集 (HWLCC)

WTS   *Wu tai shih* 五代詩 (KM)

YHCHTC   Li Chi-fu 李吉甫, *Yüan ho chün hsien t'u chih* 元和郡縣圖志 (1880 ed.)

YTCS   Wang Hsiang-chih 王象之, *Yü ti chi sheng* 輿地紀勝 (1849 ed.)

YTFY   *Yang tzu fa yen* 揚子法言 (SPPY)

YYTT   Tuan Ch'eng-shih 段成式, *Yu yang tsa tsu* 酉陽雜俎 (TSCC)

Collectanea, Encyclopedias and Anthologies

CARY   Hiraoka Takeo 平岡武夫 ed., *Chōan to Rakuyō* 長安と洛陽 (Kyoto, 1956)

CHC   Hsü Chien 徐堅, *Ch'u hsüeh chi* 初學記 (1962 ed.)

CLW   *Ch'üan Liang wen* 全梁文 (Shanghai, 1930)

CT   *Ch'u tz'u* 楚辭

CTS   *Ch'üan T'ang shih* 全唐詩 (Taipei, 1967)

CTW   *Ch'üan T'ang wen* 全唐文 (Taipei, 1972)

FYCL   Tao-shih 道世, *Fa yüan chu lin* 法苑珠林 (SPTK)

HCTY   *Hsüeh chin t'ao yüan* 學津討源

HHTTS   *Han hsüeh t'ang ts'ung shu* 漢學堂叢書

HMSY   *Hsing ming su yüan* 星命溯源 (Shanghai, 1934–1935)

HWLCC   *Han Wei Liu ch'ao po san chia chi* 漢魏六朝百三家集 (Taipei, 1963)

HWTS   *Han Wei ts'ung shu* 漢魏叢書

IWLC   Ou-yang Hsün 歐陽詢, *I wen lei chü* 藝文類聚 (1965 ed.)

KHCPTS   *Kuo-hsüeh chi-pen ts'ung shu* 國學基本叢書

KKTSCS   *Kuan ku t'ang so chu shu* 觀古堂所叢書

KM   *K'ai ming* 開明 edition

LTFH   *Li tai fu hui* 歷代賦彙 (1886 ed.)

MSTS   *Mei shu ts'ung shu* 美書叢書

PKLT   *Po K'ung liu tieh* 白孔六帖

PSNIC   *Pi shu nien i chung* 秘書廿一種

SPPY   *Szu pu pei yao* 四部備要

SPTK   *Szu pu ts'ung k'an* 四部叢刊

SSKTS   *Shou shan ko ts'ung shu* 守山閣叢書

STCN   *San tung chu nang* 三洞珠囊 (TT, 780–782)

TFYK   *Ts'e fu yüan kuei* 冊府元龜 (1960 ed.)

TPKC   *T'ai p'ing kuang chi* 太平廣記 (1846 ed.)

TPYL   *T'ai p'ing yü lan* 太平御覽 (Peking, 1963)

Bibliographies

TSCC	*Ts'ung shu chi ch'eng* 叢書集成
TT	*Tao tsang* 道藏 (Shanghai, 1924–26)
TTTS	*T'ang tai ts'ung shu* 唐代叢書 (Taipei, 1968)
WH	*Wen hsüan* 文選
WSPY	*Wu shang pi yao* 無上祕要 (TT, 768)
YCCC	*Yün chi ch'i ch'ien* 雲笈七籤 (SPTK)
YHSFCIS	Ma Kuo-han 馬國翰, *Yü han shan fang chi i shu* 玉函山房輯佚書

SECONDARY SOURCES

1974   "Ho-nan Lo-yang Pei Wei Yüan I mu tiao-ch'a," *Wen-wu* (1974), 12, 53–55.

1975a   "Ho-pei Hsüan-hua Liao pi-hua-mu fa-chüeh chien-pao," *Wen-wu* (1975), 8, 31–39.

1961   *Hsin Chung-kuo ti k'ao-ku shou-huo* (n.p., 1961).

1975b   "Liao-tai ts'ai-hui hsing-t'u shih wo-kuo t'ien-wen-shang-ti chung-yao fa-hsien," *Wen-wu* (1975), 8, 40–44.

1928   *Shōsōin gyobutsu zuroku* (Tokyo, 1928–   ).

1973   "T'u-lu-fan hsien A-szu-t'a-na—Ka-la-ho-cho ku-mu-ch'ün fa-chüeh chien-pao," *Wen-wu* (1973), 10, 7–20.

Aguilar, C. H.
1971   *Religión y Magia entre los indios de Costa Rica de origen sureño* (San José, Costa Rica, 1971).

Aldred, Cyril
1971   *Jewels of the Pharaohs; Egyptian Jewellery of the Dynastic Period* (London, 1971).

Allen, R. H.
1963   *Star Names; Their Lore and Meaning* (New York, 1963).

Barnard, Noel
1973   Review of Rutherford J. Gettens, Roy S. Clarke Jr. and W. T. Chase, *Two Early Chinese Bronze Weapons with Meteoric Iron Blades, Journal of the American Oriental Society*, 93 (1973), 639–640.

Barnes, V. E.
1969   "Progress of Tektite Studies in China; Excerpts from 'A Preliminary Survey and Study of the Tektite—Lei-Gong-Mo—from Leichow Peninsula and Hainan Island, China,' by Le Da-ming, translated by C. J. Peng, with commentary," *Transactions*, American Geophysical Union, 50 (1969), 704–708.

Bauer, Wolfgang
1956   "Der Herr vom gelben Stein (Huang shih kung); Wandlungen einer chinesischen Legendenfigur," *Oriens Extremus*, 3 (1956), 137–152.

Bede, The Venerable
1954   *Baedae opera historica* (Cambridge and London, 1954).

Bibliographies

Benhamouda, A.
1951   "Les noms arabes des étoiles (essai d'identification)," *Annales* de l'Institut d'Etudes Orientales (Faculté des Lettres de l'Université d'Alger), 9 (1951), 76–210.

Bernhard, H. J., D. A. Bennett, and H. S. Rice
1948   *New Handbook of the Heavens* (New York, Toronto, London, 1948).

Bertuccioli, Giuliano
1974   "Reminiscences of the Mao-shan," *East and West*, n.s. 24/3–4 (1974), 403–413.

Bezold, E.
1920   "Sze-ma Ts'ien und die babylonische Astrologie," *Festschrift für Friedrich Hirth zu seinem 75. Geburtstag 16. April 1920* (Berlin, 1920).

Boots, J. L.
1934   "Korean Weapons and Armor," *Transactions* of the Korea Branch of the Royal Asiatic Society, 23/2 (December, 1934), 1–37.

Botley, C. M.
1956   "Halley and the Aurora," *The Journal of the British Astronomical Association*, 66 (1956), 31–33.
1957   "Some Great Tropical Aurorae," *The Journal of the British Astronomical Association*, 67 (1957), 188–192.

Chang An-chih
1973   "Hsi Han po-hua ti i-shu ch'eng-chiu," *Wen-wu*, 9 (1973), 66–70.

Chang Hung-chao
1921   "Shih ya," *Ti-chih chuan-pao*, B, 2 (Peking, 1921).

Chao Wei-pang
1946   "The Chinese Science of Fate-Calculation," *Folklore Studies*, 5 (1946), 279–315.

Chao Yuan-jen
1917   "Chung-kuo hsing-ming-t'u k'ao," *K'o-hsüeh*, 3 (1917), 42–52, 270–308.

Chapin, H. B.
1940   *Toward the Study of the Sword as Dynastic Talisman: The Fêng-ch'êng Pair and the Sword of Han Kao Tsu* (unpublished Ph.D. dissertation, University of California, Berkeley, 1940)

Chatley, Herbert
1940   "The lunar mansions in Egypt," *Isis*, 3 (1940), 394–397.

Ch'en Jung
1957   *Chung-kuo shu-mu fen-lei-hsüeh* (Shanghai, 1957).

Chu Ch'i-feng
1934   *Tz'u t'ung* (Shanghai, 1934).

Chu Kuang-chien
1948   "Yu hsien shih," *Wen-hsüeh tsa-chih*, 3/4 (1948), 1–14.

Clerke, A. M.
1905   *The System of the Stars* (London, 1905).

Darwin, Erasmus
1825   *The Botanic Garden, A Poem in Two Parts; containing The Economy of Vegetation and the Loves of the Plants with Philosophical Notes* (London, 1825).

Dubs, H. H.
1958   "The Beginnings of Chinese Astronomy," *Journal of the American Oriental Society*, 78 (1958), 295–300.

Bibliographies

Duyvendak, J. J. L.
1947   "The Dreams of the Emperor Hsüan-tsung," *India Antiqua* (Leiden, 1947), 102–108.

Eberhard, W.
1940   "Untersuchungen an astronomischen Texten des chinesischen Tripitaka," *Monumenta Serica*, 5 (1940), 208–262.
1942   "Index zu den Arbeiten über Astronomie, Astrologie, und Elementlehre," *Monumenta Serica*, 7 (1942), 242–266.
1968   *The Local Cultures of South and East China* (Leiden, 1968).

Forbes, R. J.
1950   *Metallurgy in Antiquity: A Notebook for Archaeologists and Technologists* (Leiden, 1950).

Frankel, H. H.
1954   "The Date and Authorship of the *Lung-ch'eng lu*," *Silver Jubilee Volume* of the Zinbun-Kagaku-Kenkyusyo, Kyoto University (Kyoto, 1954), 129–149.

Frodsham, J. D.
1970   *The Poems of Li Ho (791–817)* (Oxford, 1970).

Frye, Northrop
1966   "The Archetypes of Literature," *Myth and Literature* (Lincoln, Nebraska, 1966), 87–89.

Futrell, D. S.
1972   "Indochinites: Glass from the Sky," *Mineral Digest*, 4 (Winter, 1972), 106–116.

Gettens, R. J., R. S. Clark Jr., and W. T. Chase, *Two Early Chinese Bronze Weapons with Meteoric Iron Blades* (Freer Gallery of Art, Occasional Papers, 4/1, Washington, 1971).

Graham, A. C.
1965   *Poems of the Late T'ang* (Harmondsworth, 1965).

Granet, Marcel
1925   "Remarques sur le Taoïsme ancien," *Asia Major*, 2 (1925), 146–151.
1926   *Danses et Légendes de la Chine Ancienne* (Vol. II, Paris, 1926).

Hastings, James
1962   *Encyclopaedia of Religion and Ethics* (New York, 1962).

Hawkes, David
1962   *The Songs of the South; an Ancient Chinese Anthology* (Beacon Paperback, 1962).

Ho Peng Yoke
1966   *The Astronomical Chapters of The Chin Shu; with Amendments Full Translation and Annotations* (Paris and The Hague, 1966).

Ho Ping Yü and Joseph Needham
1959   "Ancient Chinese Observations of Solar Haloes and Parhelia," *Weather*, 4 (1959), 124–134.

Hsü Dau-lin
1970   "Crime and Cosmic Order," *Harvard Journal of Asiatic Studies*, 30 (1970), 111–125.

Huang Jen-heng
1937   "Pu Liao Shih I wen chih," *Erh shih wu shih pu pien* (Shanghai, 1937).

Huber, E.
1906   "Termes Persans dans l'astrologie bouddhique chinoise," *Bulletin de l'École Française d'Extrême-Orient*, 6 (1906), 39–43.

Bibliographies

Ishii Masako
1968   "Shin-kō no seiritsu wo meguru shiryō-teki kentō," *Dōkyō kenkyū*, 3 (1968), 79-246.
Kaltenmark, Max
1953   *Le lie-sien tchouan* (Peking, 1953).
1960   "Ling-pao: note sur un terme du taoisme religieux," *Mélanges* publiés par l'Institut des Hautes Études Chinoises, 2 (1960), 559-588.
Kühnert, Franz
1901   "Über die von den Chinesen 'Te-sing' oder Tugendgestirn genannte Himmelerscheinung," *Sitzungsberichte* der Mathematisch-Naturwissenschaftlichen Classe der kaiserlichen Akademie der Wissenschaften," 110/2a (Vienna, 1901), 619-695.
Lamont, H. G.
1974   "An Early Ninth Century Debate on Heaven: Liu Tsung-yüan's *T'ienshuo* and Liu Yü-hsi's *T'ienlun*. An Annotated Translation and Introduction," *Asia Major*, 19 (1974), 37-85.
Laufer, Berthold
1928   *The Prehistory of Aviation* (Field Museum of Natural History, Publications 253; Anthropological Series, Vol. 8 No. 1, Chicago, 1928).
Lanciotti, Lionello
1955   "Sword casting and related legends in China," *East and West*, 6 (1955), 106-114.
Liu Wu-chi
1966   *An Introduction to Chinese Literature* (Bloomington, 1966).
Lundmark, Knut
1921   "Suspected New Stars Recorded in Old Chronicles and among Recent Meridian Observations," *Publications* of the Astronomical Society of the Pacific, 33 (1921), 225-238.
McCready, W. D.
1973   "Papal Plenitude Potestatis and the Source of Temporal Authority in Late Medieval Papal Hierocratic Theory," *Speculum*, 48(1973), 654-674.
Mair, G. R.
1969   *Aratus, with an English Translation* (Loeb Classical Library, Cambridge and London, 1969).
Maspero, Henri
1929   "L'astronomie chinoise avant les Han," *T'oung Pao*, 26 (1929), 267-356.
1950a   "L'astronomie chinoise dans la Chine ancienne, histoire des instruments et des découvertes," *Études Historiques* (Mélanges Posthumes, 3, Paris, 1950), 13-34.
1950b   *Le Taoïsme* (Mélanges Posthumes, 2, Paris, 1950).
Meadows, A. J.
1969   *The High Firmament; A Survey of Astronomy in English Literature* (Leicester, 1969).
Minnaert, M.
1959   *Light and Colour in the open air* (London, 1959).
Nakayama Shigeru
1966   "Characteristics of Chinese Astrology," *Isis*, 57 (1966), 442-454.
Nakayama Shigeru and Nathan Sivin
1973   *Chinese Science; Exploration of an Ancient Tradition* (Cambridge and London, 1973).

Bibliographies

Needham, Joseph and Wang Ling
1956   *Science and Civilisation in China*, 2 (Cambridge, 1956).
1958   *Science and Civilisation in China*, 3 (Cambridge, 1958).
Needham, Joseph, Wang Ling and K. G. Robinson
1962   *Science and Civilisation in China*, 4/1 (Cambridge, 1962).
Neruda, Pablo
1971   *Las Piedras del Cielo* (Buenos Aires, 1971).
Olivieri, Alexander
1897   *Pseudo-Eratosthenis Catasterismi* (Leipzig, 1897).
Pelliot, Paul
1912   Review of L. Wieger, *Taoisme*, *Journal Asiatique*, 20 (1912), 141–171.
Porkert, Manfred
1974   *The Theoretical Foundations of Chinese Medicine: Systems of Correspondence* (Cambridge and London, 1974).
Reischauer, E. O.
1955   *Ennin's Diary; The Record of a Pilgrimage to China in Search of the Law* (New York, 1955).
Santillana, Giorgio de and Bertha Von Dechend
1969   *Hamlet's Mill; An essay on myth and the frame of time* (Boston, 1969).
Santayana, George
1896   *The Sense of Beauty; being the Outlines of Aesthetic Theory* (New York, 1896).
Saussure, Leopold De
1921   "Le système cosmologique des Chinois," *Revue générale des Sciences*, 32 (1921), 729–736.
Schafer, E. H.
1952   "The Pearl Fisheries of Ho-p'u," *Journal of the American Oriental Society*, 72 (1952), 155–168.
1954   *The Empire of Min* (Tokyo, 1954).
1961   *Tu Wan's Stone Catalogue of Cloudy Forest: A Commentary and Synopsis* (Berkeley and Los Angeles, 1961).
1963a   "The Auspices of T'ang," *Journal of the American Oriental Society*, 83 (1963), 197–225.
1963b   *The Golden Peaches of Samarkand: A Study of T'ang Exotics* (Berkeley and Los Angeles, 1963).
1963c   "Mineral Imagery in the Paradise Poems of Kuan-hsiu," *Asia Major*, 10 (1963), 73–102.
1965a   "The Idea of Created Nature in T'ang Literature," *Philosophy East and West*, 15 (1965), 543–550.
1965b   "Li Te-yü and the Azalea," *Asiatische Studien*, 18/19 (1965), 105–114.
1967   *The Vermilion Bird: T'ang Images of the South* (Berkeley and Los Angeles, 1967).
1970   *Shore of Pearls* (Berkeley and London, 1970).
1972   "Two Late T'ang Poems on Music," *Literature East and West*, 16 (1972), 980–996.
1973a   *The Divine Woman: Dragon Ladies and Rain Maidens in T'ang Literature* (Berkeley, Los Angeles and London, 1973).
1973b   "Dragon-Women in Medieval Prose Literature," *Papers of the C.I.C. Far Eastern Languages Institute*, Vol. IV: Joseph K. Yamagiwa Memorial Issue (Ann Arbor, 1973), 48–62.

331

Bibliographies

1974   "The Sky River," *Journal of the American Oriental Society*, 94 (1974), 401–407.

Schipper, Kristopher M.

1965   *L'empéreur Wou des Han dans la légende taoiste; Han wou-ti nei-tchouan* (Publications de l'École Française d'Extrême-Orient. Paris, 1965).

Schlegel, Gustave

1967   *Uranographie chinoise; ou preuves directes que l'astronomie primitive est originaire de la Chine, et qu'elle a été empruntée par les anciens peuples occidentaux à la sphère chinoise* (Taipei, 1967).

Schove, D. J.

1951   "Sunspots, Aurorae and Blood Rain: The Spectrum of Times," *Isis*, 42 (1951), 133–138.

1956   "Halley's Comet and Kamienski's Formula," *Journal of the British Astronomical Association*, 66 (1956), 131–139.

1975   "Comet Chronologies in Numbers, A.D. 200–1882," *Journal of the British Astronomical Association*, 85 (1975), 401–407.

Shirley, J. W.

1974   *Thomas Harriot; Renaissance Scientist* (Oxford, 1974).

Shumaker, Wayne

1972   *The Occult Sciences in the Renaissance; A Study in Intellectual Patterns* (Berkeley, Los Angeles and London, 1972).

Sivin, Nathan

1969   *Cosmos and Computation in Early Chinese Mathematical Astronomy* (Leiden, 1969; reprinted from *T'oung Pao*, 15).

Solger, F.

1922   "Astronomische Anmerkungen zu chinesischen Märchen," *Mitteilungen der deutschen Gesellschaft f. Natur- u. Völkerkunde Ostasiens*, 17 (1922), 133–207.

Soothill, W. E.

1952   *The Hall of Light; A Study of Early Chinese Kingship* (New York, 1952).

Soothill, W. E. and L. Hodous

1970   *A Dictionary of Chinese Buddhist Terms* (Taipei, 1970).

South, Margaret T.

1972   "Po Chü-i's 'The Observatory Tower,'" *Journal of the Oriental Society of Australia*, 9/1–2 (1972–1973), 3–13.

Soymié, Michel

1954   "Le lo-feou chan; étude de géographie réligieuse," *Bulletin de l'École Française d'Extrême-Orient*, 48 (1954), 1–139.

1962   "La lune dans les religions chinoises," *La Lune; Mythes et Rites* (Sources Orientales V, Paris, 1962), 291–321.

1972   "Histoire et philologie de la Chine médiévale et moderne," *Annuaire* 1971–1972, École pratique des Hautes Études, ive section, sciences historiques et philologiques (Rapports sur les conférences) (Paris, 1972), 661–666.

Stein, R.

1942   "Jardins en miniature d'Extrême-Orient," *Bulletin de l'École Française d'Extrême-Orient*, 42 (1942), 1–104.

Straughair, Anna

1973   *Chang Hua: A Statesman-Poet of the Western Chin Dynasty* (Canberra, 1973).

Bibliographies

Strickmann, Michel
1975   "Taoism, History of," *Encyclopaedia Britannica* (Chicago, 1975).
Tate, G. H. H.
1947   *Mammals of Eastern Asia* (New York, 1947).
Thompson, D. W.
1895   *A Glossary of Greek Birds* (Oxford, 1895).
Touché-Skadding, R. J. De
1947   "The Agni-Mani—Magic Gem of the Orient," *Rocks and Minerals,* 22
        (1947), 603–609.
Vickery, J. B.
1966   *Myths and Literature; Contemporary Theory and Practice* (Lincoln, Ne-
        braska, 1966).
Waley, Arthur
1950   *The Poetry and Career of Li Po 701–762 A.D.* (London and New York, 1950).
1956   *The Nine Songs; A Study of Shamanism in Ancient China* (London, 1956).
1963   "A song from Tun-huang," *Bulletin of the School of Oriental and African
        Studies,* 26 (1963), 149–151.
Wallenquist, Åke
1966   *Dictionary of Astronomical Terms* (Garden City, New York, 1966).
Wang Ch'e and Ch'en Hsü
1974   "Lo-yang Pei Wei Yüan I mu ti hsing-hsiang-t'u," *Wen-wu,* 12 (1974), 56–60.
Wang Ling
1947   "On the invention and use of gunpowder and firearms in China," *Isis,* 37
        (1947), 160–178.
Wang, S. S.
1962   *Tsaur Jyr's Poems of Mythical Excursion* (unpublished M.A. thesis, Uni-
        versity of California, Berkeley, 1962).
Ware, J. R.
1966   *Alchemy, Medicine, Religion in the China of* A.D. 320: *The Nei P'ien of Ko
        Hung (Pao-p'u tzu)* (Cambridge and London, 1966).
Wen Shion Tsu
1934   "The Observations of Halley's Comet in Chinese History," *Popular
        Astronomy,* 42 (1934), 191–201.
Wheatley, Paul
1971   *The Pivot of the Four Quarters; A Preliminary Enquiry into the Origins
        and Character of the Ancient Chinese City* (Edinburgh, 1971).
Whicher, G. F.
1965   *The Goliard Poets; Medieval Latin Songs and Satires* (New York, 1965).
Whitelock, Dorothy
1961   *The Anglo-Saxon Chronicle; A revised translation* (London, 1961).
Willis, N. P.
1837   *Melanie and Other Poems* (New York, 1837).
Wylie, Alexander
1966   *Chinese Researches* (Taipei, 1966).
Yabuuti Kiyosi
1954   "Indian and Arabian Astronomy in China," *Silver Jubilee Volume* of the
        Zinbun-Kagaku-Kenkyusyo, Kyoto University (Kyoto, 1954), pp. 585–603.
1961   "Tō-dai ni okeru seihō temmongaku ni kan-suru ni-san no mondai," *Tsu-
        kamoto hakase shōju kinen bukkyō shigaku ronshū* (Kyoto, 1961),
        883–894.

Bibliographies

1963a   "Astronomical Tables in China from the Han to the T'ang Dynasties," *Chūgoku chūsei kagaku gijutsu-shih no kenkyū* (Tokyo, 1963), pp. 445–492.

1963b   "The Chiuchih-li—An Indian Astronomical Book in the T'ang Dynasty," *Chūgoku chūsei kagaku gijutsu-shi no kenkyū* (Tokyo, 1963), pp. 135–198.

1973   "Chinese Astronomy: Development and Limiting Factors," *Chinese Science: Exploration of an Ancient Tradition* (Cambridge and London, 1973).

Yanagisawa Takashi

1967   "Tō-hon hokuto mandara no ni-irei," *Dōkyō kenkyū*, 2 (1967), 205–236.

334

Glossaries

GLOSSARY OF HUMAN AND DIVINE NAMES*

Ch'a Nü	姹女	Cheng Ku	鄭谷
Chang Chi	張籍	Cheng Yü	鄭嵎
Chang Ch'ien	張騫	Ch'eng Yen-hsiung	成彥雄
Chang Chih-ho	張志和	Ch'i-chi	齊己
Chang Heng	張衡	Ch'i Ching Kung	齊景公
Chang Hua	張華	Ch'ien Ch'i	錢起
Chang Kuo	張果	Ch'ih-yu	蚩尤
Chang Shou-chieh	張守節	Ch'in Tsung-ch'üan	秦宗權
Chang Su-ch'ing	張素卿	Ch'in T'ao-yü	秦韜玉
Chang Wei	張為	Ching Fang	京房
Chang Yüeh	張說	Ch'ing Nü	青女
Ch'ang-hsi	常羲	Ch'iung O	瓊娥
Ch'ang-i	常義，常儀	Chou Fang	周昉
Ch'ang Kun	常袞	Chu Tzu-jung	朱子容
Ch'ang-o (*Zhang-nga)	嫦娥	Chuan-hsü	顓頊
Ch'ang Ts'an	常粲	Ch'u Kuang-hsi	儲光羲
Ch'en Kuang	陳匡	Ch'üan Te-yü	權德璵
Ch'en Shih-chen	陳碩真	Fan Ch'iung	范瓊
Ch'en Tzu-ang	陳子昂	Fang Kan	方干
Ch'en Yü	陳羽	Han Tsung	韓宗
Cheng Ch'i	鄭綮	Han Wo	韓偓
Cheng Chiao-fu	鄭交甫	Heng-o (*Gheng-nga)	姮娥
Cheng Ho	鄭和	Ho Ch'ang-shou	何長壽
Cheng Hsi-ch'eng	鄭希誠	Ho Ning	和凝

*See also "Glossary of Words and Phrases" for some personified asterisms.

335

Hsi-ch'an 棲 蟾	Li Hsiao-i 李 李 逸
Hsi-ho 羲 和	Li Hsü-chung 李 虛 中
Hsiang Szu 項 璃 斯	Li Kuei 李 軌
Hsiao Yü 蕭 璃	Li P'ing 李 永 善
Hsieh Lan 解 蘭 偓	Li Shan 李 善
Hsieh Yen 謝 偃	Li Tung 李 洞 緯
Hsien hsing yüan 仙 星 媛	Li Wei 李 緯
Hsing Fei 星 妃	Ling-hu Ch'u 令 狐 楚
Hsing O 星 娥	Ling-i 靈 一
Hsing Wu 星 婺	Liu Ch'ang-ch'ing 劉 長 卿
Hsü Ching-tsung 許 敬 宗	Liu Ching-ning 劉 靜 凝
Hsü Hsüan 徐 鉉	Liu Hsiang 劉 向
Hsü Hun 許 渾	Liu Yün-chi 劉 允 濟 賓
Hsü Nü 須 女	Lo Pin-wang 駱 賓 王
Hsü Yin 徐 夤	Lo Yin 羅 隱
Hsüan miao yü nü 玄 妙 玉 女	Lu Ch'i 盧 杞
Hsüan Nü 玄 女	Lu Ming-yüeh 盧 明 月
Hsüeh Chü 薛 舉	Lu T'ung 盧 仝
Hsüeh T'ao 薛 濤	Lu Wo 盧 渥
Hsüeh Tsung 薛 綜	Lü Ts'ai 呂 才
Hu Tseng 胡 曾	Lü Yen 呂 巖
Hua-jui Fu-jen 花 蕊 夫 人	Ma Huai-su 馬 懷 素
Huai-i 懷 義	Ma-ku 麻 姑
I-hsing 一 行	Ma Tai 馬 戴
Kao Huan 高 歡	Mao Wen-hsi 毛 文 錫
Kao Shih 高 適	Meng Chiao 孟 郊
Kao Yüeh 高 越	Meng K'ang 孟 康
K'o-p'in Yü 可 軿 瑜	Nan-kung Yüeh 南 宮 說
Kuan Lu 管 輅	Nü Kua 女 媧
Kuang Ch'eng tzu 廣 成 子	Nü Shu 女 樞
Kuo Han 郭 翰	P'ei Kuang-t'ing 裴 光 庭
Lan Ts'ai-ho 藍 采 和	P'ei Min 裴 旻
Lei Huan 雷 煥	P'ei Yüeh 裴 說
Li Chao-hsiang 李 昭 象	P'i Jih-hsiu 皮 日 休 簡
Li Ch'eng 李 憕	Po Hsing-chien 白 行 簡
Li Chi-fu 李 吉 甫	Pu-k'ung 不 空
Li Ch'i 李 頎	San yüan fu-jen 三 元 婦 人
Li Chiao 李 嶠	Sang Tao-mao 桑 道 茂
Li Ch'un-feng 李 淳 風	Shang Hsien-fu 尚 獻 甫
Li Chung 李 中	Shao Yeh 邵 謁
Li Chung-mao 李 重 茂	Shen Ch'üan-ch'i 沈 全 期
Li Hao 李 浩	Shen Pin 沈 彬

Glossaries

Shih Hu	石虎		Wei Ch'ü-mou	韋渠牟
Shu Hsi	束皙		Wei Kao	韋皋
Sŏndŏk	善德		Wei Shu-ch'ing	韋叔卿
Su Cheng	蘇拯		Wu Chih	吳質
Sun Ch'üan (king)	孫權		Wu Kang	吳剛
Sun Ch'üan	孫佺		Wu Ling-hsün	武令珣
Sun Ho	孫郃		Wu Nü	婺女
Sun T'i	孫逖		Wu Tao-hsüan	吳道玄
Sun Wei	孫位		Wu Tzu-hsü	伍子胥
Sung Chün	宋均		Wu Jung	吳融
Szu-k'ung T'u	司空圖		Wu Yüan	伍員
T'ai su san yüan chün	太素三元君		Wu Yün	吳筠
Ti huang t'ai tsun	帝皇太尊		Yang Chiung	楊炯
Ting Ling-wei	丁令威		Yang Hsi	楊羲
Ts'ai Ching	蔡經		Yang T'ing-kuang	楊庭光
Ts'ao Chih	曹植		Yao Ho	姚合
Ts'ao Sung	曹松		Yao Yüan-pien	姚元辨
Tu Kuang-t'ing	杜光庭		Yeh-lü Ch'un	耶律紇
Tu Mu	杜牧		Yen Chen-ch'ing	顏眞卿
Tung-fang Shuo	東方朔		Yen Kuang	嚴光
Wang Chien	王建		Yen Li-pen	閻立本
Wang Ch'ung	王充		Yen Li-te	閻立德
Wang I	王逸		Yen Ling	嚴陵
Wang Ts'ao	王慥		Yü Hsi	虞喜
Wang Tsun	汪遵		Yü Hsüan-chi	魚玄機
Wang Yai	王涯		Yü Shih-nan	虞世南
Wei (Lady)	韋		Yüeh O	月娥
Wei Chien-su	韋見素		Yüeh Yen-wei	欒彥璋
Wei Hua-ts'un	魏華存			

GLOSSARY OF PLACE NAMES

An-yeh	安業		Lo-fou shan	羅浮山
Ch'ang-li	昌利		Mao-shan	茅山
Hao-chou	濠州		Ming-t'ang (*hsien*)	明堂 (縣)
Heng-shan	衡山		Mu-chou	睦州
Hsing-yüan	興元		Nam-Viet	南越
Kou-ch'ü	句曲		Ning-chou	寧州
Kuang-ling	廣陵		Ping-chou	并州
Lai-shan	萊山		P'u	蒲
Lo-ch'uan	羅川		Shun-yang	順陽

337

Glossaries

T'ang ch'ang kuan 唐昌觀
T'ien chin chiao 天津橋
Wan-nien (*hsien*) 萬年 (縣)
Yang-ch'eng 陽城
Yang-p'ing 陽平

Yen-p'ing chin 延平津
Ying-chou 瀛洲
Yü jui yüan 玉蕊院
Yüeh (*Viet) 越

GLOSSARY OF OFFICIAL TITLES, OFFICES, INSTITUTIONS, AND ERAS

Chan hsing t'ai 瞻星台
Chieh tu shih 節度使
Ch'ien yüan tien 乾元殿
Fu t'ien 符天
Hou (heirgiver) 后
Hou fei szu hsing 后妃四星
Hsing kuan 星官
Hsing lang 星郎
Hua fei 華妃
Hui fei 惠妃
Hun i chien 渾儀監
Hun t'ien chien 渾天監
Lang chung 郎中
Li fei 麗妃
Ling t'ai 靈臺
Ming t'ang 明堂

Pi shu ko chü 祕書閣局
Pin (guestmaiden) 嬪
Shang shu 尚書
Szu i 司夜
Szu po 司簿
Szu t'ien t'ai 司天臺
T'ai ch'ang 太常
T'ai shih chien 太史監
T'ai shih chü 太史局
T'ai shih ling 太史令
T'ien tsun 天尊
Ts'e ching t'ai 測景臺
T'ung hsüan yüan 通玄院
Wan hsiang shen kung 萬象神宮
Yüan kuang 元光

GLOSSARY OF BOOK NAMES

Chen kao 真誥
Ch'i shih ching 起世經
Ch'i yao ching 七曜經
Ch'i yao jao
　tsai chüeh 七曜攘災訣
Chiu chih li 九執曆
Fa hsiang chih 法象志
Hsien ching 仙經
Hsing ming tsung kua 星命總括
Hsiu yao ching 宿曜經
Hsüan miao
　ching chieh 玄妙經解
Jen wang ching 仁王經

Kuan hsing yao chüeh 觀星要訣
Kuan hsing hsin ch'uan
　k'ou chüeh 觀星心傳口訣
Kuo Ch'eng wen ta 果橙問答
Ming shu 命書
Mo-teng-chia ching 摩登伽經
Po hsüeh chi 博學記
Ta chi ching 大集經
Ta T'ang k'ai yüan
　chan ching 大唐開元占經
*Tu-li ywĕt-siĕ 都利聿斯
T'ung hsüan i shu 通玄遺書
Yüeh ling 月令

338

Glossaries

GLOSSARY OF WORDS AND PHRASES

an (sun-dark) 暗

chan hsing 占星

ch'an ch'iang 欃槍

ch'an hui 蟾輝

ch'an kuang 蟾光

ch'an p'o 蟾魄

ch'ang (glory) 昌

ch'ang-ho (*ch'ang-ghap) 閶闔

ch'ang keng 長庚

chao hsia 朝霞

chao-yao 招搖

ch'ao teng chiu
    hsien wei 超登九仙位

che hsien 謫仙

*chem-zhyo 蟾蜍

chen hsing 鎮星

ch'en (chronogram) 辰

ch'en (clam monster) 蜃

ch'en hsing 辰星

cheng (zither) 箏

chi (strand) 紀

chi shih 積尸

ch'i (energizing breath, pneuma) 氣

ch'i ch'iao 乞巧

ch'i hsing chih yü 七星之旗

ch'i kuan 騎官

chia i 甲乙

chiang (come down) 降

chiang hsing 將星

chieh (kalpa) 劫

ch'ien (hypophenomenal) 潛

ch'ien (potential) 乾

ch'ien lin 騫林

ch'ien niu 牽牛

chih (substance) 質

chih fa 執法

chih nü 織女

chin huo 金火

chin p'o 金魄

ching (clear sky) 晴

ching (germ, embryo; embryonic
    essence) 精

ching (phosphor) 景

ching (warp) 經

ching yün 景雲

ch'ing shang 清商

ch'ing wei 清微

ch'iung (rose-gem) 瓊

chu (vermilion) 朱

chu ming 朱明

ch'u shih 處士

ch'ui 垂

chung k'ung 中空

chung t'ai 中台

ch'ung hsüan 沖玄

ch'üan (authority) 權

erh ching 二景

fa hsing 罰星

fan tou ch'a 犯斗槎

fen yeh 分野

feng po 風伯

fu (rhapsody) 賦

fu (sustainer) 輔

*ghung 虹

ha-ma 蝦蟇

hao t'ien shang ti 昊天上帝

heng 恆 姮

ho ku (river drum) 河鼓

ho ku (drum bearer) 荷鼓

hou kung 后宮

hsiang (counterpart) 象

hsiang (minister) 相

hsiao-yao shen
    ming yü 逍遙神明域

hsiao-yao yu 逍遙遊

hsien (transcendent) 仙

hsien ch'a 仙槎

hsien ch'an 仙蟾

hsin hsing 新星

hsing (form) 形

Glossaries

hsing (activity) 行

hsing ch'a 星槎

hsing ch'en 星辰

hsing chï 星紀

hsing pien 星變

hsing t'an 星壇

hsiu (lodging, nakshatra) 宿

hsü (a Jupiter station) 戌

hsüan (suspend) 懸

hsüan hsiang 玄象

hsüan yeh 宣夜

hsüan-yüan 軒轅

huang tao yu i 黃道遊儀

hui (hsing) 慧(星)

hun (cloud-soul) 魂

hun ching yang ming 魂精陽明

hun-lun 渾淪

hun ming k'ung ling 魂明空靈

hun shen 魂神

hun t'ien 渾天

hun-t'un 渾沌

hun yüan 混元

hung fan 洪範

jen wu 人物者

jih che 日者

jui hsing 瑞星

kai t'ien 蓋天

kang (mainstay) 綱

k'o hsing 客星

kou ch'en 句陳

kou ch'ien 鉤鈐

kua hsing ch'a 挂星槎

kuan so 貫索

kuan yüeh ch'a 貫月槎

kuang (light) 光

k'un (latent) 埍

k'ung-tung, kung-tung 空洞, 崆峒
  空桐, 空峒, 空同

k'ung-tung chih
  hsiao t'ien 空洞之小天

*kwa (snail) 蝸

*k'wĕt 堀窟

*k'ywāt 闕

lei kung mo 雷公墨

li (almanac) 歷

li (plough) 犂

li chu 離珠

ling (numinous) 靈

ling ch'a 靈槎

ling pao 靈寶

liu hsing 流星

liu-li 琉璃

lo 落

lo hsing 落星命

lu ming 祿命

mang (awn) 芒

mao (yaktail) 旄

meng hsing 濛星

meng-hung 濛鴻

*miet (mihr) 蜜

ming (luminous) 明

ming ch'i 明器

ming hsia 明霞

ming hsing 命星

nei p'ing 內屏

*ngei 蜺

*ngywāt ywĕk 月域

niu 紐

niu yu kung 紐幽宮

nü chen 女貞

nü kuan tzu 女冠子

o (*nga "fairy") 娥

pa yüeh ch'a 八月槎

pao chien 寶劍

pao tao 寶刀

pao ting 寶鼎

pei ch'en 北辰

pei chi 北極

pen hsing (running star) 奔星

pen hsing (spurting star) 賁星

pi (straightener) 弼

pi-li fu 霹靂斧

piao (manifesto) 表

po (*bwĕt) 孛

340

Glossaries

po yün hsiang　白雲鄉

p'o (*p'ăk "protopsyche")　魄

p'o (new-born moon)　霸

p'o ching　魄精

pu chih jan erh jan i　不知然而然矣

p'u (unhewn)　朴

san ch'ing　三清

san t'ai　三台

shang (sky god)　商

shang ch'ing　上清

shao yang　少陽

shen (spirit)　神

shen (triaster)　參

shen ming　神明

sheng jen　聖人

shih ch'ing　石青

shu shu　術數

shui ching　水精

shuo (conjunction)　朔

so chi　莎雞

su kuang　素光

su o　素娥

su yüeh　素月

sui hsing　歲星

szu kuai　司怪

ta chiang chün　大將軍

ta ch'ih　大赤

ta lo　大羅

ta ming　大明

t'ai chi　太極

t'ai ch'u　太初

t'ai hsia　太霞

t'ai hsü　太虛

t'ai i (grand interchangeability)　太易

t'ai i (grand monad)　太一

t'ai po　太白

t'ai shang chen jen　太上真人

t'ai shih　太始

t'ai su　太素

t'ai tzu　太子

t'ai wei　太微

t'ai wei kung　太微宮

t'ai yang　太陽

tao shih　道釋

ti (god-king, theocrat)　帝

ti hou　地候

ti hsien　地仙

t'i ch'i ch'iao wen　題乞巧文

tien pao　典寶

t'ien ch'an　天攙

t'ien chiang　天江

t'ien feng　天鋒

t'ien fu　天桴

t'ien hsien　天仙

t'ien huang ta ti　天皇大帝

t'ien jan tsai hsüan hsü　恬然在玄虛

t'ien kuan　天官

t'ien shih　天使

t'ien (chih) shu　天(之)樞

t'ien ta chiang chün　天大將軍

t'ien ti t'an　天帝壇

t'ien wai　天外

t'ien wen　天文

to (drop)　隓

tou　斗

ts'ang (gray)　蒼

ts'ang (watchet)　滄

ts'ang-ts'ang　蒼蒼

tsao hua　造化

tsao wu che　造物者

t'ui ming　推命

tung (grotto)　洞

tung (tube)　筒

tung chün　東君

tung fan　東藩

tung t'ien　洞天

tzu jan　自然

tz'u (Jupiter station)　次

wa (frog)　蛙

wa (name of Nü Kua)　娃

wang (full moon)　望

wang liang　王良

wang shih　枉矢

wei (weft)　緯

341

Glossaries

wei (tail) 尾

wen ch'ang 文昌

wen hsing 文星

wu (*myu "minx") 婺

wu chu hou 五諸侯

wu ch'ü 武曲

wu pu 巫步

wu shan i tuan yün 巫山一段雲

wu ti tso 五帝座

wu wei 五緯

yang ming 陽明

yao (mythological god-king) 堯

yao (turquoise) 瑤

yao ch'an 妖蟾

yao hsing 妖星

yao ling 耀靈

yao p'o pao 耀魄寶

yao t'u 瑤兔

yeh jih fen ko ying 野日分戈影

yeh kuang yün 夜光雲

yin (tiger sign) 寅

yin huo 陰火

yin p'o 陰魄

yin-yün 氤氳

ying (finest bloom) 英

ying huo 熒惑

ying kuang 螢光

yu wu ch'iung che 游無窮者

yü ch'ing 玉清

yü kuei 輿鬼

yü lin 羽林

yü pu 禹步

yü shih 雨師

yü yü 禹餘氣

yüan ch'i 元氣

yüan chün 元君

yüan p'o 圓魄

yüan yu 遠遊

yüeh (marchmount) 嶽

yüeh (battleaxe) 鉞

yüeh-chih 月氏 月支

yüeh kuei 月桂

yüeh k'u 月窟

yüeh p'o 月魄

yün (nimbus) 暈

yün (fall) 隕

Index

Note: Some words and concepts pervade the book, constituting its fabric. Accordingly they have not been indexed. They are the following: asterism, constellation, counterpart, divination, eclipse, five activities (wu hsing), god, goddess, heaven, imagery, metaphor, meteorology, myth, omen, pacing the void, poem, poetry, religion, simulacrum, T'ang, yin and yang. In addition to these, words which occur as the titles of chapters and sub-chapters have no corresponding entries in the index. These include astral, astrology, comet, cosmogony, meteor, moon, nova, planet, sky, sun.

Index

Index

Index

Index

Index

348

Index

Index

351